Group Theory in China

Mathematics and Its Application (China Series)

Managing Editor:

M. HAZEWINKEL

Centre for Mathematics and Computer Science,
Amsterdam, The Netherlands

Group Theory in China

Edited by

Zhe-xian Wan
The Institute of System Science of the Chinese Academy of Sciences
Beijing, People's Republic of China

and

Sheng-ming Shi
Capital Normal University
Beijing, People's Republic of China

This book serves as the 35th volume of the Series on the Pure and Applied
Mathematics published by Science Press

Science Press
New York/Beijing

Springer Science+Business Media, LLC

Library of Congress Cataloging in Publication Data

ISBN 978-94-010-6294-7   ISBN 978-94-011-5454-3 (eBook)
DOI 10.1007/978-94-011-5454-3

'Et moi, ..., si j'avait su comment en revenir, je n'y serais point allé.'

JulesVerne

The series is divergent; therefore we may be able to do something with it.

O.Heaviside

One service methematics has rendered the human race. It has put common sense back where it belongs, on the topmost shelf next to the dusty canister labelled 'discarded non-secse'.

Eric T. Bell

Mathematics is a tool for thought. A highly necessary tool in a world where both feedback and nonlinearities abound. Similarly, all kinds of parts of mathematics serve as tools for other parts and for other sciences.

Applying a simple rewriting rule to the quote on the right above one finds such statements as: 'One service topology has rendered mathematical physics ...'; 'One service logic has rendered computer science ...'; 'One service category theory has rendered mathematics ...'. All arguable true. And all statements obtainable this way form part of the raison d'être of this series.

This series, *Mathematics and Its Applications,* ttarted in 1977. Now that over one hundred volumes have appeared it seems opportune to reexamine its scope. At the time I wrote

"Growing specialization and diversification have brought a host of monographs and textbooks on increasingly specialized topics. However, the 'tree' of knowledge of mathematics and related fields does not grow only by putting forth new branches. It also happens, quite often in fact, that branches which were thought to be completely disparate are suddenly seen to be related. Further, the kind and level of sophistication of mathematics applied in various sciences has changed drastically in recent years: measure theory is used (non-trivially) in regional and theoretical economics; algebraic geometry interacts with physics; the Minkowsky lemma, coding theory and the structure of water meet one another in packing and covering theory; quantume fields, crystal defects anf mathematical programming profit from homotopy theory; Lie algebras are relevant to filtering; and prediction and electrial engineering can use Stein spaces. And in addition to this there are such new emerging subdisciplines as "experimental methematics', 'CFD', 'completely integrable systems', 'chaos, synergetics and large-scale order', which are almost impossible to fit into the existing classification schemes. They draw upon widely different sections of mathematics."

By and large, all this this still applies today. It is still true that at first sight mathematics seems rather fragmented and that to find, see, and exploit the deeper underlying interrelations more effort is needed and so are books that can help mathematicians and scientists do so. Accordingly MIA will continue to try to make such book available.

If anything, the description I gave in 1977 is now an understatement. To the examples of interaction areas one should add string theory where Riemann surfaces, algebraic geometry, modular functions, knots, quantum field theory, Kac-Moody algebras, monstrous moonshine (and more) all come together. And to the examples of things which can be usefully applied let me add the topic 'finite geometry'; a combination of words which sounds like it might not even exist, let alone be applicable. And yet it is being applied: to statistics via designs, to radar/sonar detection arrays (via finite projective planes), and to bus connections of VLSI chips (via difference sets). There seems to be no part of (so-called pure) mathematics that is not in immediate danger of being applied. And, accordingly, the applied mathematician needs to be aware of much more. Besides analysis and numerics, the traditional workhorses, he may need all kinds of combinatorics, algebra, probability, and so on.

In addition, the applied scientist needs to cope increasingly with the nonlinear world and the extra mathematical sophistication that this requires. For that is where the rewards are. Linear models are honest and a bit sad and depressing: proportional efforts and results. It is in the nonlinear world that infinitesimal inputs may result in macroscopic outputs (or vice versa). To appreciate what I sm hinting at; if electronics were linear we would have no fun with transistors and computers; we would have no TV; in fact you would not be reading these lines.

There is also no safety in ignoring such outlandish things as nonstandard analysis, superspace and anticommuting integration, p-adic and ultrametric space. All three have applications in both electrical engineering and physics. Once, complex numbers were equally outlandish, but they frequently proved the shortest path between 'real' results. Similarly, the first two topics named have already provided a number of 'wormhole' paths. There is no telling where all this is leading-fortunately.

Thus the original scope of the series, which for various (sound) reasons now comprises five subseries: white (Japan), yellow (China), red (USSR), blue (Eastern Europe), and green (everything else), still applies. It has been enlarged a bit to include book treating of the tools from one subdiscipline which are used in others. Thus the series still aims at books dealing with:

– a central concept which plays an improtant role in several different mathematical and/or scientific specialization areas;
– New applications of the results and ideas from one area of scientific endeavour into another;
– influences which the results, problems and concepts of one field of enquiry have, and have had, on the development of another.

The present volume, one of the first in the 'Chinese subseries' of MIA, also appropriately enough, one dealing with fundamental issues: interrelations between logic and computer science. The advent of computers has sparked off revived interest in a host of fundamental issues in science and mathematics such as computability, recursiveness, computational complexity and automated theorem proving to which latter topic ths author has made seminal contributions for which he was awarded the ATP prize in 1982.

It is a pleasure to welcome this volume in this series.

The shortest path between two truths in the real domain passes through the complex domain

J. Hadamard

La physique ne nous donne pas seulement l'occasion de rèsoudre des problèmes ... elle nous fait presentir la solution.

H. Poincaré

Never lend books, for no one ever returns them; the only books I have in my library are books that other folk have lent me.

Anatole France

The function of an expert is not to be more right than other people, but to be wrong for more sophisticated reasons.

David Butler

Bussum, August 1989

Michiel Hazewinkel

Preface

Professor Hsio-Fu Tuan is a famous Chinese mathematician, who has made important contributions to the theories of both finite groups and Lie groups. He has also had a great influence on the development of algebra and, in particular, group theory in China. On the occasion of Professor Tuan's eighty second birthday, we publish the present book GROUP THEORY IN CHINA and dedicate it to him.

This book consists of a collection of essays on various aspects of group theory and some related areas, written by Professor Tuan's former students and colleagues, who are known by their own work in these aspects respectively. Each essay contains the main results, in particular, the recent results of one aspect, obtained by Chinese mathematicians. We hope the book will help the mathematicians in the world to know how group theory in China has been developed, what main results in the field have been obtained by Chinese mathematicians and what problems of this field Chinese mathematicians are interested in.

Editors
February, 1996

Dedicated to Professor Hsio-Fu Tuan
for His Eighty Second Birthday

Hsio-Fu Tuan was born on July 29th, 1914. He graduated from the Tsinghua University and got his B. S. degree in 1936. In 1940 he entered the Toronto University as a graduate student and got his M. S. degree next year. Later, Tuan transferred to the Princeton University and, under the supervision of R. Brauer, got his Ph. D. degree in 1943. Then he stayed at the Department of Mathematics, Princeton University as a research associate for two years. There, under the guidance of Brauer and C. Chevalley and finally together with them, Tuan completed important work on modular representation theory of finite groups and algebraic Lie groups and algebras. During that time Tuan also visited E. Artin for four months. 1945 – 1946 Tuan worked at the Institute of Advanced Study, Princeton, as an assistant of H. Weyl.

In autumn of 1946, Tuan returned to China and became a professor of the Tsinghua University. Since 1952 Tuan is a professor of the Peking University and the Dean of the Department of Mathematics till 1981. Tuan is a Member of the Chinese Academy of Sciences and a Member of the Standing Council of the Chinese Mathematical Society (1950–1987).

The most important mathematical contribution of Tuan is on modular representation theory of finite groups and its application to finite simple groups and linear groups. His joint work with Chevalley about algebraic Lie algebra and algebraic Lie group is the beginning of the modern theory of algebraic groups. In his study of p-groups, he successfully generalized Kulakoff's Ahzahl Theorem. Since the 1970's he began to study the applications of finite group theory to certain combinatorial problems and achieved important results. Tuan has also made important contributions to the development of mathematics and the improvement of education in China.

Contents

Generalized Simple Groups of Lie Type

Chen Chong Hu (Chen Zhonghu)

Department of Mathematics, Xiangtan University
Xiangtan, Hunan, 411105, P. R. of China

For each simple Lie algebra L over $\mathbf{C}$ and arbitrary field K, Chevalley showed in [1] how to construct the simple groups $L(K)$ which are called the Chevalley groups of type L over K. Clearly. The Chevalley groups are analogues of the complex simple Lie groups over arbitrary field. Using the diagram automorphisms of L and certain automorphisms of K whose orders are the same, Steinberg in [2] and Ree in [3] showed how to costruct the twisted groups over K which can be interpreted as the generalization of the Chevalley groups. The simple groups of Lie type which are Chevalley groups or twisted groups are very important in the theory of simple groups, especially, in the theory of finite simple groups. Chevalley suggested five open problems in [1] which deal with the theory of real simple Lie groups, of classical groups and of the simple algebraic groups. Clearly, the five open problems suggested by Chevalley in [1] are very important in group theory.

(1) The fifth problem suggested by Chevalley in [1] was how to generalize [1] so that the generalization of [1] can suit the real forms of simple Lie algebras over $\mathbf{C}$ which always called the real simple Lie algebra, and Chevalley in [1] indicated that the fifth problem is important and difficult possibly. Clearly, this fifth problem is equivalent to ask how to construct certain simple groups over K which are analogues of the real simple Lie groups over arbitrary field K. In [4, 5], the author made use of the Satake diagram of a real form of simple Lie algebra L over $\mathbf{C}$ to determine an automorphism of L corresponding to the real form, moreover, using the automorphism of L corresponding to the real form of L and a certain automorphism of K whose orders are the same, the author constructed certain simple groups over K which are analogues of the real simple Lie groups and are called the generalized simple groups of Lie type. Clearly, the methods of the constructions used in [4, 5] are the generalization of

the methods of the constructions used in [1, 2], but the automorphisms of L corresponding to the real forms of L are often not diagram automorphisms of L (in fact, are often the compositions of inner automorphisms and diagram automorphisms of L), so, the constructions in [4, 5] are often difficult and complicated.

(2) When Chevalley suggested the first problem in [1], he indicated that for classical types, the simple groups over K constructed in [1] only include the orthogonal and unitary groups corresponding to the forms whose Witt index is sufficiently large. Carter in [6] indicated that the orthogonal and unitary groups which are not the simple groups of Lie type of the classical types can be interpreted as "non-split" groups of Lie type. Therefore, one problem left open was how to construct the "non-split" groups of Lie type. By the methods which are analogues of the methods used in [4, 5], the orthogonal and unitary groups corresponding to the forms whose Witt index is not sufficiently large and is not zero have been constructed in [7]. Then all classical groups can be constructed uniformly by the generalized Chevalley's methods.

(3) The third problem suggested by Chevalley in [1] was to ask about the relation between the simple groups constructed in [1] and the simple algebraic groups. This relation was discussed in [8]. Satake in [9] , Humphreys in [10] and Springer in [11] indicated an important fact that the construction of the Chevalley groups (resp. the simple groups of Lie type) over the algebraically closed field $\mathbf{K}$ (resp. finite field $\mathbf{F}$) gives a uniform proof of existence theorem of simple algebraic groups defined over $\mathbf{K}$ (resp. $\mathbf{F}$), i. e. existence theorem of Chevalley. Moreover, the simple groups of the $\mathbf{K}$-rational (resp. $\mathbf{F}$-rational) points of the simple algebraic groups defined over $\mathbf{K}$ (resp. $\mathbf{F}$) can be described by the simple geoups of Lie type over $\mathbf{K}$ (resp. $\mathbf{F}$). For a local field k, Bruhat and Tits in [12] gave a description of the simple groups of the k-rational points of the simple algebraic groups defined over k by deeper geometric theory. In [13], the author made use of the methods in [9] , [8] and [4, 5], to give a uniform proof of existence theorem of the simple algebraic groups defined over a p-adic field p. In [14], [15] and [16], by a similar construction of the generalized simple groups of Lie type given in [4, 5], the author gave the decriptions of the simple groups of the K-rational points of the simple algebraic groups defined over K where K is a p-adic field p, a formally real field $\mathbf{R}$ or a number field in [14], [15] or [16] respectively.

References

[1] Chevalley C., Sur certain groupes simple, Tohoku Math. J., 7(1955), 15–66.

[2] Steinberg R., Variations on a theme of Chevalley, Pacific. J. Math., 9(1959), 875–891.

[3] Ree. R, A family of simple groups asscoiated with the simple Lie algebra of type F_4; (G_2), Amer. J. Math., 83(1961), 401–420; (432–462).

[4] Chen Chong Hu, Some simple groups of Lie type constructed by inner automorphism, Chin. Ann. Math., 1(1980), 161–176.

[5] Chen Chong Hu, A family of simple groups associated with the Satake diagrams, J. Austral. Math. Soc. (Series A), 41(1986), 13–43.

[6] Carter. R., Simple groups of Lie type, Wiley, London, New York, 1972.

[7] Chen Chong Hu, Classical groups and generalized simple groups of Lie type, J. Austral. Math. Soc. (Series A), 47(1989), 53–78.

[8] Steinberg. R, Lectures on Chevalley groups, mimeographed lecture notes, New Haven, Yale Univ. Math. Dept., 1968.

[9] Satake. I, Classification theory of semisimple algebraic groups, Dekker, New York, 1971.

[10] Humphreys. J., Linear algebraic groups, Springer-Verlag, New York, Berlin, 1975.

[11] Springer. T., Linear algebraic groups, Birkhauser, Boston, 1980.

[12] Bruhat. F and Tits. J, Groups reductifs sur un corps local, I, II, Publ. Math. I. H. E. S., 41(1972), 5-252; 60(1984), 197–376.

[13] Chen Chong Hu, On existence theorem of simple algebraic groups over p-adic field, Acta. Math. Sinica, 36 (1993), 233–244, (in Chinese).

[14] Chen Chong Hu, A note on simple groups over p-adic field, Proceedings of the Conference on Order structures and algebra of computer Languages, World Scientific, Singapore, London, 1–17.

[15] Chen Chong Hu, On the construction of simple groups over formally real fields, Chin. Ann. Math., 8(B) (1987), 170–180.

[16] Chen Chong Hu, A note on the construction of simple groups over number field, SEA Bull. Math., 15 (1991), 1–12.

Abstract Group Theory

Z. M. Chen

Department of Mathematics, Southwest Teachers University
Beibei, Chongqing, 630715, P. R. China

A series of reseach on abstract group theory has been carried out in P. R. China over the last 15 years. Among these, there are several aspects playing a special role in the development. We divide them in the following to state briefly the main results relating to them.

I. The solvable groups

II. The supersolvable groups

III. The nilpotent group

IV. The influence of some subgroups, some factor-groups, or some automorphism groups on finite groups

V. Formations

VI. The infinite groups

I. The solvable groups

Under what conditions, a group G must be solvable? This is a quite interesting question, and many efforts from different directions have been made to answer it by lots of persons. We state the following results which give us a good report about the reseach in P. R. China on this question.

Theorem 1 ([1] and [15]) *A finite group having exactly two conjugacy classes of maximal subgroups is solvable.*

Theorem 2 ([12]) *A finite group G having two same order classes of maximal subgroups is solvable, or $G/S(G) \cong PSL\,(2,\,7)$, where $S(G)$ is the maximal normal solvable subgroup of G.*

Theorem 3 [9] *If a non-solvable group G has exactly two same order classes of non-normal maximal subgroups, then $G/S(G) \cong PSL\,(2,\,7)$.*

Instead of the maximal subgroups, one considers the elements with the same order in G and we get the following:

Theorem 4 (Suskin's problem, [18]) *In a finite group G, if any two p-elements of the same order are conjugate for any prime p, then $G \cong 1$, Z_2, or S_3.*

Theorem 5 [10] *If $Aut(G)$ acts double transtively on elements of the same order of a group G, then $G \cong Z_3$, S_3, or an elementary Abelian 2-group whose order is greater than 2.*

By inverstigating the stucture of minimal non-solvable groups with an acting group one proves in [3] the well-known conjecture for solvability of a group which possesses a fixed-point-free automorphism group in broad sense.

Theorem 6 [3] *Suppose that a group A acts on a group G, $(|A|, |G|) = 1$. If the fixed-point subgroup C_G (A) of G has no sections isomorphic to S_3, A_4 , or $Sz(2) = 5 : 4$, then G is solvable. Especially G is solvable when C_G (A) is of odd order, or nilpotent.*

The following results are examples that the solvability of a group G follows from some conditions on its some special subgroups.

Theorem 7 (Extension of Thompson's Theorem, [13] and [16]) *Let M be a maximal subgroup of a group G and $P \in Syl_2 M$. If M is nilpotent and $\Omega_2(P) = \le Z(P) \le Z(P)$, then G is solvable.*

Theorem 8 [16] *Assume a group A acts on a group G and G has a maximel A-invariant subgroup M which is nilpotent. Let $P \in Syl_2 M$. If each non-cyclic subgroup P_1 of P containing $Z(P)$ is A-invariant and normal in P and one of the following conditions is satisfied, then G is solvable:*

(1) *A is solvable,*

(2) *$(|A|, |G : M|) = 1$, or*

(3) *G has an A-invariant Sylow q-subgroup $Q \neq 1$ for some prime $q \in \pi(G) - \pi(M)$.*

Theorem 9 (Extension of Laffey Theorem, [2] and [6])*If every p'-subgroup of a group G is 2-closed for some p and one of the following conditions holds, then G is solvable:*

(1) *$p \ge 5$, or*

(2) *$p = 3$ and G has no sections isomorphic to $PSL(2, 7)$, or $PSL(3, 3)$.*

Theorem 10 [7] *If every non-maximal proper subgroup of a group G is 2-closed and G is not isomorphic to A_5, or $SL(2, 5)$, then G is solvable.*

Theorem 11 [8] *Suppose group $G = AB$, A, B are subgroups of G and (a) A has a 2-decomposable normal subgroup H such that A/H is an elementary abelian group;*

(b) *B has a nilpotent subgroup K of odd order such that $|B : K| \leq 2$; and*

(c) *$(|A|, |K|) = 1$. Then G is solvable.*

The subgroup H of a group G is called an r-minimal subgroup if there exists a subgroup chain

$$1 = H_0 < H_1 < \cdots < H_r = H,$$

where H_i is maximal in H_{i+1}. G is said to be an SQN-r-group if every r-minimal subgroup of G is S-quasinormal in G.

Theorem 12 (Extension of Sastry and Deskins' Theorem, [11] and [17]) *An SQN-r-group ($r = 1, 2, 3$) is solvable and its fitting length is less than 3.*

Theorem 13 [19] *Suppose that the Sylow p-subgroups of a group G is a T. I. set and G has no section $PSL(2, 2^3)$ when $p = 3$; G has no section $Sz(2^5)$ when $p = 5$. If there exists normal subgroup M of G such that $p \mid (|M|, |G/M|)$, then G is p-solvable.*

Deskins has defined the normal index of maximal subgroup M in a finite group G as the order of a chief factor H/K of G where H is minimal in the set of normal supplements to M in G. A maximal subgroup M of G is called c-maximal, if $|G : M|$ is composite.

Theorem 14 [4] *Let p be the large prime factor dividing the order of a group G. Then G is p-solvable if and only if $[\eta(G : M)]_p = 1$ for each non-nilpotent c-maximal subgroup M of G with $[G : M]_p = 1$, where $\eta(G : M)$ denotes the normal index of M in G.*

At the end of this section, we mention two results within the sovable groups.

Theorem 15 (Extension of Hall's Theorem, [5] and [14]) *Let G be a solvable group. If the derived subgroup of every Sylow subgroup of G is contained in the centre of G and G has no sections isomorphic to a Schmidt group with normal extraspecial Sylow subgroup, then $\gamma_\infty(G) \cap Z(G) = 1$.*

Let π be a set of primes and H be a π-nilpotent subgroup of a group G. If H contains a Hall π'-subgroup of G and $N_G(H) = H$, then H is said to be a π-Carter subgroup of G.

Theorem 16 ([20] and [21]) *Let G be a π-solvable group. Then*

(a) G has π-Carter subgroups,

(b) if K_1 and K_2 are any two π-Carter subgroups, then K_1 and K_2 are conjugate in G,

(c) if $G = AB$, $A \lhd G$, A, B are π-nilpotent, then B is contained in a certain π-Carter subgroup, and

(d) let $K_\pi(G)$ be the π-nilpotent residual of G and $H_\pi(G)$ be the π-hypercentre of G. Then $(K_\pi(G), H_\pi(G)) \leq 0_{\pi'}(G)$.

References

[1] S. Adnan, *On groups having exactly 2 conjugacy classes of maximal subgroups*, Lincei-Rend. Sc. fis. mat. e net, 66 (1979), 175–178.

[2] Z. M. Chen, *On inner-p-closed groups*, (in Chinese) Adv. Math., 15:4 (1986), 385–388.

[3] Z. M. Chen and Y. M. Wang, *Minimal non-solvable groups with an acting group*, (in Chinese) Koxue Tong Bao, 34:22 (1989), 1691–1693.

[4] X. Y. Guo, *The normal index of maximal subgroups in finite groups*, (in Chinese) Acta Math. Sin., 34 (1991), 208–212.

[5] P. Hall, J. Math., 182 (1940), 206–214.

[6] T. J. Laffey, *Solubility theorems for finite groups*, Proc. Cambri. Phil. Soc., 73 (1973), 1–6.

[7] S. R. Li, *Finite groups in which every non-maximal proper subgroup of even order is 2-closed*, (in Chinese) Chin. Ann. of Math., 4B (1983), 199–206.

[8] S. R. Li and S. Y. Li, *A theorem on the solvability of finite factorizable groups*, (in Chinese) Acta Math. Sin., 29:3 (1986), 413–416.

[9] S. R. Li, *Finite groups having exactly two same order classes of non-invariant maximal subgroups*, (in Chinese) Acta Math. Sin., 33:3 (1990), 388–392.

[10] X. H. Li, *On a problem of finite groups*, (in Chinese) J. Southwest-China Teachers Univ., 15:1 (1990), 144–146.

[11] N. S. Sastry and W. E. Deskins, *Influence of normality conditions on almost minimal subgroups of a finite group*, J. Alg., 52 (1978), 364–377.

[12] W. J. Shi, *Finite groups with two same order classes of maximal subgroups*, (in Chinese) Chin. Ann. of Math., 10A:5 (1989), 532–537.

[13] J. G. Thompson, *Normal p-complements of finite groups*, J. Alg., 1 (1964), 43–46.

[14] R. J. Wang, *On generalized A-groups*, (in Chinese) J. Math. Res. and Exposition, 9:4 (1989), 509–518.

[15] M. Y. Xu, *Another proof for the solvability of finite groups with at most two conjugacy classes of maximal subgroups*, (in Chinese) Chin. Ann. of Math., 6B:2 (1985), 211–213.

[16] G. X. Zhang, *On two theorems of Thompson*, Proc. Amer. Math. Soc., 98:4 (1986), 579–582.

[17] J. P. Zhang, *Influence of S-quasinormality condition on almost minimal subgroups of a finite group*, (in Chinese) Acta Math. Sin., (New Series) 3:2 (1987), 125–132.

[18] J. P. Zhang, *On Syskin's problem of finite groups*, (in Chinese) Chin. Sci. Bull. A:2 (1988), 124–128.

[19] J. P. Zhang, *On the p-solvability of the finite groups with a T. I. Sylow p-subgroup*, (in Chinese) Chin. Sci. Bull., 34:3 (1989), 177–179.

[20] Z. R. Zhang, *On Carter's Theorem*, (in Chinese) J. of Southwest-China Teachers College, 2 (1984), 6–9.

[21] Y. N. Zhen, Π-*properties of* Π-*solvable groups*, (in Chinese) J. Math., 6:3 (1986), 297–366.

II. The supersolvable groups

For supersolvable groups, there are lots of well-known theorems about the structure and the criterion. Some of these, proved by foreign mathematicians, have a good generalization. We list them in the following.

Theorem 1   (Extension of Mclain's Theorem, [3] and [9]) *A group G is supersolvable if and only if G has two chains of subgroups:*

$$G = G_0 > G_1 > \cdots > G_r > E, \ and$$
$$G = H_0 > H_1 > \cdots > H_r > E,$$

such that the indices of one chain are primes from small to large and the indices of another chain are primes from large to small.

Theorem 2   (Extension of Doerk's Theorem, [4] and [6]) *If the indices of subgroups A_i, $i = 1, 2, 3, 4$, of a group G are relatively prime and A_i, $i = 1, 2, 3$, are supersolvable, A_4 is meta-nilpotent, then G is supersolveble.*

Theorem 3   (Extension of Asaad's Theorem [1] and [4]) *Suppose that the indices of subgroups A, B, C of a group G are relatively prime and square free. If A, B are supersolvable and C is meta-nilpotent, then G is supersolvable.*

Theorem 4   (Extension of Humphrey's Theorem [8] and [14]) *Assume that G is a QCLT group. G is supersolvable if and only if G has no sections isomorphic to S_4.*

Before we state some new results, we give two definitions.

Definition 1   A solvable group G is called general nilpotent, if there exists a Sylow system $P_1, \cdots, P_k$ of G such that $\langle a_i \rangle \langle a_j \rangle = \langle a_j \rangle \langle a_i \rangle$ for every element $a_i \in P_i$ and $a_j \in P_j$.

Definition 2   Assume that H is a subgroup of a group G. Let $P_G(H) = \langle x \in G \, ; \, \langle x \rangle H = H \langle x \rangle \rangle$. We call that G satisfies the permutizer condition if $H < P_G(H)$ for every proper subgroup H of G.

Theorem 5   [4] *Let G be an odd order group and $G = AB = BC = CA$. If A, B are general nilpotent and C is supersolvable, then G is supersolvable.*

Theorem 6   ([7], [12] and [15]) *If solvable group G stisfies the permutizer condition, then*

(1) *G is supersolvable if and only if G has no S_4 section;*

(2) *G is p-supersolvable for every odd prime p, [15]; and*

(3) *G is supersolvable if and only if G' satisfies permutizer condition.*

There are a series of theorems for supersolvability stated by normality of some subgroups. To weaken normality may derive certain analogous theorems. Let G be a group and H a subgroup of G.

Quasinormal: *H is called quasinormal in G, if $HK = KH$ for every subgroup K of G.*

S-quasinormal: *H is called S-quasinormal in G, if $HK = KH$ for every Sylow subgroup K of G.*

Weakly S-normal: *H is called weakly S-normal in G, if for each p there exists a Sylow p-subgroup S_p, such that $HS_p = S_pH$. [15]*

Semi-normal: *H is called semi-normal in G, if $HK = KH$ for every subgroup K with $(|H|, |K|) = 1$. If furthermore we restrict K to the Sylow subgroups, then H is called S-seminormal in G. [5]*

Half-normal: *H is called half-normal, if there exists a subgroup B such that $G = HB$ and $HK = KH$ for every subgroup K of B. [10]*

Theorem 7   (Extension of Buckley's Theorem, [2] and [13]) *Suppose that $N \triangleleft G$ and G/N is supersolvable. Assume one of the following conditions is satisfied:*

(1) *the subgroups of prime order of N are all contained in the hyper-generalized-centre of G; or*

(2) *the subgroups of prime order of N are all S-quasinormal in G. Then G is supersolvable if and only if G has no D_{2q} section, where D_{2q} is a Schmidt group whose Sylow 2-subgroup is normal and has no generators of order 2.*

Theorem 8   (Extension of Srinivasan's Theorem [5] and [11]) *If every maximal subgroup of every Sylow subgroup of a group G are all S-seminormal in G, then G is supersolvable.*

Finally, we mention one more result.

Theorem 9   [15] (1) *For any factor d of the order of a supersolvable group there exists a weakly S-normal subgroup of order d;*

(2) *A group G is an X group if and only if all primitive subgroups of G are weakly S-normal; and*

(3) *A group G is a Y group if and only if all subgroups of G are weakly S-normal.* [5]

References

[1] M. Asaad, *A note on supersolvable finite groups*, Proc. Math. Phys. Soc. Egypt, 46 (1978), 25–27.

[2] J. Buckley, *Finite groups whose minimal subgroups are normal*, Math. Zeit., 116 (1970), 15–17.

[3] Z. M. Chen, *Some characterizations of finite supersolvable groups*, (in Chinese) Chinese Ann. Math., 3:5 (1982), 561–566.

[4] Z. M. Chen, *Several theorems of finite supersolvable groups*, (in Chinese) J. Southwest-China Teachers Univ., 2 (1986), 1–6.

[5] Z. M. Chen, *An extension of Srinivasan's theorem*, (in Chinese) J. Southwest-China Teachers Univ., 1 (1987).

[6] K. Dorek, *Minimal nicht Über auflösbare andliche Gruppen*, Math. Zeit., (1966), 198–205.

[7] X. Y. Guo, *The necessary and sufficient conditions for a finite groups to be super-solvable*, (in Chinese) 9 (1989), 161–164.

[8] J. F. Humphreys, *On groups satisfying the converse of Lagrange's theorem*, Proc. Cambri. Phil. Soc., 75 (1974), 25–32.

[9] D. H. Mclain, *The existence of subgroups of given order in finite groups*, Proc. Cambri. Phil. Soc., 53:2 (1957), 278–285.

[10] X. Y. Shu, *Half-normal subgroups of finite groups*, (in Chinese) J. Math., 8:1 (1988), 5–9.

[11] S. Srinivasan, *Two sufficient conditions for supersolvability of finite groups*, Isreal J. Math., 35:3 (1980), 210–214.

[12] J. P. Zhang, *On finite groups which satisfy the permutizer condition*, (in Chinese) Koxue Tong Bao, 14 (1985), 1048–1049.

[13] J. P. Zhang and L. W. Zhang, *The supersolvability of a class of finite groups*, (in Chinese) Acta Math. Sin., 30:5 (1987), 622–625.

[14] J. P. Zhang, *On the supersolvability of QCLT-groups*, Acta Math. Sin., 31:1 (1988), 29–32.

[15] Y. N. Zhen, *Weakly S-normal subgroups*, (in Chinese) J. Math., 10:1 (1990), 33–38.

III. The nilpotent group

A group G is said to be nilpotent if G has a finite series

$$1 = Z_0(G) < Z_1(G) < \cdots < Z_n(G) = G$$

such that each $Z_i(G)$ is normal in G and each $Z_{i+1}(G)/Z_i(G)$ is the centre of $G/Zi(G)$.

Theorem 1   ([13] and [14]) *If $N_G(P)/C_G(P)$ is a p-group for every p and every Sylow p-subgroup P of a group G, then G is nilpotent.*

Theorem 2   (Extension of Doerk's Theorem, [2] and [4]) *Suppose that A, B, C are nilpotent subgroups of a group G and their indices in G are relatively prime. If the classes of B and C are less than k, then G is nilpotent and the class is less than k. Especially, G is Abelian when B and C are Abelian.*

Theorem 3   [7] *The hypercentre of a group G is the intersection of all maximal nilpotent subgroups of G.*

A generalization of nilpotence is p-nilpotence which means a group G has a normal subgroup K such that $|K|$ and $|G/K|$ are coprime and $p \mid |G/K|$ for the prime p.

Theorem 4   [6] *If the derived group G' of a group G is p-nilpotent and $[x, y^{p-1}]$ is a p'-element for every p-element $x \in G'$ and every p'-element $y \in G$, then G is p-nilpotent.*

Theorem 5   [1] *Assume P is a Sylow p-subgroup of a group G. If each p'-element of $N_G(P_1)$ commutes with elements of P_1, then G is p-nilpotent, where P_1 is any subgroup of P with class k and satisfies that $Z_k(P) \le P_1$.*

Theorem 6   ([3], [5] and [11]) *Assume P is a Sylow p-subgroup of a group G and $N_G(P)$ is p-nilpotent. If one of the following conditions is satisfied, then G is p-nilpotent:*

(1) *p is odd and $\Omega_1(P) \le Z(P)$,*

(2) *$p = 2$ and $\Omega_2(P) \le Z(P)$,*

(3) *the rank of abelian sections of every intersection of any two Sylow p-subgroups is less than p, or*

(4) *every term $Z_i(P)$ in the upper central series of P is weakly closed in P with respect to G.*

From the study of p-nilpotence, it is natural to consider the existence of a normal π-complement in G for a Hall π-subgroup H of G, where π is a set of primes.

Theorem 7   [8] *Let H be a Hall π-subgroup of G. Then there is a normal π-complement of H in G if and only if the following condition is satisfied: If $H_1 \le H$ and $N_G(H_1) \ne G$, then $N_G(H_1) = N_{H_1}(H)O_{\pi'}(N_G(H_1))$.*

Theorem 8   [8] *Let H be a Hall π-subgroup of G. Then there is a* *normal π-complement of H in G if and only if G is a π-solvable, and $|(H - 1)^{G,\pi}| = |G : H||H - 1|$, where $(H - 1)^{G,\pi}$ is the union of all π-sections of G which involve $H - 1$.*

Theorem 9 (Extension of Brauer-Suzuki's Theorem, [9] and [10]) *Following properties of a group G are equivalent:*

(1) *The π-Hall subgroup H of G has a normal π-complement in G;*

(2) (a) *Any two elements of H which are conjugate in G must be conjugate in H, and* (b_1) *$C_H(x)$ is a π-Hall subgroup of $C_G(x)$ if $x \in H_1$;*

(3) (b_2) *$C_H(x)$ is a π-Hall subgroup of $C_G(x)$ and $\gamma_\pi(C_G(x)) = |C_G(C_G(x)/C_H(x))|$ if $x \in H - 1$, where $\gamma_\pi(C_G(x))$ denotes the number of π-elements in $C_G(x)$, and* (c) *$|(H - 1)^{G,\pi}| = |G : H||H - 1|$;*

(4) (b_3) *$O_{\pi'}(C_G(x)) = C_G(x)C_H(x)$ if $x \in H$ and $C_G \neq G$, and $(C_1)|(H - H_0)^{G,\pi}| = |G : H\,||H - H_0|$ if $H_0 = \{x \in H; C_G(x) = G\}$;*

(5) (b_4) *$N_H(H_1)$ is a π-Hall subgroup of $N_G(H_1)$, $\gamma_\pi(N_G(H)) = |N_G(H_1)/N_H(H_1)|$ if $H_1 \leq H$ and $N_G(H_1) \neq G$, and* (c) *in (3);*

(6) (b_5) *$\gamma_\pi(C_G(x)) = |C_G(x)|_{\pi'}$ if $x \in H$ and $C_G(x) \neq G$, and* (c) *in (3);*

(7) (d) *Every π-element of G is contained in a conjugate of H, and* (b_5) *in (6); and*

(8) (b_6) *$\gamma_\pi(N_G(H_1)) = |N_G(H_1)|_{\pi'}$ if H_1 is a cyclic subgroup of H with $N_G(H_1) \neq G$, and* (c) *in (3).*

For convenience, let (G, H, H_0, π) denote the following configuration: G is a finite group with subgroups H and H_0 such that $H_0 \triangleleft H$ and $\pi = \pi(H/H_0)$.

Theorem 10 [10] *If (G, H, H_0, π) satisfies the conditions (A_0), (B_1), (D) and $(|G : H|, |H : H_0|) = 1$, then there is a unique relative normal complement G_0 of H over H_0 in G, where conditions (A_0), (B_1) and (D) are as follows:*

(A_0) *If two π-elements h_1 and h_2 of $H - H_0$ are conjugate in G, then $h_1 H_0$ and $h_2 H_0$ are conjugate in H/H_0;*

(B_1) *Let h be a π-element of $H - H_0$ and $p \in \pi$. If p does not divide the order of h, then p does not divide the index $|C_G(h) : C_H(h)|$; and*

(D) *For each $p \in \pi$, any p-element of H_0 cannot be conjugate to any element of $H - H_0$.*

It is clear that the condition (A_0) are a necessary condition for the existence of the relative normal complement of H over H_0.

Theorem 11 [10] *Suppose that (G, H, H_0, π) satisfies the conditions*

(B$_2$), (C), (D) and $(|G:H|,|H:H_0|) = 1$. Then there is a unique relative normal complement of H over H_0 in G, where conditions (B$_2$) and (C) are as follows:

(B$_2$) For every π-element $x \in H - H_0$, $|C_G(x) : C_H(x)|$ is a π'-number, $C_G(x)$ exactly contains $|C_G(x)|_{\pi'}$-elements, and $C_H(x)$ exactly contains $|C_H(x)|_{\pi'}$-elements; and

(C) $|(H - H_0)^{G,\pi}| = |G/H||H - H_0|$.

As a corollary of the above theorem, we have

Theorem 12 [10] Let H be a π-Hall subgroup of $\dot{G}$. Then there exists a normal π-complement of H in G if and only if the following two conditions hold:

(ā) If two π-elements of $H - 1$ are conjugate in G, then they are conjugate in H; and

(b̄) For every $h \notin H$ and $h \in Z(G)$, if p does not divide the order of h, then p does not divide the index $|C_G(h) : C_H(h)|$.

At the end of this section, we mention three more results which should belong to this section.

Theorem 13 [5] Let $\pi = \{2, p, q, \cdots\}$, $p, q, \cdots$ are primes of form $4k + 1$. Then a group G is π'-closed if and only if G is π-homogeneous, that is $N_G(K)/C_G(K)$ is a π-group for each π-subgroup K of G.

Theorem 14 [12] Suppose that $N_G(P)/C_G(P)$ is a π-group for each p-subgroup P of a group G with prime p in π. Then G is π'-closed if one of the following statements is satisfied:

(1) Each π-subgroup of G is 2-closed; or

(2) Each π-subgroup of G is $2'$-closed.

Theorem 15 [12] If $N_G(Z(J(P)))$ is π'-closed for a Sylow p-subgroup P of a group G with prime p in π, then so is G.

References

[1] Z. M. Chen, *A theorem for finite groups having normal p-complements*, (in Chinese) Chin. Adv. Math., 11:4 (1982), 318–320.

[2] Z. M. Chen, *Several theorems of finite supersolvable groups*, (in Chinese) J. Southwest-China Teachers Univ., 2 (1986), 1–6.

[3] Z. M. Chen, *On B_p-groups*, (in Chinese) Acta Math. Sin., 32:6 (1989), 834–840.

[4] K. Dorek, *Minimal nicht Über auflösbare endliche Gruppen*, Math. Zeit., (1966), 198–205.

[5] Z. W. Du, *On the π-homogeneity and the π'-closure of finite groups*, (in Chinese) Chin. Ann. Math., 10A:1 (1989), 51–53.

[6] Y. Fan, *On the p-supersolvability*, (in Chinese) Chin. Adv. Math., 15:2 (1986), 201–204.

[7] Y. Fan, *A note on hypercentres and hyper-generalized centres*, (in Chinese) J. Math., 6:2 (1986), 215–220.

[8] X. Y. Guo, *A note on the normal π-complement*, (in Chinese) J. of Northeast-China Math., 4 (1988), 446–452.

[9] X. Y. Guo, *On the relative normal complement in finite groups*, (in Chinese) Chin. Ann. Math., 10A (1989), 699–704.

[10] X. Y. Guo and B. L. Zhang, *Normal π-complements in finite groups*, Comm. in Alg., 17:7 (1989), 1601–1606.

[11] Y. C. Ren, *Several remarks for p-nilpotent groups*, (in Chinese) J. Sichuan Univ., 26:1 (1989), 35–38.

[12] X. F. Wang, *On inner-π′-closed groups and normal π-complements*, (in Chinese) Chin. Ann. Math., 10B:3 (1989), 323–331.

[13] Y. M. Wang, *On Zassenhaus' conjecture*, (in Chinese) Koxue Tong Bao, 5 (1991).

[14] W. J. Xiao, *On Zassenhaus' conjecture*, (in Chinese) Koxue Tong Bao, 4 (1989), 244–246.

IV. The influence of some subgroups, some factor-groups, or some automorphism groups on finite groups

Let σ be a group property. Then a σ-group means the group possesses the the property σ. Suppose G is a finite non-σ group. G is called an inner-σ group if every proper subgroup of G is a σ group; G is called an out-σ group if every proper factor-group of G is a σ group; G is called a minimal non-σ group if G is both inner-σ group and out-σ group.

The study of the above three kinds of groups leads to the classification of some finite groups. Also it provides a useful method for such research.

The following groups have been studied quite well and their structure is clear.

1. *Inner-p-closed groups;* [3] 2. *Inner-π′-closed groups;* [15]

3. *Inner-(π, π')-closed groups;* [6]

A group G is called π-closed if G has a normal Hall π-subgroup. G is called (π, π')-closed if G is π-closed or π'-closed.

4. *Inner-Bp-groups;* [5]

A group G is called a B_p-group if "$N_G(P)$ is p-nilpotent which implies that G is p-nilpotent, where P is a Sylow p-subgroup of G."

5. *Inner-$\gamma_k - pn$ groups;* [4]

A group G is called a $\gamma_k - pn$ group if the k-th term $\gamma_k(G)$ in the lower central series of G is p-nilpotent.

6. *Odd group G with p-rank $\gamma_p(G) > 2$, but the p-rank of every proper subgroup of G is not greater than 2;* [7]

7. *Inner-regular p-groups of order p^{p+1} with an abelian subgroup of order p^p;* [16]

8. *Non-2'-closed groups each of whose 2-maximal subgroups is 2'-closed;* [8]

9. *Non-3-closed groups each of whose 2-maximal subgroups is 3-closed;* [9]

10. *Non-solvable 3d-groups each of whose 2-maximal 3d-subgroups is supersolvable;* [11]

A group G is called a *pd-group* if $p \mid |G|$.

11. *Non-solvable groups each of whose 2-maximal pd-subgroups is nilpotent, where p is the smallest odd prime divisor of the order of the group;* [10]

12. *Finite simple groups each of whose 2-maximal subgroups is PQN group;* [14]

A group G is called a PQN group if the minimal subgroups and the cyclic subgroups of order 4 are all quasinormal in G.

13. *Non-solvable groups each of whose solvable subgroups is a Schmidt group, p-decomposable, 2-closed, or 2'-closed, where p is the smallest odd prime factor of the order of the group;* [13]

14. *Non-2-closed groups each of whose supersolvable subgroups is 2-closed or a Schmidt group;* [12]

15. *Groups whose automorphism group is of order pqr;* [1]

16. *Groups whose automorphism group is a Schmidt group.* [2]

References

[1] G. Y. Chen, *Finite groups with the automorphism group having order a product of three distinct primes*, Proc. R. Ir. Acad., 90A:1 (1990), 57–62.

[2] G. Y. Chen, *Finite groups with the Schmidt group as the automorphism group*, (in Chinese) Chin. Ann. Math., 13B:1 (1992), 105–109.

[3] Z. M. Chen, *On inner-p-closed groups*, (in Chinese) Chin. Adv. Math., 15:4 (1986), 385–388.

[4] Z. M. Chen, *Outer-σ groups of finite order*, (in Chinese) Chin. Ann. Math., 8B:1 (1987), 109–119.

[5] Z. M. Chen, *On B_p-groups*, (in Chinese) Acta Math. Sin., 32:6 (1989), 834–840.

[6] Z. M. Chen, *Inner-(π, π')-closed groups and Isaacs' theorem*, (in Chinese) Chin. Ann. Math., 11A:5 (1990), 576–582.

[7] Y. Fan, *On groups of odd order and p-rank ≤ 2*, Arch. Math., 52 (1989), 12-18.

[8] S. R. Li, *Finite groups in which every non-maximal proper subgroup of even order is 2'-closed*, (in Chinese) J. Math., 3:1 (1983), 1–6.

[9] S. R. Li and S. Y. Li, *Finite groups in which every non-maximal proper subgroup is 3-closed*, (in Chinese) Acta Math. Sin., 29:4 (1986), 498–503.

[10] S. R. Li, *A class of finite non-solvable groups*, (in Chinese) Chin. Adv. Math., 16:3 (1987), 289–293.

[11] S. R. Li, *Finite non-solvable groups with supersolvable second maximal 3d-subgroups*, (in Chinese) Chin. Ann. Math., 9A:1 (1988), 32–37.

[12] S. R. Li, *Finite groups in which every supersolvable subgroup is either 2-closed or a Schmidt group*, (in Chinese) Acta Math. Sin., 31:3 (1988), 341–347.

[13] S. R. Li and Y. Q. Zhao, *Some finite non-solvable groups characterized by their solvble subgroups*, (in Chinese) Acta Math. Sin., (N. S) 4:1 (1988), 5–13.

[14] Y. C. Ren, *Finite groups in which every minimal subgroup and cyclic subgroup of order 4 of 2-maximal subgroups are all quasinormal*, (in Chinese) Acta Math. Sin., 33:6 (1990), 798–803.

[15] X. F. Wong, *On inner-π'-closed groups*, (in Chinese) Chin. Ann. Math., 10B:3 (1989), 323–331.

[16] D. G. Zhu, *On inner-regular p-groups*, (in Chinese) J. Southwest-China Teachers Univ., 2 (1987).

V. Formations

Let F be a formation locally defined by $\{f(p)\}$ and F_p is a formation defined by

$$\begin{cases} g(q) = \{f(p), q = p, \\ \text{class of finite groups, } q \neq p. \end{cases}$$

Theorem 1 *If every $f(p)$ is subgroup closed, then so is F; if each $f(p)$ is full and integrated and group closed, then so is F.*

Analogous results hold for the product of normal subgroups. [2]

Theorem 2 *Suppose that group A acts on a p-solvable group G and $A \geq InnG$. Let $G_p \in Syl_pG$. If $N_A(P)$ is F_p-stable on P for any p-subgroup P with $Z(G_p) \leq P \triangleleft G_p$. Then A is F_p-stable on G.* [2]

Theorem 3 *Let f be a formation function and $\Phi f(G) = \cap\{H < G; H \text{ is } f\text{-abnorma}\}$. Suppose that the formation function f satisfies $S_n f(p) = f(p)$ (that is, $f(p)$ is closed under taking the normal subgroups) for all $p \in P$ (the set of prime numbers), and that the set defined by $\pi = \{p \in P; f(p) \neq \phi\}$ is non-empty. Then*

(1) $O_\pi(\Phi f(G)) \in LE(f)$, and

(2) $\Phi f(G/O\pi(\phi f(G)) = \phi(G/O\pi(\phi f(G))) \leq O\pi'(G/O\pi(\phi f(G)))$. [5]

Theorem 4   *Let G be a group and $\Phi p(G) = \cap\{H; H$ is a maximal subgroup with pd-complement chief factor$\}$. Then*

(1) $\Phi p(G) \geq Op'(G)$, and

(2) $\Phi p(G)/Op'(G) = \Phi(G/Op'(G))$. [3]

[4] discusses the operator on a formation F to derive an equivalent condition for F being a saturated formation, and using the result to the solvable groups of rank ≤ 2 to prove that $H_1(G, V) = 0$ for all irreducible module V with dimension > 2 if and only if G is solvable and the rank of $G/\Phi(G)$ is less than 2.

Theorem 5   *Assume that F contains the class of nilpotent groups, and that the group G has a normal subgroup N such that G/N inF. If $N_G(P)/C_G(P)$ is a p-group for every prime $p \mid |G|$ and every Sylow p-subgroup P of G, then $G \in F$.* [1]

Theorem 6   *Assume that G has the Sylow tower property and G has a normal subgroup such that $G/N \in F$. If $N_G(P)/C_G(P) \in f(p)$ for every $p \mid |G|$ and every Sylow p-subgroup P of G, then $G \in F$.* [1]

Theorem 7   *Assume that $G(m) > G(n)$ are two terms of the series of derived groups of G and $G(m)$ is p-nilpotent. If $G/G(m) \in F_p$, then $G \in F_p$.* [1]

References

[1] Z. M. Chen, *A way extending theorem in group theory*, (in Chinese) Chin. Sci. Bull., 38:11 (1993), 884–887.

[2] Y. Fan, *The F-stability and the F-criticality*, (in Chinese) Acta Math. Sin., 29:1 (1986), 117–126.

[3] Y. Fan, *On three problems of maximal subgroups*, (in Chinese) Acta Math. Sin., 29:5 (1986), 628–631.

[4] Y. Fan, *Centralities, normalities and the properties of cohomology concerning formations*, (in Chinese) Chin. Sci. Bull., 11 (1987), 1130–1138.

[5] Y. Q. Teng and B. L. Zhang, *Frattini subgroups relating to formation functions*, J. of Pure and Applied Alg., 64 (1990).

VI. The infinite groups

A few years ago, some algebraists in P. R. China began to consider a number of questions related to infinite groups and have got a series of results. Some of these results are similar to that in finite group theory, but some are different and special. We list parts of them for indicating the several aspects.

1. Minimal non-P-groups and just non-P-groups

Let P be a group theoretical property and G be a group. If all its proper subgroups have the property P, but G itself does not, then G is said to be a minimal non-P-group (or an inner-Σ-group as in the finite case); if all its proper quotients have the property P, but G itself does not, then G is called a just non-P-group (or an out-Σ-group as above). The structure of minimal non-P-groups and just non-P-groups has been studied by several authors for a variety of properties P. One motivation for such investigations is that an understanding of the structure of groups that "minimal fail" or "just fail" to have a property can provide insight into the nature of the property itself. Another reason is that interesting problems about modules or other algebraic systems are sometimes involved.

The following types of minimal non-P-groups are investigated.

Theorem 1.1 [1]   *Let G be an abelian group.*

(1) *If G is a torsion-free minimal non-P-group where property P is preserved under isomorphisms, then G is a direct sum of isomorphic copies of Q;*

(2) *If G is directly indecomposable minimal non-P-group, then G is isomorphic with Q or $Z_p\infty$.*

Theorem 1.2   [1] *Let G be an Abelian group. Then the following statements about G are equivalent:*

(1) *G is a minimal non-(being a direct product of cyclic groups)-group;*

(2) *G is a minimal non-supersolvable-group; and*

(3) *G is isomorphic with $Z_p\infty$.*

Theorem 1.3   [1] *Let G be a solvable group. Then G is a minimal non-(finitely generated)-group if and only if G is isomorphic with $Z_p\infty$.*

Theorem 1.4   [1] *Let G be an infinite group with $G' < G$. Then the following statements about G are equivalent:*

(1) *G is a minimal non-finite-group;*

(2) *G is a minimal non-cyclic-group; and*

(3) *G is isomorphic with $Z_p\infty$.*

How to describe and construct just non-P-groups for certain kinds of properties P seems to be an interesting problem. Of course a simple group which is not a P-group is a just non-P-group. In order to avoid consideration of arbitrary simple groups most studies have dealt only with just non-P-groups having a non-trivial Abelian normal subgroup, i.e., nontrivial Fitting subgroup.

JNFA-groups

Groups that are just not finite-by-Abelian are called $JNFA$-groups. They were studied in a joint work of D. J. S. Robinson and Z. R. Zhang (see [10]). There is a natural dichotomy of $JNFA$-groups with non-trivial Fitting subgroup into those with trivial centre and those with non-trivial centre. The two types exhibit markedly different behaviour.

The following result provides a characterization of $JNFA$-groups with non-trivial centre.

Theorem 1.5 [10] (i) *If G is a $JNFA$-group with non-trivial centre, then G is torsion-free nilpotent of class 2. Furthermore, G' is infinite cyclic and $Z(G)$ is isomorphic with a subgroup of the additive group of rational numbers.*

(ii) *Conversely, a group G with the structure described in (i) is a $JNFA$-group with non-trivial centre.*

Proposition 1.6 [10] *Let Q be a non-trivial torsion-free Abelian group and let C be any non-trivial subgroup of the rationals. Then there is a $JNFA$-group with centre C and central quotient group Q if and only if Q admits a non-degenerate alternating (bilinear) form $\langle\ \rangle : Q \times Q \to Z$.*

It appears to be difficult to decide in general which Abelian group Q admits such a form, but a partial answer is given as following.

Theorem 1.7 [10] (i) *If Q is a non-trivial abelian group admitting a non-degenerate alternating form $\langle\ \rangle : Q \times Q \to Z$, then Q is isomorphic with a subgroup of a product of $|Q|$ copies of Z;*

(ii) *Let Λ be any non-empty set. For each λ in Λ let $\langle a_\lambda \rangle$ and $\langle b_\lambda \rangle$ be infinite cyclic groups. Choose a subgroup S_1 of $\prod_\lambda \langle a_\lambda \rangle$ containing $\coprod_\lambda \langle a_\lambda \rangle$ and put $S_2 = \coprod_\lambda \langle b_\lambda \rangle$. Then $Q = S_1 \oplus S_2$ has a non-degenerate alternating form $\langle\ \rangle : Q \times Q \to Z$;*

(iii) *A countable torsion-free Abelian group Q has a non-degenerate alternating form $\langle\ \rangle : Q \times Q \to Z$ if and only if it is free Abelian with rank being not equal to an odd integer.*

The following is fundamental to the theory of $JNFA$-group with trivial centre and non-trivial Fitting subgroup.

Theorem 1.8 [10] (i) *Let G be a $JNFA$-group with trivial centre and non-trivial Fitting subgroup A. Then every non-trivial G-admissible subgroup A_0 of A is a faithful just infinite module for the non-trivial finite-by-*

Abelian group G/A (By a just infinite module we mean an infinite module whose proper quotient modules are all finite);

(ii) *Conversely, let A be a faithful just infinite module for a non-trivial finite-by-Abelian group Q. Then every extension of A by Q inducing the given module structure is a $JNFA$-group with trivial centre and Fitting subgroup being equal to A.*

Theorem 1.9 [10] *Let G be a $JNFA$-group with Fitting subgroup F. Then there is a subgroup X such that $X \cap F = 1$ and $|G : XF|$ is finite.*

Via Theorem 1.9, it is easy to see that the study of just infinite modules is central to the theory of $JNFA$-groups. A reduction techuique (see [10, (3.7)]) reveals that not too much is lost if we restrict attention to such modules over nilpotent groups.

Theorem 1.10 [10] *Let Q be a non-trivial nilpotent group with finite derived subgroup, and let K be the centre of Q. Then Q has a faithful just infinite module of characteristic $p \geq 0$ if and only if (I) the torsion-subgroup of K is a locally cyclic p'-group, and one of the following is true:*

(II) *the torsion-free rank of K is infinite;*

(III) *Q is finitely generated and infinite;*

(IV) *Q is an infinite torsion group and $p > 0$; or*

(V) *Q is finite and $p = 0$.*

We can also say when a simple module of this type exists. A special case of interest is when Q is abelian.

Theorem 1.11 [10] (i) *Let R be an infinite Noetherian domain in which every non-zero prime ideal has finite index. Let Q be a subgroup of the group of units of R that generates R as a ring and view R as a Q-module via the ring multiplication. Then R is a faithful just infinite Q-module;*

(ii) *Conversely, if Q is any abelian group with a faithful just infinite module, then all such modules arise as finite essential extensions of modules of the type described in (i).*

Relating to the study of $JNFA$-groups, we have studied groups that are just not centre-by-finite ($JNCF$-groups). Since a centre-by-finite group has finite derived subgroup, a $JNCF$-group is either finite-by-Abelian or a $JNFA$-group. It turns that the $JNCF$-groups with finite derived subgroup occur as subgroups of wreath products of nilpotent just non-Abelian p-groups and finite groups. Then, by applying the results on $JNFA$-groups, the characterization of $JNCF$-groups with non-trivial

Fitting subgroup that are not finite-by-Abelian is obtained.

$JNFN_c$-groups

A generalization of $JNFA$-groups is the $JNFN_c$-groups.

By a group G being an FN_c-group if it is finite-by-nilpotent of class $\leq c$ (or finite-by-N_c). A group is called a $JNFN_c$-group if all its proper quotients are FN_c-groups, but G itself is not. Taking $c = 1$ we get $JNFA$-groups.

We have studied $JNFN_c$-groups with non-trivial Fitting subgroup for arbitrary c, and naturally divided them into those with non-trivial centre and those with trivial centre [21].

It turns out that $JNFN_c$-groups with non-trivial centre are just the torsion-free nilpotent groups of class $c + 1$ such that $\gamma_{c+1}G$ is infinite cyclic and $Z(G)$ is a subgroup of rationals. For a $JNFN_c$-group G with trivial centre and non-trivial Fitting subgroup A, it is shown that A is Abelian and then A is a faithful just infinite module for the non-trivial FN_c-group $Q =: G/A$. The investigation also shows that it is significantly harder to describe $JNFN_c$-groups with $c \geq 2$ than that for $c = 1$ in [10]; some interesting but complicated problems about groups, modules and cohomology and many new techniques are involved.

In [21], faithful just infinite modules over nilpotent FN_c-groups are meanwhile considered. For this purpose a well-know theorem of P. Hall is extended in the following form: if G is an FN_c-group, then $G/C_G(\gamma_c G)$ is residually finite and of finite exponent. Moreover, $G/Z_{2c-2}(G)$ is of finite exponent, provided $c \geq 2$. Also a result on faithful just infinite modules is given: let Q be a nilpotent group with centre K. If Q has infinite torsion-free rank and the torsion-subgroup of K is locally cyclic p'-group ($p \geq 0$), then Q has a faithful just infinite module of characteristic p that is simple. By using these results, necessary and sufficient conditions are obtained for a non-trivial nilpotent FN_c-group Q to have a faithful just infinite module or a simple faithful just infinite module. These conclusions are markedly different from those of $JNFA$-groups.

JN min FO-groups

Groups that possess only a finite number of elements of each order, including ∞ , are said to be FO-groups. An FO-group with min (or a min FO-group) is just an extension of a finite group by direct product of finitely many quasicyclic groups. By using the results on $JNFA$-groups, groups that are just not min FO-groups (JN min FO-groups) with non-trivial Fitting subgroup are completely described.

Theorem 1.12   [19] *Let G be an Abelian group.*

(a) *If G is a JN min FO-group, then G is an essential extension of an infinite cyclic group by direct product of finitely many quasicyclic groups and cyclic groups of prime-power order;*

(b) *Conversely, a group G with the structure described in (a) is a JN min FO-group.*

Theorem 1.13   [19] (a) *Let G be a JN min FO-group. Then G is not Abelian if and only if the centre $Z(G)$ of G is trivial;*

(b) *Let G be a non-Abelian JN min FO-group with non-trivial Fitting subgroup A. Then A is a faithful just infinite module of characteristic 0 for the non-trivial finite group G/A, or a simple faithful just infinite module of characteristic p (p being a prime) for the infinite min FO-group G/A;*

(c) *Conversely, let A be a faithful just infinite module of characteristic 0 for a non-trivial finite group Q, or a simple faithful just infinite module of characteristic p (p being a prime) for an infinite min FO-group Q. Then every extension of A by Q inducing the given module structure is a non-Abelian JN min FO-group with Fitting subgroup being equal to A.*

2. The P-radical of a group and its applications

Let P be a group theoretical property. A group is a P-group if it has P. Denote the lattice which consists of all normal subgroups of a group G by $L(G)$. A member of $L(G)$ is P-element if it has P. If P satisfies the following conditions:

(a) *an image of a P-group is a P-group,*

(b) *in the lattice $L(G)$ induced by any group G, there exists a P-element N, which includes each P-element of $L(G)$, and*

(c) *the quotient lattice $L(G)/N$ contains no non-trivial P-elements,*

then P is the radical property of groups and N is the P-radical of G (or of $L(G)$). And if $N = 0$, then G is P-semisimple. For avoiding confusion, the radical of G is signified by $N_P(G)$.

By studying the properties of the radical of groups of centain classes, these groups may be described to some extent. This method is often used in the theory of rings. For describing P-semisimple the following notations are needed. A normal subgroup Q of a group G is said to be a prime normal subgroup of G if for any normal subgroups A and B of G, $[A,B] \leq Q$ implies that $A \leq Q$ or $B \leq Q$. A group G is called a prime group if $\{e\}$ is a prime normal subgroup of G with e the unity of G.

Solvable-semisimple groups

The solvable-radical of groups is first considered.

Theorem 2.1 [14] *Suppose that G is a solvable-semisimple group. Then the intersection of all prime normal subgroup of G is trivial.*

Theorem 2.2 [14] *A group G is solvable-semisimple if and only if G is the subdirect product of some prime groups.*

Theorem 2.3 [14] *Let G be a group with max-n, the maximal condition on normal subgroups. Then*

(i) *the solvable-radical $N_S(G)$ exists, and*

(ii) *$N_S(G)$ is just the intersection of all prime normal subgroups of G.*

Perfect subgroups and hypoabelian groups

A group D is said to be a perfect group if $D = [D, D]$.

Theorem 2.4 [17] (i) *In any group G there exists a unique maximal perfect (normal) subgroup $D(G)$ (called the perfect-radical of G) containing every perfect (normal) subgroup of G. Therefore $D(G)$ is a characteristic subgroup of G, and*

(ii) *$D(G/D(G))$ is trivial.*

As an application of the perfect-radical, hypoabelian groups are described.

Theorem 2.5 [17] *In any group G, the perfect-radical $D(G)$ and the transfinite derived subgroup coincide. Therefore, G is hypoabelian if and only if its perfect-radical $D(G)$ is trivial, that is, there is no non-trivial perfect subgroup in G.*

Theorem 2.6 [17] *Let $\{A_i; i \in I\}$ denote the set of all normal subgroups of G such that G/A_i is hypoabelian. Then the perfect-radical $D(G) = \cap_{i \in I} A_i$.*

Theorem 2.7 [17] *Suppose that $G = A \times B$. Then $D(G) = D(A) \times D(B)$. Therefore, G is hypoabelian if and only if A and B are hypoabelian.*

K-subgroups and hypocentral groups

A subgroup H of a group G is said to be a K-subgroup of G if $H = [H, G]$.

Theorem 2.8 [18] *In any group G, there exists a unique maximal K-subgroup $K(G)$ (called the K-radical) containing all K-subgroups of G. Therefore, $K(G)$ is a characteristic subgroup of G. Furthermore, $K(G/K(G))$ is trivial.*

It should be mentioned that the results of Theorem 2.8 are given in the case that G has min, the minimal condition on subgroups, in [22].

Theorem 2.9   *In any group G, the K-radical $K(G)$ and the hypocentre coincide. Therefore, G is hypocentral if and only if G has no non-trivial K-subgroup.*

Further applications of the radical of groups will occur in the next section.

3. FC-groups

Groups with finite conjugacy classes are called FC-groups. They have been well studied.

The decomposition of local principal factor blocks in periodic FC-groups and its application

Let $\sigma(G)$ be the set of all prime factors of the orders of elements in a periodic FC-group G. Let F be a finite normal subgroup of G. Then a factor of any G-admissible series of F is said to be a local principal factor of G.

For any $p, q \in \sigma(G)$ with $p \neq q$, p and q are adjacent if some local pricipal factor of G can be divided by the product pq of p and q; p is adjacent to itself if some local principal factor can be divided by p. We define a equivalence relation "$\sim$" among the set $\sigma(G)$ in the following way: $p \sim q$ holds if and only if there exists k primes of $\sigma(G)$ $p_1 = p$, $p_2, \cdots, p_i, \cdots, p_{k-1}$ and $p_k = q$ with k a positive integer such that, for $i = 1, \cdots, k-1, p_i$ and p_{i+1} are adjacent. Then $\sigma(G)$ is the disjoin union

$$\sigma(G) = \bigcup_{i \in I} \pi_i \qquad (*)$$

of equivalence classes π_i, where I is some index set. π_i is called a local principal factor block of G and the formula $(*)$ is called a decomposition of local principal factor blocks of G.

Of course, in finite groups local principal factor blocks are just principal factor blocks. From the related results in finite groups, a local conjugacy theorem of periodic FC-group is obtained.

Theorem 3.1   [16] *Suppose that π is a local principal factor block or union of local principal factor blocks of a periodic FC-group. Then any two Sylow π-subgroups of G are locally conjugate in G. Therefore, they are isomorphic.*

Via Theorem 3.1, Sylow theory of periodic FC-groups (see, for example, [11]) may be extended to the local principal factor blocks case without any difficulty.

Equivalence among some kinds of generalized nilpotencies in FC-groups

By the properties of the K-radical of groups in Theorem 2.8 and Theorem 2.9 we can discuss equivalence among some kinds of generalized nilpotencies in some special types of FC-groups.

Theorem 3.2 [18] *Let G be a BFC-group, a group with boundedly finite conjugacy classes. Then G is hypocentral if and only if it is locally nilpotent.*

Theorem 3.3 [18] *Let G be an X-group, a group with finite classes of conjugate subgroups. Then the following statements about G are equivalent:*

(a) *G has only normal maximal subgroups,*

(b) *G is locally nilpotent,*

(c) *G satisfies the normalizer condition,*

(d) *G is a hypercentral group,*

(e) *G is a hypocentral group, and*

(f) *G has no non-trivial K-subgroups.*

Remark   Via a well-known result of B. H. Neumann, X-groups are just centre-by-finite groups.

The locally solvable redical of FC-groups and its application

By studying properties of the locally solvable radical of FC-groups, a structure description is obtained.

Theorem 3.4 [20] *In any FC-group G there is a unique maximal normal locally solvable subgroup $N_{LS}(G)$ (called the LS-radical) including all normal locally solvable subgroup of G such that $G/N_{LS}(G)$ is (locally solvable)-semisimple.*

Theorem 3.5 [20] *Let G be an FC-group. Then G is (locally solvable)-semisimple if and only if G is a subcartesian product of some primary FC-groups.*

Summing up the results of these two theorems, we have:

Theorem 3.6 [20] *In any FC-group G, the locally solvable radical exists.*

Therefore, any FC-group is an extension of a locally solvable group by a subcartesian product of some primary FC-groups.

4. Other results on infinite groups

A generalization of the Burnside Basis Theorem

The Burnside Basis Theorem in finite p-groups is an elementary result of the theory of p-groups. By studying the properties of the Frattini sub-group $\Phi(P)$ in p-group P of the following type, this theorem is extended. Let G be a group. If, for any proper set H of elements in G, there exists a maximal subgroup M of G such that $H \leq M < G$, then G is said to satisfy the weakly maximal condition.

Theorem 4.1 [15] *Let P be a non-trivial p-group satisfying the weakly maximal condition. Denote $D = \Phi(P)$. If P has only normal maximal subgroups, then*

(a) *$\overline{P} := P/D$ is an elementary Abelian group,*

(b) *if $\overline{S}$ is a basis of $\overline{P}$, then S is a minimal set of generators of P,*

(c) *conversely, if S is a minimal set of generators of P, then $\overline{S}$ is a basis of $\overline{P}$, and*

(d) *from a set T of generators, a minimal set S of generators may be selected such that $S \leq T$.*

Normal series of D_2-groups

A group G is said to be a D_2-group if G has a normal series of finite length in which each factor is an Abelian group generated by at most 2 elements.

Theorem 4.2 [7] *If G is a D_2-group, then*

(1) *G has a normal series*

$$1 = G_0 < G_1 < \cdots < G_s < G_{s+1} < \cdots < G_{s+f} < \cdots < G_n = G,$$

in which each factor is either an elementary Abelian p-group (p may be different) of rank ≤ 2 or a free Abelian group of rank ≤ 2; further, if $i < s$ then G_{i+1}/G_i is a p_i-group and $p_{i+1} > p_i > 3$; if $s \leq i < s + f$ then G_{i+1}/G_i is a free group; and if $s + f \leq i < n$ then G_{i+1}/G_i is either a 2-group or a 3-group.

(2) *G has a normal series*

$$1 = G_0 < G_1 < \cdots < G_f < G_{f+1} < \cdots < G_{f+s} \leq G,$$

in which the factors below G_f are all free Abelian groups of rank ≤ 2, G_{f+s}/G_f is a finite group possessing a Sylow tower, and

$$G/G_{f+s} \cong (A_4 \times, \cdots, \times A_4) \times E,$$

where A_4 is the alternating group of degree 4 and E is an elementary Abelian 2-group which acts faithfully on $A_4 \times \cdots \times A_4$ and each A_4 is E-invariant (maybe $G/G_{f+s} = 1$).

Corollary 4.3 [7] *All $\{2,3\}'$-elements of a D_2-group G form a torsion subgroup of G.*

Corollary 4.4 [7] *A D_2-group G has a Sylow tower if and only if G is independent of A_4.*

SD_2 -groups

Inducing from the study of D_2-groups, one inverstigates the structure of SD_2-groups, which are the groups whose proper subgroups are all generated by at most 2 elements, and has got the following

Theorem 4.5 [9] *If G is an infinite solvable SD_2-group, then*

(1) *G is not finitely generated if and only if $G \cong Z_p\infty$, or*

(2) *G is finitly generated if and only if G is generated by at most 2 elements. Further, the Hirsch length $l(G)$ of G is either 1 or 2, and if $l(G) = 1$ then G is supersolvable as well as metabelian with G being an extension of $Z \times T$ by a group of order at most 2, where T is a finite group all of whose Sylow-subgroups are cyclic; if $l(G) = 2$, then G is the natural extension of the free abelian group $Z + Z$ by a finite subgroup of $GL(2, Z)$.*

Weakly supersolvable groups

By a group G being weakly supersolvable if G has a normal series

$$1 = G_0 \leq G_1 \leq \cdots \leq G_t = G$$

in which each subgroup of the factor G_{i+1}/G_i is G-invariant under the natural conjugate action of G on G_{i+1}/G_i, $i = 0, 1, \cdots, t$.

It is clear that nilpotent groups and supersolvable groups are all weakly supersolvable.

The following three results are fundamental.

Theorem 4.6 [8] *A weakly supersolvable group is solvable.*

Theorem 4.7 [8] *The class of weakly supersolvable groups is closed under forming subgroups, factor-groups, and direct product. Further, if G/M and G/N are both weakly supersolvable, then so is $G/M \cap N$.*

Theorem 4.8 [8] *Any pricipal factor of a weakly supersolvably group is of prime order while any maximal subgroup of a weakly supersolvable group has a prime index.*

Some deeper results on weakly supersolvable groups are also proved.

Theorem 4.9 [8] *For any prime $p > 2$, all of the π_p-elements of a weakly supersolvable group G form a cheracteristic subgroup in G, where πp is a set of primes not less than p.*

Theorem 4.10 [8] *If G is a weakly supersolvably group, then the following conditions are equivalent:*

(1) *G is supersolvable,*

(2) *G is finitely generated,*

(3) *G satisfies the maximal condition on subgroups, and*

(4) *G satisfies the maximal condition on normal subgroups.*

Theorem 4.11 [8] *Suppose G is a Cernikov group, then G is weakly supersolvably if and only if each finite subgroup of G is supersolvable and P is nilpotent for any p-subgroup P of G, where p is a prime.*

AT-groups and IT-groups

Let G be a group. G is called an *AT*-group (or an *IT*-group) if each abelian (or infinite) subgroup of G is normal.

Theorem 4.12 [12] *Suppose G is either an AT-group or an IT-group. Then the following are equivalent:*

(1) *G is solvable,*

(2) *$C_G(F) = \xi(F)$, where F is the Fitting subgroup of G,*

(3) *$C_G(F)$ is solvable,*

(4) *there exists a solvable subnormal subgroup H such that $C_G(H)$ is solvable, and*

(5) *G is a hypercyclic group.*

Theorem 4.13 [12] *Finitely generated solvable IT-groups are either finite or Abelian.*

Theorem 4.14 [12] *Finitely generated solvable AT-groups are either finite, abelian, or non-torsion and non-abelian with the centralizer of the derived subgroup in the group being also non-torsion.*

The decomposition of modules over hyperfinite groups

Let G be a group, A a ZG-module. If $A = A^f + A^{\overline{f}}$ in which each irreducible ZG-factor of the ZG-submodule A^f of A as an abelian group is finite while the ZG-submodule $A^{\overline{f}}$ of A has no factors being such type, then A is said to have an f-decomposition. In 1986, D. I. Zaitsev proved that: if G is a hyperfinite locally solvable group, then any artinian ZG-module A has an f-decomposition [13]. Dual to this result, we have

Theorem 4.15 [2] *If G is a hyperfinite locally solvably group, then any noetherian ZG-module A has an f-decomposition.*

Further, by the joint work of Z. Y. Duan and M. J. Tomkinson, there is

Theorem 4.16 [3] *If G is a hyperfinite locally solvable group, then any minimax ZG-module A has an f-decomposition.*

For the structure of A^f and $A^{\overline{f}}$, we have

Theorem 4.17 [4] *If G is a hyperfinite locally solvable group and A is a noetherian ZG-module, then A^f as a group is finitely generated and $|G/C_G(A^f)| < \infty$.*

Theorem 4.18 [4] *If G is a periodic Abelian group with $\pi(G)$ being finite, where $\pi(G) = \{p; p$ is a prime and, for some $x \in G$, x is of order $p\}$, and if A is a Noetherian ZG-module. Then A^f as an Abelian group is torsion and has a finite exponent as well as a ZG-composition series of finite length.*

Extensions of Abelian by hyper-(cyclic or finite) groups

Suppose A is a normal subgroup of a group E and B is a subgroup of A. E is said to split conjugately over A modulo B if $E = AE_1$ with $A \cap E_1 = B$ for some subgroup E_1 of E and all such E_1 are conjugate in E modulo B.

Theorem 4.19 [5] *If G is a hyper-(cyclic or finite) locally solvable group and A a periodic artinian ZG-module, then any extension E of A by G splits conjugately over A modulo A^f.*

Theorem 4.20 [6] *If G is a hyper-(cyclic or finite) locally solvable group and A a Noetherian ZG-module, then any extension E of A by G splits conjugately over A modulo A^f.*

References

[1] Z. M. Chen and Y. Xiao, *Infinite minimal non-σ-groups and the group Z_{p^∞}*, (in Chinese) J. of Southwest-China Teachers College (Natural Science Edition), No. 2 (1979).

[2] Z. Y. Duan, *Noetherian modules over hyperfinite groups*, Ph. D. thesis, Univ. of Glasgow, (1991).

[3] Z. Y. Duan and M. J. Tomkinson, *The decomposition of minimax modules over hyperfinite groups*, Arch. Math., 61 (1993) 340–343.

[4] Z. Y. Duan, *The structure of noetherian modules over hyperfinite groups*, Math. Proc. Cambri. Phil. Soc., 112 (1992); 21–28.

[5] Z. Y. Duan, *Extensions of abelian by hyper-(cyclic or finite) groups I*, Comm. Alg., 20:8 (1992), 2305–2321.

[6] Z. Y. Duan, *Extensions of abelian by hyper-(cyclic or finite) groups II*, Rend. Sem. Mat. Univ. Padova, 89 (1993), 113–126.

[7] Y. Fan, *On the normal series of D_2-groups*, (in Chinese) Chinese Annals of Mathematics, 8A:5 (1987), 626–631.

[8] Y. Fan, *Weakly supersolvable groups*, (in Chinese) Chinese Annals of Mathematics, 9A:6 (1988), 711–723.

[9] H. G. Liu, *Solvable SD_2-groups of infinite order*, (in Chinese) Acta Math. Sinica, 35:3 (1992), 339–349.

[10] D. J. S. Robinson and Z. R. Zhang, *Groups whose proper Quotients have finite derived subgroups*, J. of Alg., 118:2 (1988), 346–368.

[11] M. J. Tomkinson, *FC-groups, Pitman publishing Limited*, London, 1984.

[12] M. Q. Xu, *Solvability of AT-groups and infinite IT-groups*, (in Chinese) Acta Math. Sinica, 31:5 (1988), 663–670.

[13] D. I. Zaitsev, *Splitting of extensions of abelian groups*, AN USSR. Inst. Mat., Kiev. (1986), 21–31.

[14] Z. R. Zhang, *A structure theorem of solvable-semisimple groups*, (in Chinese) Journal of Southwest-China Teachers College (Natural Science Edition), No. 2, (1985).

[15] Z. R. Zhang, *A generalization of the Burnside Basis Theorem*, (in Chinese) Journal of Mathematics, 7:1 (1987), 70–74.

[16] Z. R. Zhang, *The decomposition of local principal factor blocks in periodic FC-groups and its applications*, (in Chinese) Mathematics in Practice and Theory, 3 (1987), 82–86.

[17] Z. R. Zhang, *D-subgroups and hypoabelian groups*, (in Chinese) Journal of Chengdu Institute of Meteorology, 2 (1988), 84–88.

[18] Z. R. Zhang, *On equivalence relations among some kinds of generalized nilpotencies in X-groups*, (in Chinese) Journal of Southwest-China Teachers University (Natural Science Edition), 14:3 (1989), 14–20.

[19] Z. R. Zhang, *Groups whose proper quotients are FO-groups with min*, (in Chinese) Acta Math. Sinica, 33:3 (1990), 344–347.

[20] Z. R. Zhang, *A structure theorem of FC-groups*, (in Chinese) Acta Math. Sinica, 34:3 (1991), 324–328.

[21] Z. R. Zhang, *Groups whose proper quotients are finite-by-nilpotent*, (in Chinese) Archiv der Mathematik, 57 (1991), 521–530.

[22] Y. l. Zheng, *K-subgroups and its application*, (in Chinese) Journal of Wuhan University, (in Chinese) Special Issue for Mathematics II, (1989), 99–103.

Several Questions About Finite Group Representations and Their Applications *

Fan Yun

Department of Mathematics, Hubei University

Wuhan, 430062

The research of finite group representations and their applications have a long history and are a vast area. As early as the beginning of this century, the systematic theory of finite group representations has been founded by Burnside, Frobenius, Schur and others, the core of the theory is the characters. Not long afterwards, it had been well-known that this theory is essetially still valid for the case when the representation fields, i.e. the coefficient fields, are algebraicly closed and of characteristic prime to the order of the groups. However, the situations are far more different and becoming very complicated when one considers the representations of a finite group over a field of characteristic p which divides the order of the group. From the point of view of the algebraic structures, in the former case the group algebras are semi-simple, in the later case, whereas, the group algebras have non-zero radicals. In the thirties of this century R. Brauer connected the case of characteristic p and the case of characteristic 0 with so-called p-modular systems, and created a series of theories of Brauer characters, block theory etc. This is just the so-called modular representation theory of finite groups. For the above historical comments please see [39] and [30].

Since the 1970s, several ideas, methods and theories such as local representations, structural researches were further developed, and are developing, please consult [1], [33], [6] etc.

In this paper we will state several quetions which we were and are studying. We always assume that F is a field of characteristic p where p is a prime number, and G is a finite group, and all algebras and modules are finite dimensional.

---

* Partially supported by NSFC

1. The change of ground-fields and the actions of Galois groups

The modular theory was developed mainly in the case of splitting fields. It has been shown by an important work due to R. Brauer that the ground field F is splitting for the group G if F contains a $(|G|_{p'})'th$ primitive root of unity. In this case there are many beautiful theories. Some of the theories appear nearly the same in the non-splitting case, but some of them appear not. In order to avoid the troubles of small ground-fields many authors often assume that the ground-field is "large enough", see [16, §17].

About the representations over a field not large enough, there have been some important knowleges such as the theory of Schur index, the Witt-Berman's thoerem on the number of irreducible (ordinary, modular) characters([16]), in which the action of a suitable Galois group has been involved. Later, M. Broué developed the ideas on the action of a Galois group in an unpublished paper [8].

Our primitive observation for the question is that the formulae of the change of ground-field about an algebra and its modules are dual to the Mackey's Subgroup Theorem when the Galois Group is considered (see [18]).

Theorem. *Let K and L be two finite extensions of the field F with K separable, and let N be the nomal closure of L and K, and G be the Galois group of N over F. Assume A is a K-algebra and V is a righgt A-module. Then*

(i) $(A \downarrow_F) \uparrow^L \cong \bigoplus_{\alpha \in G_K \backslash G/G_L} ((A^\alpha)^{K^\alpha L})_L;$

(ii) $(V \downarrow_F) \uparrow^L \cong \bigoplus_{\alpha \in G_K \backslash G/G_L} ((V^\alpha)^{K^\alpha L})_L,$ *where* $((V^\alpha)^{K^\alpha L})_L$ *is an*

$((A^\alpha)^{K^\alpha L})_L$-*module.*

Such understanding provides us an efficient research angle for some more questions.

The first question is about the Cartan invariants over arbitrary ground-fields. We extended the results of [32] by P. Landrock, see [27].

Next we found that the Cartan invariants over arbitrary ground-fields is in deed dependent on the Galois group $\Gamma = \text{Gal}(F(\varepsilon)/F)$, where ε is a $(|G|_{p'})'th$ primitive root of unity (after that I was informed that the result had been obtained in Broué's work [8]). Furthermore, we observed that the Scott coefficients(lower defect multiplicities) and related invariants and other block invariants are all determined by the action of $G \times \Gamma$ on G. Of course, some modifications of related concepts should be made with the action of the Galois group. In this way we gave reasonable connections of the several kinds of invariants and showed a structural understanding of them, see [20]. For example, for the modified multiplicity $m_{FG,B}^{(u)}(P, B_P)$ we showed

Theorem. *If $P = D(B_P)$ is a defect group of B_P where (P, B_P) is a B-Brauer pair, then*

$$m_{FG,B}^{(u)}(P, B_P) = |\mathrm{Cl}(G_u \cap Z(P)/N(B_P) \times \Gamma)|$$

If F is splitting for G, then the action of Γ on G is trivial and all the results return to the same as those in the case of large enough fields, including the work [14] by D. Burry, [9] by M. Broué, and [12] by M. Broué and J.B. Olsson.

In addition, it was succeeded to use the ideas on change of ground-fields and the action of the Galois groups to the study on permutation modules and p-permutation modules in the Green rings of finite groups.

The p-permutation modules are also called trivial source modules. The notation comes from an observation in [11] due to M. Broué: an FG-module V is a trivial source module if and only if its restriction to every p-subgroup is a permutation module, this is a natural result.

For the Green algebra of a group G, the subalgebra generated by all p-permutation modules plays an important role, see [6]. However, several structures of the Green algebra described by the p-permutation modules are in fact determined just by the subalgebra generated by the permutation modules, e.g. see [40]. So a natural question is: how far are the two subalgebras different from each other? It is well-known that the p-permutation modules are determined by the Conlon species of the Green algebra, this is just like that the irreducible modules are determined by their Brauer characters; and the Conlon species coincide with the values on conjugacy classes of irreducible Brauer characters if $p = \mathrm{char} F \dagger |G|$, see [6]. An observation for the values of Conlon species suggested us to guess: it is reasonable that the subalgebra generated by permutaion modules is just the fixed point set of the Galois group $T = \mathrm{Gal}(\mathbf{Q}(\varepsilon)/\mathbf{Q})$ acting suitablly on the subalgebra generated by p-permutaiton modules, where $\mathbf{Q}$ is the rational field and ε is a $(|G|_{p'})'th$ primitive root of unity. This asertion was proved in [22]. The action of T on the subalgebra generated by p-permutation modules is defined by means of the action of T on the Conlon spscies, and we proved in fact the following result on integral Green ring $a(FG, Q)$ of Q-projectives:

Theorem. *Notations are as above. Then*

$$|G| \cdot a(FG, Triv, Q)^T \subseteq a(FG, Perm, Q) \subseteq a(FG, Triv, Q)^T$$

This implies that the above guess is true if $|G|$ is an invertible element in the coefficient ring. As a byproduct we extended the Berman-Witt Theorem on irreducible characters to the Conlon species([22]).

Another byproduct is that the dimensions of the subspaces of fixed points of cyclic subgroups of G can partially determine the subalgebra generated by permutation modules, see [21].

2. The change of subgroups and the local structural methods for blocks

One of the most important notions due to Brauer is the blocks. From the point of view of algebraic structures, a block ideal and its block idempotent of an algebra are just an indecomposable algebraic summand of the algebra and its central idempotent. The block defined by Brauer is the set of all characters belonging to the block ideal, or more widely, of all modules belonging to the block ideal. It is natural to reduce the study of the representations of the algebra to the study of its blocks. The distinguishing features of the group representations are the change of subgroups and the character theory. In the following we abbreviate "block ideal" to "block" too for convenience.

The most important one of the changes of subgroups is the change from a group to its local subgroups. Brauer introduced the Brauer's map, and defined the Brauer correspondence of blocks from subgroups to the group in terms of central characters, and founded three basic Main Theorems on blocks, see [30]. Later on, Broué and Puig extended the Brauer's map to a very general situation, and stressed the role of its dual, i.e. the trace map. This is still one of the most useful ideas and tools upto now.

To the end of the seventies, from the point of view of module theory, J.L. Alperin and D.W. Burry introduced the concept of block covers and module covers with which another Brauer correspondence of blocks is defined, and they showed that this Bauer correspondence coincides with the Brauer correspondence in the sense of Brauer in a classical case, cf. [4]. It was known soon after that the two Brauer correspondences are in fact the same provided they are both defined ([41]). However, it is easy to show examples in which they are not both defined.

The first interest to us is that there may be some very close relationships of block covers and the corresponding module covers. In the case that the subgroup contains the center of the defect group of the block we got the connection soon in [19]. It was more fascinating that the following is true.

Theorem. *Let B and $\tilde{B}$ be an FG- and FH-block resp., where $H \leq G$, and let $\tilde{D}$ be a defect group of $\tilde{B}$. Then $\tilde{B}$ is covered by B, if and only if for every $\tilde{B}$-module U there is a B-module which covers U, if and only if an indecomposable $\tilde{B}$-module U of vertex $\tilde{D}$ is covered by a B-module. And it is also true if B and $\tilde{B}$ in the statement are interchanged with each other.*

This was proved when we got several local characterizations of block covers in terms of Brauer maps, Brauer pairs etc. in [24], one of them is as follows:

Theorem. *B and $\tilde{B}$ and $\tilde{D}$ are the same as above. Let e and $\tilde{e}$ be the block idempotents of B and $\tilde{B}$ resp. Then $\tilde{B}$ is covered by B if and only if*

$$\mathrm{Br}_{C_G(\tilde{D})}(e\tilde{e}) = \mathrm{Br}_{C_G(\tilde{D})}(e) \cdot \mathrm{Br}_{C_G(\tilde{D})}(\tilde{e}) \neq 0$$

Using these characterizations, not only the above connection of block covers and the corresponding module covers are easy to derived as corollary, and many classical results such as Nagao's theorem, the description of Brauer pairs can be extended as well, and a local characterization of the existence of the Brauer correspondence in the sense of Alperin-Burry is also obtained. For the above see [24].

This research was by means of analysis of local structures, especially, of analysis of algebraic structures and module structures, with less character analysis. As a byproduct, we gave another result on the distribution into blocks of components of a reduced module in [23].

With the ideas and methods we analysed the Brauer correspondence in the sense of Brauer (i.e. the one defined in terms of central characters).

Lemma. *Let D be a defect group of an FG-block idempotent b and $\overline{C_G(D)} = C_G(D)/Z(D)$, let $\tau : F\overline{C_G}(D) \to F\overline{C_G(D)}$ be the canonical algebra homomorphism. Denote $\overline{b_0} = \tau(\mathrm{Br}_{C_G(D)}(b))$. Let λ_b be the central character associated with b. Then for $c \in ZFG$,*

$$\tau(\mathrm{Br}_{C_G(D)}(c)) \cdot \overline{b_0} = \lambda_b(c) \cdot \overline{b_0}$$

This provides a local structural way to consider the Brauer correspondence in the sense of Brauer even if for non-splitting ground-fields. With it we treated the related questions more easily and got some new results. Particularly, for nomal subgroups, we obtained relations among the existence of the Brauer correspondence in the sense of Brauer, the local properies in terms of the Brauer maps and of the trace maps, and the existence of the Brauer correspondence in the sense of Alperin-Burry. One of them is

Theorem. *Let H be a normal subgroup of a group G, let $\tilde{b}$ and b be blocks of FG and FH resp. Then the Brauer correspondence of $\tilde{b}$ in the sense of Brauer exists and equals b if and only if b is the unique weakly regular FG-block which covers $\tilde{b}$ and the center of a defect group of b is contained in H.*

For the above see [25]. H. Blau gave another proof for some results of this work with characters later ([7]).

3. Related questions on nilpotent blocks

One of the central theories of modular representation theory is the block theory, main aim is to obtain the structures and the properties of blocks and to make classifications. The most successful structural theories may be two kinds: cyclic blocks and nilpotent blocks. The works on the cyclic blocks were completed by R. Brauer, E. Dade and others (cf. [30, Ch. VII]). The nilpotent block is introduced by M. Broué and L. Puig from the point of view of local theories. Through their cooperative works, it reached a beautiful theoretic top which is due to Puig in terms of the so-called source algebras.

The famous Frobenius Theorem on finite p-nilpotent groups is put in a frame of the structure of p-subgroups formed by the Sylow Theorem. With respect to the modular representations of finite groups, [3] provides a similar frame in terms of Brauer block pairs. From this point of view, in [13] of M. Broué and L. Puig, a FG-block b is said to be nilpotent if the quotient group $N_G(P, b_P)/C_G(P, b_P)$ is a p-group for every b-Brauer pair (P, b_P). Their work [13] revealed that such blocks have very good properties; and, in some cases, the structures of such blocks were also obtained. What about it in general cases? Luis Puig came to a new idea.

L. Puig created a frame of the so-called pointed groups, he refined the usual local structures and introduced the local categories. Moreover, he defined the source algebra of a pointed group which is a striking idea. A systematic theory which is called Puig's theory sometimes was formed, see [35], [36], [37], [38]. One of the remarkable successes of Puig's theory is to give the structure of the source algebra of a nilpotent FG-block b when the groud-field is algebraically closed. He showed that the block b is nilpotent in the sense of Frobenius (defined as above) if and only if its local category is equivalent to the local catgory of its defect pointed group; and proved that, roughly speaking, b is nilpotent if and only if its source algebra is an extension of its defect group algebra by a matrix algebra, see [37].

What about it when the ground-field F is not algebraic closed? Puig guessed that his result is still ok if F contains a $(|G|_{p'})$'th root of unity. Encouraged by M. Broué and Puig, we paid a lot of efforts to study the question. We found that the defining condition in the sense of Frobenius is no longer equivalent to the defining condition in terms of local categories when the ground-field F is not large enough. From the point of view of local categories we give

Definition *The F-block b of G is called an F-nilpotent block if its local category is equivalent to the local category of its defect pointed group.*

And we proved

Theorem *Notations are as above. Assume P is a defect group of b, and $\widehat{F}$ is the extension of F by adding the all values of the all absolutely irreducible Brauer characters belonging to b. Then the F-block b of G is F-nilpotent if and only if there is an interior FP-algebra S which as F-algebra is just the matrix algebra $M_n(\widehat{F})$ such that the source algebra of b is isomorphic to $S \otimes_F FP$ as interior FP-algebras.*

It turns out Puig's result if $F = \widehat{F}$. This also shows that Puig's guess is true. In fact, we also pointed out when the three equivalent conditions for nilpotent blocks obtained by Puig for the algebraically closed fields are still equivalent to each other. These were done in [26].

We remark that all the works stated in this section are done in a somewhat general case, i.e. for a p-modular system not only for a field F. Furthermore questions are being in considerations.

4. Others

The group graded rings may be regarded as an extension of group algebras. Some works about group algebras, e.g. Clofford's theory, can be extended to group graded rings. E. Dade did a lot of deep results, e.g. [17].

Using Dade's theory, M.E. Harris extended results [5], [15] of J.L. Alperin and others to a more general case (see [31]):

Let $N \trianglelefteq G$ and U be an FG-module with $U \downarrow_N$ absolutly irreducible. Let V be an $F\overline{G}$-module where $\overline{G} = G/N$. Then there is a filtration of the FG-projective cover of $V \otimes_F U$ which is well connected with a filtration of the $F\overline{G}$-projective cover of V.

Our research showed that the result of Alperin and others is in fact valid for general fully group graded rings:

Theorem. *Let $N \trianglelefteq G$ and R be a fully G-graded ring and R_N be the subring generated by N-components. Let I_N and K_N be two G-invariant ideals of R_N, and set $I := I_N R$ and $\overline{R} = R/I$. Let P and $\overline{P}$ be a projective R-module and an $\overline{R}$-module resp. Then the following statements (as R-isomorphisms) are equivalent to each other:*

(i) $P/PI \cong \overline{P} \otimes_{R_1} (R_N/(K_N + I_N))$;

(ii) $PI^i/PI^{i+1} \cong \overline{P} \otimes_{R_1} (I_N^i/(K_N I_N^i + I_N^{i+1})), \quad i = 0, 1, 2, \ldots$

We also extended the Harris's result to some extent, though it can not be extended to group graded rings. It is interesting that these results can be proved in an elementary module-theoretical way, see [29].

Another application of representations of finite groups we have done is about the decompositions of tensor spaces. Let V be a finite dimensional F-space, and let $W = \overset{n}{\otimes} V$ be the n'th tensor product space of V and G be a permutation group of degree n. Then G acts on W by permuting the tensor factors of W. When F is the complex field, S. Pierce [34] gave the G-invariant decomposition of W by means of irreducible characters of G. Pierce's method is a classical method in the analysis of symmetric tensors. From the angle of representations of finite groups we saw that the Pierce decompositions are essentially dependent on the block idempotents of group algebras, and the block idempotent decompositions can be gotten in any coefficient fields, but the expressions of block idempotents in different fields may have different forms. We realized this observation in [28].

References

[1] J.L. Alperin, Local Representation Theory, Cambridge Univ. Press, Cambridge, 1986.

[2] J.L. Alperin, The Green correspondence and normal subgroups, J. Algebra, **104** (1986), 74–77.

[3] J.L. Alperin, M. Broué, Local methods in block theory, Ann. Math., **110** (1979), 142–157.

[4] J.L. Alperin, D.W. Burry, Block theory with modules, J. Algebra, **65** (1980), 225–233.

[5] J.L. Alperin, M.J. Collins, D.A. Sibley, Projective modules, filtrations and Cartan invariants, Bull. L. M. S., **16** (1984), 416–420.

[6] D. Benson, Modular Representation Theory: New Trends and methods, LNM 1081, Springer, Berlin, 1984.

[7] H.I. Blau, On block induction II, J. Algebra, to appear.

[8] M. Broué, On projective relative, bloc, groups de defaut, These, A L'université de Paris-VII, 1975.

[9] M. Broué, Brauer coefficients of p-subgroups associated with a p-block of a finite group, J. Algebra, **56** (1979), 256–383.

[10] M. Broué, Remarks·on blocks and subgroups, J. Algebra, **51** (1978), 228–232.

[11] M. Broué, On Scott modules and p-permutation modules, Proc. A. M. S., **93** (1985), 401–408.

[12] M. Broué, J.B. Olsson, Subpair multiplicities in finite groups, J. Reine Angew. Math., **371** (1986), 125–143.

[13] M. Broué, L. Puig, A Frobenius theorem for blocks, Invent. Math., **56** (1980), 117–128.

[14] D.W. Burry, Scott modules and lower defect groups, Comm. Algebra, **10** (1982), 1855–1872.

[15] J.F. Carlson, M.F. Collins, Filtrations for projective modules, Bull. L. M. S., **18** (1986), 591–592.

[16] C.W. Curtis, I. Reiner, Methods of Representation Theory I, John Wiley and Sons Inc., New York, 1982.

[17] E.C. Dade, Group–graded rings and modules, Math. Z., **174** (1980), 241–262.

[18] Y. Fan, Two theorems on change of ground-fields of algebras, Chinese Science Bull.(Kexue Tongbao), **32** (1987), 1304–1307.

[19] Y. Fan, Block covers and module covers of finite groups, Group Theory Proc. 1987 Singapore Conference, Walter de Gruyter, Berlin, 1989, 357–366.

[20] Y. Fan, On Scott coefficients and block invariants: An approach for non-splitting case, Comm. Algebra, **18** (1990), 2199–2242.

[21] Y. Fan, Fixed point sets of modules and permutation modules (Chinese), J. Hubei Univ., **12** (1990),281–287.

[22] Y. Fan, Permutation modules, p-permutation modules and Conlon species, Science in China (Ser. A), **34** ((1991), 1290–1301.

[23] Y. Fan, On the components of a reduced module, Chinese Science Bull., **37** (1992), 881–884.

[24] Y. Fan, Local characterizations of block covers and their applications, J. Algebra, **152** (1992), 397–416.

[25] Y. Fan, Remarks on Brauer correspondences, J. Algebra, **157** (1993), 213–223.

[26] Y. Fan, The souce algerbas of nilpotent blocks over arbitrary ground-fields, J. Algebra, to appear.

[27] Y. Fan, Xiaopei Li, On Cartan matrices of Frobenius algebras (in Chinese), J. of Math.(PRC) **8** (1988), 333–338.

[28] Y. Fan, Qiang Huang, Decomposition of tensor spaces and block idempotents of group rings, Linear and Multil. Algebra, **26** (1990), 299–305.

[29] Y. Fan, Borong Zhou, Some remarks on filtrations for projective modules, Bull. L. M. S., **24** (1992), 431–436.

[30] W. Feit, Representation Theory of Finite Groups, North-Holland Publishing Company, Amsterdam, 1982.

[31] M.E. Harris, Clifford theory and filtrations, J. Algebra, to appear.

[32] P. Landrock, The Cartan matrix of a group algebra modulo any power of its radical, Proc. A. M. S., **88** (1983), 205–206.

[33] P. Landrock, Finite Group Algebras and their Modules, Cambridge Univ. Press, Cambridge, 1983.

[34] S. Pierce, Orthogonal decompositions of tensor spaces, J. Res.(U.S. Nat. Bur. Standards), **74B** (1970), 41–44.

[35] L. Puig, Pointed groups and construction of characters, Math. Z., **176** (1981), 265–292.

[36] L. Puig, Local fusion in block source algebras, J. Algebra, **104** (1986), 358–369.

[37] L. Puig, Nipotent blocks and their source algebras, Invent. Math., **93** (1988), 77–116.

[38] L. Puig, Pointed groups and construction of modules, J. Algebra, **116** (1988), 7–129.

[39] J.-P. Serre, Linear Representations of Finite Groups, Springer, Berlin, 1977.

[40] J. Thévenaz, Permutaton representations arising from simplicial complexes, J. of Combin. Ser. A, **46** (1987), 121–155.

[41] A. Watanabe, Relations between blocks of a finite group and its subgroup, J. Algebra, **78** (1982), 282–291.

Recent Progress on Classical Groups and Algebraic K-Theory in China*

Li Fu-an

Institute of Mathematics, Academia Sinica, Beijing

You Hong

Department of Mathematics, Northeast Normal University, Changchun

Initiated by the late Professor L.K.Hua in the late 1940s, Chinese mathematicians used the matrix method to study the structure and automorphisms of classical groups, and made a lot of contributions in the 1950s and early 1960s. These results on classical groups were well known by Western and Russian mathematicians, and compiled in the monograph *Classical Groups* [16] by L.K.Hua and Z.X.Wan which appeared in 1963. After the "Cultural Revolution", when coming back to this field, we found that the Western and Russian mathematicians had made considerable progress on classical groups, and built up and developed algebraic K-theory. We tried to renew our study and to catch up with the research trend on these fields. In the recent fifteen years we have made some progress, including solving some important open problems, some of which will be sketched below.

1. Isomorphisms of Linear Groups over Division Rings

Let K be a division ring, and let $GL_n(K), SL_n(K), PGL_n(K)$ and $PSL_n(K)$ denote, respectively, the general linear group, the special linear

* Supported in part by the National Natural Science Foundation of China

group, the projective general linear group and the projective special linear
group of dimension n over K. The automorphisms and isomorphisms of
these groups were completely determined in the early fifties except the
difficult case $n = 2$. Z.X. Wan, H.S. Ren and X.L. Wu did a lot of work
for the case $n = 2$ (cf. [46, 48, 49, 73]), and finally solved this problem in
1987 (see [50, 51, 64]), hence ended the study of the automorphisms and
isomorphisms of these groups over division rings. They obtained

Theorem 1. *The automorphisms of* $GL_n(K), SL_n(K), PGL_n(K)$ *and*
$PSL_n(K)$ *for* $n \geq 2$ *are all standard. For instance, any automorphism of*
$PSL_n(K)$ *is either of the form*

$$\overline{A} \mapsto \overline{PA^\sigma P}^{-1} \quad \text{for all } \overline{A} \text{ in } PSL_n(K),$$

where P *is an element in* $GL_n(K), \overline{P}$ *denotes the projective image of* P
in $PGL_n(K)$, *and* σ *is an automorphism of* K, *or of the form*

$$\overline{A} \mapsto \overline{P(A^\tau)'^{-1}\overline{P}^{-1}} \quad \text{for all } \overline{A} \text{ in } PSL_n(K),$$

where τ *is an anti-automorphism of* K.

Theorem 2. *Suppose that* K *and* K' *are division rings,* n *and* n' *are*
integers ≥ 2. *Then the following statements are equivalent:*

(0) $n = n'$, *and* K *is isomorphic or anti-isomorphic to* K'.

(1) $GL_n(K) \simeq GL_{n'}(K')$.

(2) $SL_n(K) \simeq SL_{n'}(K')$.

(3) $PGL_n(K) \simeq PGL_{n'}(K')$.

If we rule out the exceptional isomorphisms $PSL_2(\mathbf{F}_4) \simeq PSL_2(\mathbf{F}_5)$ *and*
$PSL_2(\mathbf{F}_7) \simeq PSL_3(\mathbf{F}_2)$, *then the above statements are also equivalent to*

(4) $PSL_n(K) \simeq PSL_{n'}(K')$.

2. Automorphisms of the Commutator Subgroups of Orthogonal Groups over Fields

Let F be a field of characteristic $\neq 2$, f a non-singular symmetric
bilinear form in n indeterminates over F. Let $O_n(F, f)$ be the orthogonal group over F with respect to f, and $\Omega_n(F, f)$ the commutator subgroup of $O_n(F, f)$. The automorphisms of $\Omega_n(F, f)$ were determined by

O.T.O'Meara [40] using his residual space method for $n \geq 7$ and $n \neq 8$. The cases $n = 5, 6$ and $n = 8$ were settled by A.A. Johnson [18] and A. Hahn [14] respectively. When $n = 3, 4$ with f isotropic, $P\Omega_n(F, f)$ is isomorphic to $PSL_3(F), PSL_3(F) \times PSL_3(F)$ or $PSL_3(K)$ where K is a quadratic separable extension of F, their automorphisms being known. The determination of automorphisms of $\Omega_3(F, f)$ and $\Omega_4(F, f)$ with f anisotropic was proposed as an open problem by E.A. Connors and A. A. Johnson in 1979 (cf. [17]). This problem has been settled by Z. X. Li [34, 35].

Theorem 3. *Every automorphism of $\Omega_3(F, f)$ with f anisotropic is standard, i.e. induced by a semilinear transformation which preserves orthogonality. Hence*

$$\text{Aut}\Omega_3(F, f) \simeq P\Gamma O_3(F, f).$$

Theorem 4. *Let F and f be as above. Denote by $O_4'(F, f)$ the spinor subgroup of $O_4(F, f)$. Then there is a split exact sequence $1 \rightarrow Aut\Omega_3(F, f_1) \times Aut\Omega_3(F, f_1) \rightarrow AutPO_4'(F, f) \overrightarrow{\leftarrow} S_2 \rightarrow 1$, where f_1 is the restriction of f on a certain 3-dimensional subspace, and S_2 is the symmetric group on two letters. Furthermore, every automorphism of $P\Omega_4(F, f)$ is induced by an automorphism of $PO_4'(F, f)$.*

By means of the above split exact sequence, Z. X. Li found an exceptional automorphism of $PO_4'(F, f)$ for $F = \mathbf{Q}(\sqrt{2})$. Note that, however, the existence of exceptional automorphisms of $PO_4'(F, f)$ and $P\Omega_4(F, f)$ depends heavily on the underlying field F. For instance, if $F = \mathbf{Q}$ or $\mathbb{R}$, then all automorphisms of $PO_4'(F, f)$ and $P\Omega_4(F, f)$ are standard.

F.A.Li [20, 21] discussed the non-defective orthogonal groups over a field F of characteristic 2 with $F \neq \mathbf{F}_2$. He proved that every automorphism of Ω_n and $O_n'(n \geq 6$ with $n \neq 8)$ is standard, i.e. induced by a semilinear transformation which preserves the quadratic structure. This improved a result of E.A.Connors [9]. F. A. Li also settled the exceptional behaviour of $Aut\Omega_8$ and $AutO_8'$, which is closely related to the triality principle, and established a result similar to that of A.Hahn [14] for the case of characteristic $\neq 2$.

3. Isomorphisms of Linear Groups over Commutative Rings

Let R be a commutative ring with 1. $GL_n(R), SL_n(R)$ and $E_n(R)$ are, respectively, the general linear group, the group of matrices with determinant 1, and the elementary group of rank n over R. V. M. Petechuk [43] proved that, when $n \geq 4$, the automorphisms of $E_n(R), SL_n(R)$ and $GL_n(R)$ are all standard. In [42] he determined the automorphisms of $SL_3(R)$ and $GL_3(R)$ for R a commutative local ring, and found a new type of automorphisms φ_α^σ. F. A. Li and Z. X. Li [30] derived the general form of the isomorphisms of GL_3 over arbitrary commutative rings, and found a new type of isomorphisms Φ_α^σ which is a generalization of Petechuk's φ_α^σ.

Theorem 5. *Let R and R' be commutative rings. Then every isomorphism $\Lambda : E_3(R) \to E_3(R')$ can be expressed as $\Lambda = \Psi \cdot \Phi_\alpha^\sigma$ where Ψ is a standard automorphism of $E_3(R')$. Moreover, Λ is exceptional if and only if $\alpha \neq 1$.*

Theorem 6. *Let R and R' be commutative rings. Let $\Lambda : GL_3(R) \to GL_3(R')$ be an isomorphism. Then $\Lambda | GE_3(R) = \Psi \cdot \Phi_\alpha^\sigma$ where $GE_3(R)$ is the subgroup of $GL_3(R)$ generated by $E_3(R)$ and all diagonal matrices, and Ψ is a standard automorphism of $GE_3(R')$.*

Combining the above two theorems with Petechuk's result, we have

Theorem 7. *Let R and R' be commutative rings, and let V and V' be free modules of ranks $n, n' \geq 3$ over R and R' respectively. Suppose that $GL(V) \simeq GL(V')$. Then $n = n'$. If $n \geq 4$, or $n = 3$ and 2 is a unit in R (or more generally, card $(R/M) > 2$ for all maximal ideals M of R), or $n = 3$ and R is a domain, then R and R' are isomorphic. However, there do exist examples in which R and R' are not isomorphic, and hence, not Morita equivalent.*

F. A. Li and Z. X. Li also exhibited an example in which the rings $\mathbb{Z}_4$ and $\mathbf{F}_2[x]/(x^2)$ are not Morita equivalent, but they have isomorphic 3-dimensional linear groups. This gives a counterexample to an open problem proposed by O.T.O'Meara and L.N.Vaserstein (cf. [17]).

The exceptional isomorphism Φ_α^σ, which can not be explained by Morita theory, is induced by a non-trivial system $(\varphi, \alpha, \sigma)$ from R to R'. It is interesting to explore more deeply the nature of Φ_α^σ and $(\varphi, \alpha, \sigma)$. F. A. Li [23] pointed out that every system $(\varphi, \alpha, \sigma)$ is intrinsically "glued" by some local ones, and $(\varphi, \alpha, \sigma)$ is non-trivial if and only if at least one local system is non-trivial. In [27] he also discussed the nature of $(\varphi, \alpha, \sigma)$ from a view point of homology.

For two-dimensional linear groups over commutative rings, F. A. Li and H. S. Ren [32] determined the automorphisms of $E_2(R)$ and $GE_2(R)$ under the hypothesis that $2, 3$ and 5 are units. See [62, 66, 67, 71] for other results.

4. Automorphisms and Isomorphisms of Classical Groups over Rings

Let R be an associative ring with 1, and let $Sp_{2n}(R), O_n(R), O_n^+(R)$ and $\Omega_n(R)$ denote, respectively, the symplectic group, the orthogonal group, the subgroup of matrices in $O_n(R)$ with determinant 1, and the commutator subgroup of $O_n(R)$.

A lot of works on classical groups over semilocal rings, Φ-surjective rings and some other rings were done in the early eighties by H. Q. Zhang, X. P. Tang and their students. H. Q. Zhang and L. Q. Wang [88], and X. P. Tang and Z. Z. Lin [55] considered the isomorphisms of symplectic groups over Φ-surjective rings. But V. M. Petechuk [44] completely determined the isomorphisms of symplectic groups of dimension $2n \geq 6$ over arbitrary commutative rings. H. You [76] proved that all isomorphisms of the 4-dimensional symplectic groups over commutative rings are standard, provided that the rings have no factor rings of perfect fields with characteristic 2. (There exist exceptional automorphisms of $Sp_4(F)$ where F is a perfect field with characteristic 2. See [16].) For arbitrary commutative rings, the isomorphism problem of 4-dimensional symplectic groups is now still open.

H. You and R. F. Wang [84], and Y. Z. Zhang [89] studied the automorphisms of $O_n(R)$ and $O_n^+(R)$ over semilocal rings. X. P. Tang and J. B. An [54] used O'Meara's residual spece method to determine the isomorphisms of $PO_n(R), PO_n^+(R)$ and $P\Omega_n(R)$ over local rings when the Witt indeces ≥ 6. A. Hahn and Z. X. Li [15] dealt with the isomorphisms of the hyperbolic unitary groups by using hermitian Morita theory. For other results refer to [3, 36, 47, 63].

5. Structure of Classical Groups over Rings

Many authors discussed the structure of classical groups over rings in recent years. See [2, 8, 22, 25, 53, 61, 65, 70, 74, 80, 87].

We now consider a more general case. Assume an anti-automorphism* is defined on an associative ring R such that $a^{**} = \varepsilon a \varepsilon^*$ for some unit $\varepsilon = \varepsilon^{*-1}$ and all a in R. Let $R_\varepsilon = \{a - a^*\varepsilon | a \in R\}$ and $R^\varepsilon = \{a \in R | a = -a^*\varepsilon\}$. Fix an additive subgroup Λ of R with the following properties:

$$r^*\Lambda r \subseteq \Lambda \quad \text{for all } r \in R, \quad \text{and}$$

$$R_\varepsilon \subseteq \Lambda \subseteq R^\varepsilon.$$

Let $O_{2n}(R)$ be the pseudo-orthogonal group of rank $2n$ over R (which depends on *, ε and Λ). Note that, when $\Lambda = R$ (resp. $\Lambda = 0$) , then $O_{2n}(R)$ is just the symplectic group $Sp_{2n}(R)$ (resp. the ordinary orthogonal group). See [5,6, 39] for the definitions of $EO_{2n}(R), EO_{2n}(J), EO_{2n}(R, J)$, and $O_{2n}(R, J)$, where J is a form ideal of R.

Let C denote the center of R, and C_* the subring of C generated by all cc^* with $c \in C$. L.N. Vaserstein and H. You [61] described the structure of the pseudo-orthogonal group $O_{2n}(R)$:

Theorem 8. *Assume that $n \geq 3$, and for every maximal ideal M of C_*, there exists a multiplicative set $S \subseteq C_* - M$ such that the absolute stable rank of $S^{-1}R \leq n - 2$. Then for any form ideal J of R,*

$$EO_{2n}(R, J) = [EO_{2n}(R), EO_{2n}(J)] = [EO_{2n}(R), EO_{2n}(R, J)]$$

$$= [EO_{2n}(R), O_{2n}(R, J)] = [O_{2n}(R), EO_{2n}(R, J)].$$

Moreover, evrey subgroup H of $O_{2n}(R)$ is normalized by $EO_{2n}(R)$ if and only if there is a unique form ideal J such that

$$EO_{2n}(R, J) \subseteq H \subseteq O_{2n}(R, J).$$

The authors of [61] showed that, if R is a finitely generated C-module, then the ring $(C_* - M)^{-1}R$ is semilocal for every maximal ideal M of C_*. Thus, their result covers many previous ones in [1, 6, 8, 11, 22, 25, 59, 87].

F.A. Li and M. L. Liu [31] proved for any commutative ring R and any $n \geq 3$ that, if H is a subgroup of $GL_n(R)$ normalized by an elementary congruence subgroup $E_n(R, J)$, there is an ideal L of R such that

$$E_n(R, L) \subseteq H \subseteq GL_n(R, (L : J^{40})).$$

(L. N. Vaserstein reduced the power from 40 to 4, cf. [60].) Furthermore, L is uniquely determined up to a certain equivalence relation on the set of ideals of R.

S.Z. Li and J. G. Zha discussed maximal subgroups of classical groups, and derived a number of nice results. See the survey article of S. Z. Li in this book.

6. Expressing a Matrix as a Product of Matrices of a Special Nature

Given a class of matrices, one studies products of matrices from the class, and asks about the minimal number of factors in various factorizations. Classes considered here include the class of transvections, the class of involutions, the class of commutators and many others.

A classical factorization result is that, for a field F, every matrix A in $SL_n(F)$ is a product of at most $resA + 1$ transvections (see [41]), where $resA = \text{rank}(A - I)$. F. Zhou and L. Li [90] extended this result to the case of commutative local rings.

H. You and J. Z. Nan [83] discussed the factorization of matrices in $SL_n(F)$ over a field F by 2-involutions. They showed that every matrix A in $SL_n(F)$ is a product of at most $[(resA + 1)/2] + 2$ 2-involutions when $n \geq 3$ and $char F \neq 2$, or $n \geq 4$ and $char F = 2$. They also gave an example to illustrate that the number $[(res A + 1)/2] + 2$ is the least bound for the factorization.

Y.X. Wang [72] gave a complete answer for the least length of the factorization of orthogonal groups over fields by isometries. He proved that the least length is $res A + 4$ for $A \in O_n(F)$.

B.C. Yuan [85, 86] discussed expressions of $O_n(V)$ and $\Omega_n(V)$ over a F with characteristic $\neq 2$ by Eichler transformations and 2-transvections, respectively, under the hypothesis that the Witt index $\neq 0$. He proved that every $\sigma \in O_n(V)$ is a product $\sigma = \tau\tau_1 \cdots \tau_k$ where $res \ \tau \leq 2$, and the τ_i are Eichler transformations, and the least number for k is $[res \ \sigma/2]$. When the Witt index ≥ 2, each element σ in $\Omega_n(V)$ is a product of at most $res \ \sigma + 2$ 2-transvections.

For any group G, let $c(G)$ be the least integer s such that every element in the commutator subgroup of G is a product of at most s commutators. H. You [81] proved $c(Sp_{2n}(R)) \leq 4$ for any commutative ring R with stable rank 1 and $n \geq 3$. F. Arlinghaus, L.N. Vaserstein and H. You [4] generalized this result to the pseudo-orthogonal group $O_{2n}(R)$, provided that R satisfies the Λ-stable condition, and showed for the ordinary orthogonal group that $c(O_{2n}(R)) \leq 3$ where R is a commutative ring with absolute stable rank 1 and $n \geq 3$.

7. Steinberg Groups over Rings

The Steinberg group $St_n(R)$ over a ring R is the group defined by the generators $x_{ij}(a)(1 \leq i, j \leq n, \ i \neq j, \ a \in R)$ subject to the Steinberg relations. Let $\varphi : St_n(R) \to E_n(R)$ be the canonical homomorphism sending $x_{ij}(a)$ to $e_{ij}(a)$. Denote $K_{2,n}(R) = Ker \ \varphi$.

F.A. Li [24] discussed the isomorphisms of Steinberg groups, and obtained

Theorem 9. *Let R and R' be commutative rings and $n \geq 4$.*

(1) Every isomorphism $\Lambda : St_n(R) \to St_n(R')$ induces an isomorphism $\lambda : E_n(R) \to E_n(R')$, and Λ is uniquely determined by λ.

(2) *If $St_n(R) \simeq St_n(R')$, then $K_{2,n}(R) \simeq K_{2,n}(R')$.*

(3) *Every isomorphism $\lambda : E_n(R) \to E_n(R')$ can be uniquely lifted to an isomorphism from $St_n(R)$ to $St_n(R')$.*

For the case $n = 3$, if $St_3(R)$ and $St_3(R')$ are central extensions of $E_3(R)$ and $E_3(R')$ respectively, then (1) and (2) in Theorem 9 hold. In [29] F. A. Li studied the isomorphisms of stable Steinberg groups and established a result similar to Theorem 9.

Let L (resp. U) denote the subgroup of $St_n(R)$ generated by all $x_{ij}(a), i > j$ (resp. $i < j$), and use W to denote the subgroup of $St_n(R)$ generated by all $w_{ij}(u) = x_{ij}(u)x_{ji}(-u^{-1})x_{ij}(u)$. F. A. Li [28] gave a necessary and sufficient condition for the existence of a normal form for elements of Steinberg groups:

Theorem 10. *Let R be a commutative ring. Then the following statements are equivalent.*

(1) *R satisfies the first stable range condition.*

(2) *$E_n(R) = LWLU$ for all $n \geq 2$.*

(3) *$E_n(R) = LWLU$ for some $n \geq 2$.*

(4) *$St_n(R) = K_{2,n}(R)LWLU$ for all $n \geq 2$.*

(5) *$St_n(R) = K_{2,n}(R)LWLU$ for some $n \geq 2$.*

H. You [78] extended Theorem 10 to the case of non-commutative rings, replacing W by another suitable subgroup.

8. K-groups over Some Rings

H. S. Li, M. Van den Bergh and F. Van Oystaeyen [33] proved that, if R is a Zariski ring with associated graded ring $G(R)$ of finite global dimension then there is an injection $K_0(R) \to K_0(G(R))$ mapping $[R]$ to $[G(R)]$.

F. G. Wang [68] discussed the relationship between the Grothendieck groups of a primitive ring with socle and its quotient ring, and obtained

Theorem 11. *Let R be a primitive ring.*

(1) *If $A = soc(R) \neq R$, then $K_0(R) \to K_0(R/A)$ is surjective, and its kernel is a cyclic subgroup of $K_0(R)$.*

(2) R is a homomorphic image of a certain primitive ring R' with socle, and R and R' have a same faithfully irreducible module. Furthermore, $K_0(R')$ is the direct sum of $K_0(R)$ and a cyclic group.

A ring R is called an IBN ring (ring with invariant basis number) if $R^m \simeq R^n$ implies $m = n$. R is called an IBN_1 ring (resp. IBN_2 ring) if $R^m \simeq R^n \oplus M$ implies $m \geq n$ (resp. $R^m \simeq R^m \oplus M$ implies $M = 0$). It is clear that $IBN_2 \subset IBN_1 \subset IBN$. W. T. Tong [58] studied IBN rings by means of the pre-ordering on $K_0(R)$, which was introduced by K.R.Goodearl [12], i.e. $x \leq y$ if there is a finitely generated projective left R-module P such that $y - x = [P]$. W. T. Tong characterized IBN_1 and IBN_2 rings, and gave a sufficient condition for which the pre-ordering becomes a total ordering. On the other hand, if the pre-ordering is a partial ordering, then either R is an IBN_1 ring, or $K_0(R) = 0$. In [56] W. T. Tong discussed the Euler characteristic χ of IBN rings. In particular, he showed

Theorem 12. *Let R be a commutative ring.*

(1) *For any stably free R-modules M and N,*

$$\chi(M \otimes N) = \chi(M)\chi(N).$$

(2) *If $f : K_0(R) \to \mathbb{Z}$ is a ring isomorphism, and M and N are finitely generated projective R-modules, then*

$$f([M]) = \chi(M),$$

$$\chi(M \otimes N) = \chi(M)\chi(N).$$

W.T. Tong [57] proved that, if a commutative ring R has the unimodular column property and G is any abelian group, then $K_0(RG) \simeq K_0(R)$.

M. L. Liu and F. A. Li [37, 38] gave a formula for calculation of the G_0 and G_1 groups for a class of group rings $R(\pi \rtimes \Gamma)$.

J. Z. Wang [69] described the automorphism groups $Aut_R(P)$ of projective modules over commutative rings. He introduced a normal subgroup $T^1 G_R(P)$ of $Aut_R(P)$, which plays a role similar to that of the elementary subgroup $E_n(R)$ in $GL_n(R)$. For commutative noetherian R with $J - \dim(R) = d$, he derived that $Aut_R(P \oplus R^n) = Aut_R(P)T^1 G_R(P \oplus R^n)$

if rank $P \geq d+1$. Some results on the automorphism groups of projective modules over polynomial rings were also given in [69].

Let $U_{2n}^\varepsilon(R)$ be the unitary group over an associative ring R equipped an involution. H. You [77] studied the prestabilization of $U_{2n}^\varepsilon(R)$ over a class of rings. In [82] he also discussed the stabilization of the unitary groups over polynomial rings in case $\varepsilon = -1$, and obtained some results similar to those on linear groups [52] and on symplectic groups [13, 19]. Especially, $U_{2n}^{-1}(A) = U_{2n}^{-1}(R)EU_{2n}^{-1}(A)$ for all $n \geq 3$, and $K_1U^{-1}(A) = K_1U^{-1}(R)$, where R is the ring of algebraic integers in a quadratic field $\mathbf{Q}(\sqrt{d})$, and $A = R[x_1, \cdots, x_m]$. For other results on K_1 and K_1U^ε, see [69, 79].

F.A.Li [26] proved, for a finitely generated commutative $\mathbb{Z}$-algebra A with Krull dimension d and an arbitrary finite group π, that $St_n(A\pi)$ is finitely presented whenever $n \geq 4$. If, in addition, $n \geq d+3$, and $K_1(A\pi)$ and $K_2(A\pi)$ are finitely generated, then $E_n(A\pi)$ and $GL_n(A\pi)$ are finitely presented.

Let F be a number field, and O_F the ring of algebraic integers in F. K.Q. Feng [10] derived that the Birch-Tate conjecture holds for a class of cyclic number fields:

Theorem 13. *Let F be a cyclic number field with odd prime degree p. Suppose that the following three conditions are satisfied:*

(1) *the multiplicative order of 2 mod p is even;*

(2) *2 is inertial in F; and*

(3) *the class number of F is odd.*

Then the Birch-Tate conjecture holds for F. Moreover, in this case, the even part of $K_2(O_F)$ is $(\mathbb{Z}/2\mathbb{Z})^p$, and generated by Steinberg symbols $\{-1, \varepsilon_i\}(1 \leq i \leq p)$, where $\{\varepsilon_1, \cdots, \varepsilon_{p-1}\}$ is a fundamental system of units in O_F, and $\varepsilon_p = -1$.

H.R. Qin [45] gave a necessary and sufficient condition for the existence of elements of order 4 in $K_2(O_F)$, provided that the S-class number of F is odd where S consists of all prime spots except for dyadic and Archimedean ones. He also proposed a method to determine the Sylow 2-subgroup of $K_2(O_F)$ for a class of real quadratic fields F. He calculated the K_2-group of O_F for some imaginary quadratic fields F. In particular,

$$K_2(\mathbb{Z}[\sqrt{-6}]) = 0 \quad \text{and} \quad K_2(\mathbb{Z}[(1 + \sqrt{-35})/2]) \simeq \mathbb{Z}/2\mathbb{Z} \oplus \mathbb{Z}/3\mathbb{Z}.$$

This gives a possitive answer for a conjecture of Browkin and a negative answer for another conjecture of Browkin [7].

H. You [75] gave a set of generators of $K_2(R, I)$ where R is a commutative ring with the unit 1-stable range condition, and I is an ideal of R.

References

[1] Abe, E., Normal subgroups of Chevalley groups over rings, *Contemp. Math.*, **83** (1989), 1–17.

[2] An Jianbei, The structure of symplectic groups and two-dimensional linear groups over some type of commutative rings, *Chinese Ann. Math.*, **5** A: 6 (1984), 665–680. (in Chinese)

[3] An Jianbei, Automorphisms of symplectic groups over semilocal rings of characteristic 2, *Acta Math. Sinica*, **27**: 6 (1984), 824–829. (in Chinese)

[4] Arlinghaus, F. A., Vaserstein, L. N. and You Hong, Commutators in pseudo-orthogonal groups, preprint.

[5] Bak, A., On modules with quadratic forms, *Lecture Notes in Math.*, **108** (1969), 55–66.

[6] Bass, H., Unitary algebraic K-theory, *Lecture Notes in Math.*, **343** (1973), 57–265.

[7] Browkin, J., Conjectures on the dilogarithm, *K-Theory*, **3** (1989), 29–56.

[8] Cao Chongguang and Wang Luqun, On normal subgroups of the symplectic group over a ring with one in its stable range, *Acta Math. Sinica*, **29** : 3 (1986), 323–326. (in Chinese)

[9] Connors, E. A., Automorphisms of orthogonal groups in characteristic 2, *J. Number Theory*, **5** (1973), 477–501.

[10] Feng Keqin, On the Birch-Tate conjecture for cyclic number fields, *J. Pure Appl. Algebra*, **48** (1987), 223–228.

[11] Golubchik, I., On normal subgroups of linear and unitary groups over associative rings, in *Spaces over Algebras and Questions of Network Theory* , Ufa, 1985, pp.122–142. (in Russian)

[12] Goodearl, K. R., Partially ordered Grothendieck groups, *Lecture Notes in Pure and Appl. Math.*, **91** (1984), 71–90.

[13] Grunewald, F., Mennicke, J. and Vaserstein, L.N., On symplectic groups over polynomial rings, *Math. Z.*, **206** (1991), 35–56.

[14] Hahn, A. J., Cayley algebras and the automorphisms of $PO_8'(V)$ and $P\Omega_8(V)$, *Amer. J. Math.*, **98** (1976), 953–987.

[15] Hahn, A. J. and Li Zunxian, Hermitian Morita theory and hyperbolic unitary groups, *J. Algebra*, **97** (1985), 30–52.

[16] Hua Luogeng and Wan Zhexian, *Classical Groups*, Shanghai Science and Technology Press, Shanghai, 1963. (in Chinese)

[17] James, D., Waterhouse, W. and Weisfeiler.B, Abstract homomorphisms of algebraic groups: problems and bibliography, *Commun. Algebra*, **9** (1981), 95–114.

[18] Johnson, A. A., The automorphisms of the orthogonal groups $\Omega_n(V)$, $n \geq 5$, *J. Reine Angew. Math.*, **298** (1978), 112–155.

[19] Kopeiko, V., The stabilization of symplectic groups over polynomial rings, *Math. USSR Sb.*, **34** (1978), 655–669.

[20] Li Fu-an, The automorphisms of non-defective orthogonal groups in characteristic 2, *Chinese Ann. Math.*, **6** B: 3 (1985), 363–373.

[21] Li Fu-an, The automorphisms of non-defective orthogonal groups $\Omega_8(V)$ and $O_8'(V)$ in characteristic 2, *Chinese Ann. Math.*, **7** B: 1 (1986), 1–13.

[22] Li Fu-an, The structure of symplectic groups over arbitrary commutative rings, *Acta Math. Sinica* (New Ser.), **3**: 3 (1987), 247–255.

[23] Li Fu-an, Local behaviour of systems $(\varphi, \alpha, \sigma)$, *Kexue Tongbao*, **33**: 19 (1988), 1445–1447. (in Chinese)

[24] Li Fu-an, Isomorphisms of Steinberg groups over commutative rings, *Acta Math. Sinica* (New Ser.), **5** : 2 (1989), 146–158.

[25] Li Fu-an, The structure of orthogonal groups over arbitrary commutative rings, *Chinese Ann. Math.*, **10** B: 3 (1989), 341–350.

[26] Li Fu-an, Finite presentability of Steinberg groups over group rings, *Acta Math. Sinica* (New Ser.), **5**: 4 (1989), 297–301.

[27] Li Fu-an, Homological meaning of systems $(\varphi, \alpha, \sigma)$, *Acta Math. Sinica* (New Ser.), **7** : 4 (1991), 348–353.

[28] Li Fu-an, Decomposition of Steinberg groups, *Chinese Science Bull.*, **37** :15 (1992), 1244–1247.

[29] Li Fu-an, Isomorpisms of stable Steinberg groups, *Chinese Ann. Math.*, **14** B: 2 (1993), 183–188.

[30] Li Fu-an and Li Zunxian, Isomorphisms of GL_3 over commutative rings, *Scientia Sinica* (Ser. A), **31** : 1 (1988), 7–14.

[31] Li Fu-an and Liu Mulan, A generalized sandwich theorem, *K-Theory*, **1**: 2 (1987), 171–183.

[32] Li Fu-an and Ren Hongshuo, The automorphisms of two-dimensional linear groups over commutative rings, *Chinese Ann. Math.*, **10** B: 1 (1989), 50–57.

[33] Li Huishi, Van den Bergh, M. and Van Oystaeyen, F., Note on the K_0 of rings with Zariskian filtration, *K-Theory*, **3** (1990), 603–606.

[34] Li Zunxian, The automorphisms of the orthogonal groups $\Omega_3(V)$, *Scientia Sinica* (Ser. A), **25** : 7 (1982), 693–701.

[35] Li Zunxian, Quaternion algebras and automorphisms of $PO_4^+(V)$, $PO_4'(V)$ and $P\Omega_4(V)$, *Scientia Sinica* (Ser. A), **30**: 11 (1987), 1121–1132.

[36] Lin Zongzhu, The isomorphisms of symplectic groups over local rings, *Acta Math. Sinica*, **27** : 6 (1984), 730–748. (in Chinese)

[37] Liu Mulan, $G_1(R\pi)$ for π a finite abelian group, *J. Pure Appl. Algebra*, **24** : 3 (1982), 287–291.

[38] Liu Mulan and Li Fu-an, G_0 and G_1 of a class of group rings $R(\pi \rtimes \Gamma)$, *Science Bull.*, **31** : 11 (1986), 721–724.

[39] Magum, B., Van der Kallen, W. and Vaserstein, L., Absolute stable rank and Witt cancellation for non-commutative rings, *Inv. Math.*, **91** (1988), 525–543.

[40] O'Meara, O.T., The automorphisms of the orthogonal groups $\Omega_n(V)$ over fields, *Amer. J. Math.*, **90** (1968), 1260–1306.

[41] O' Meara, O. T., *Lectures on Linear Groups*, Amer. Math. Soc., Providence, R. I., 1974.

[42] Petechuk, V. M., Automorphisms of the groups $SL_3(K), GL_3(K)$, *Math. Notes*, **31** (1982), 335–340.

[43] Petechuk, V. M., Automorphisms of matrix groups over commutative rings, *Math. USSR Sb.*, **45** (1983), 527–542.

[44] Petechuk, V. M., Isomorphisms of symplectic groups over commutative rings, *Algebra and Logic*, **22** : 5 (1983), 397–405.

[45] Qin Hourong, On K_2 and algebraic number theory, Thesis, Nanjing Univ., 1992. (in Chinese)

[46] Ren Hongshuo, Automorphisms of $SL_2(K)$ over a class of skew fields, *Acta Math. Sinica* , **24** : 4 (1981), 566–577. (in Chinese)

[47] Ren Hongshuo, Partition rings and a type of automorphisms, *Acta Math. Sinica*, **26** : 5 (1983), 538–546. (in Chinese)

[48] Ren Hongshuo and Wan Zhexian, Automorphisms of $PGL_2(K)$ over any skew field K, *Acta Math. Sinica*, **25** : 2 (1982), 208–218.

[49] Ren Hongshuo and Wan Zhexian, Automorphisms of $PSL_2^{\pm}(K)$ over any skew field K, *Acta Math. Sinica*, **25** : 4 (1982), 484–492.

[50] Ren Hongshuo, Wan Zhexian and Wu Xiaolong, Automorphisms of $PSL(2, K)$ over skew fields, *Acta Math. Sinica* (New Ser.), **3**: 1 (1987), 45–53.

[51] Ren Hongshuo, Wan Zhexian and Wu Xiaolong, Automorphisms and isomorphisms of linear groups over skew fields, *Proc. Symposia in Pure Math.*, **47** (1987), 473–476.

[52] Suslin, A. A., On the structure of the special linear group over polynomial rings, *Izv. Akad. Nauk*, Ser. Mat., **41**: 2 (1977), 235–252. (in Russian)

[53] Tang Xiangpu and An Jianbei, The Structure of symplectic groups over semilocal rings, *Acta Math. Sinica* (New Ser.), 1:1 (1985), 1–15.

[54] Tang Xiangpu and An Jianbei, The isomorphisms of orthogonal groups over local rings, *Acta Math. Sinica* (New Ser.), **2**: 4 (1986), 281–291.

[55] Tang Xiangpu and Lin Zongzhu, The isomorphisms of symplectic groups over $X_0 - \Phi$ – surjective rings, *Acta Math Sinica*, **29** : 4 (1986), 477–480. (in Chinese)

[56] Tong Wenting, On Euler characteristic of modules, *Chinese Ann. Math.*, 10B: 1 (1989), 58–63.

[57] Tong Wenting, PF-rings and the Grothendieck groups of group rings, *J. of Math. Research and Exposition*, **10**: 2 (1990), 157–162. (in Chinese)

[58] Tong Wenting, *IBN* rings and orderings on Grothendieck groups, to appear.

[59] Vaserstein, L. N., On normal subgroups of Chevalley groups over commutative rings, *Tohoku Math. J.*, **38** (1986), 219–230.

[60] Vaserstein, L. N., The subnormal structure of general linear groups over rings, *Math. Proc. Camb. Phil. Soc.*, **108** (1990), 219–229.

[61] Vaserstein, L. N. and You Hong, Normal subgroups of classical groups over rings, preprint.

[62] Wan Zhexian and Ren Hongshuo, Automorphisms of two-dimensional linear groups over local rings in characteristic 2, *Chinese Ann. Math.*, **4** A: 4 (1983), 419–434. (in Chinese)

[63] Wan Zhexian and Ren Hongshuo, The two-dimensional unitary groups with non-zero Witt index, *Chinese Ann. Math.*, **4** A: 5 (1983), 587–599. (in Chinese)

[64] Wan Zhexian, Ren Hongshuo and Wu Xiaolong, The isomorphisms of linear groups over skew fields, *Acta Math. Sinica*, **30** : 2 (1987), 187–194. (in Chinese)

[65] Wan Zhexian and Wu Xiaolong, On the second commutator subgroup of $PGL_2(\mathbb{Z})$, *C. R. Math. Rep. Acad. Sci. Canada*, **2** : 6 (1980), 303–308.

[66] Wan Zhexian and Yang Jinggeng, Automorphisms of the quaternion unimodular group of dimension 2, *Chinese Ann. Math.*, **2** A: 3 (1981), 387–397. (in Chinese)

[67] Wan Zhexian and Yang Jinggeng, Automorphisms of the projective quaternion unimodular group of dimension 2, *Chinese Ann. Math.*, 3A: 3 (1982), 395–402. (in Chinese)

[68] Wang Fanggui, Grothendieck groups of primitive rings, *Acta Math. Sinica*, **34** : 5 (1991), 645–652. (in Chinese)

[69] Wang Jingzhou, The automorphism groups of projective modules and representations of stably free modules, Thesis, Nanjing Univ., 1991. (in Chinese)

[70] Wang Luqun, On the standard form of normal subgropus of linear groups over Φ-surjective rings, *Chinese Ann. Math.*, **5** A: 2 (1984), 229–238. (in Chinese)

[71] Wang Luqun and Zhang Yongzheng, Two-dimensional linear groups over Φ-surjective rings, *Acta Math. Sinica* , **27** : 6 (1984), 860–864. (in Chinese)

[72] Wang Yangxian, On the length problem in orthogonal groups, *Acta Math. Sinica*, **24** : 2 (1981), 291–302. (in Chinese)

[73] Wu Xiaolong, A remark on "Automorphisms of $SL_2(K)$ over a class of skew fields", *Acta Math. Sinica*, **27** : 3 (1984), 289–292. (in Chinese)

[74] Wu Xiaolong, Two–dimensional linear groups over the Gaussian domain, *Acta Math. Sinica*, **28** : 6 (1985), 731–746. (in Chinese)

[75] You Hong, $K_2(R, I)$ of unit 1-stable rings, *Chinese Science Bull.*, **35** : 19 (1990), 1590–1595.

[76] You Hong, Isomorphisms of four-dimensional symplectic groups over commutative rings, *Northeast Math. J.*, **7** : 4 (1991), 406–416.

[77] You Hong, Prestabilization for $K_1 U^\varepsilon$ of Λ-2-fold rings, *Chinese Science Bull.*, **37** : 5 (1992), 357–361.

[78] You Hong, Some remarks on decomposition of Steinberg groups, *Chinese Science Bull.*, **37** : 24 (1992), 2032–2037.

[79] You Hong, K_1 of the unitary groups over the quaternion group, *Advances in Math.*, **22** : 1 (1993), 69–73. (in Chinese)

[80] You Hong, The subgroups of GL_2 which are normalized by elementary matrices over a class of non-commutative rings, to appear in *Chinese Ann. Math.*

[81] You Hong, Commutators and unipotents in symplectic groups, to appear in *Acta Math. Sinica*.

[82] You Hong, The stabilization of unitary groups over polynomial rings, preprint.

[83] You Hong and Nan Jizhu, Decomposition of matrices into 2-involutions, to appear in *Lin. Algebra and Its Appl.*

[84] You Hong and Wang Renfa, Automorphisms of orthogonal groups over semilocal rings, *Acta Math. Sinica*, **28** : 2 (1985), 279–288. (in Chinese)

[85] Yuan Bingcheng, A generator theorem of the orthogonal group $O_n(V)$, *Chinese Ann. Math.*, **7** A: 1 (1986), 29–32. (in Chinese)

[86] Yuan Bingcheng, On the length theorem of decomposition of the orthogonal group $O_n(V)$ and its commutator subgroup $\Omega_n(V)$, *Acta Math. Sinica*, **31** : 4 (1988), 523–539. (in Chinese)

[87] Zhang Haiquan and Wang Luqun, The normal subgroups of symplectic groups over Φ-surjective rings, *Acta Math. Sinica*, **28** : 2 (1985), 270–278. (in Chinese)

[88] Zhang Haiquan and Wang Luqun, Isomorphisms of symplectic groups over Φ-surjective rings, *Chinese Ann. Math.*, **6** B : 2 (1985), 157–170.

[89] Zhang Yongzheng, Automorphisms of O_n^+ over semilocal rings, *Chinese Ann. Math.*, **9** A : 3 (1988), 311–320. (in Chinese)

[90] Zhou Fang and Li Li, A generator theorem of linear groups over local rings, *Kexue Tongbao*, **26**: 14 (1981), 893.

Some Work on the Connection Between Groups and Combinatorics

Li Huiling

Department of Applied Mathematics
Zhejiang University

After the completion of the classification of finite simple groups, some new direction in finite group theory appears and some direction becomes more active. Also some work has been done on these directions in China. The purpose of this article is to survey these studies. It consists of three parts: In Section one we describe the work on the amalgam done by Huang Jianhua and others. Section two is about the works on the connection between group theory and some combinatorial structures. The work of Qiu Weisheng on multiplier conjecture is described in section three. The work on the connection between groups and graphs will not be discussed in this article. A special paper will deal with them.

1. Some work on amalgam

"Amalgam" is a new field in the group theory, which combines the local group theoretic analysis and graph theory in order to study the local structure of groups. Because several significant results have been got recently in this field, people expect the method of the amalgam to play an important role in process to simplify the proof of the classification of finite simple groups.

Let P_1, P_2, B be three groups and ψ_1 and ψ_2 are homomorphisms from B to P_1 and P_2 respectively, then $(P_1, P_2; B, \psi_1, \psi_2)$ is called an amalgam. Let G be a group and ϕ_1, ϕ_2 be homomorphisms of P_1, P_2 into G. If

$$\psi_1\phi_1 = \psi_2\phi_2,$$

then G is called a completion of the amalgam $(P_1, P_2; B, \psi_1, \psi_2)$. Every amalgam $(P_1, P_2; B, \psi_1, \psi_2)$ possesses an universal completion $\hat{G}$ such that every completion G of this amalgam is a homomorphoim image of $\hat{G}$. It is easily seen that $\hat{G}$ is an amalgamated product of P_1, P_2 with amalgamated subgroup B. We often identify P_1, P_2, B with their images in $\hat{G}$. Thus $B \leq P_1 \cap P_2$. We say that the amalgam is primitive if there is a subgroup $K \leq B$ suth that $B = N_{P_i}(K)$.

Remark The above mentioned amalgams are indeed the amalgams of rank 2. Generally we can define the amalgam of high rank.

Now we consider a group of Lie type of rank 2. We fix a Borel subgroup $B = U \rtimes H$, where U is a unipotent subgroup and H a Carter subgroup. Then there are exactly two parabolic subgroups P_1, P_2 such that $P_1 \cap P_2 = B$. Thus if we let ψ_i be the inclusion map, then $(P_1, P_2; B, \psi_1, \psi_2)$ is an amalgam and G is a completion of it. Clearly at this time $B = N_{P_i}(U)$. Thus the parabolic system of G corresponds to an amalgam. When G is finite, then U is a Sylow p-group, where p is the characteristic of G, and P_1, P_2 are the.local subgroups of G. Every group of Lie type of rank 2 has such parabolic system. Also many sporadic simple groups have a similar "parabolic system". Hence we can use the amalgam to study the local structure of simple groups.

A fundamental work on the amalgam was done by D.Goldschmidt. In [8] Goldschmidt considered such a problem: Let G be a group generated by two subgroups X_1 and X_2, $B = X_1 \cap X_2$. Suppose that $|X_i : B| = 3$ for $i = 1$, 2, and suppose that B contains no nontrivial normal subgroup of G, and ask what can be said about the structure of the groups X_1 and X_2. Using the graph theory consideration, Goldschmidt completely answered this question. He proved that $|X_i| \mid 3 \cdot 2^7$ and determined all 15 possibilities for the structure of X_1 and X_2.

In order to study the local structure of groups the general question about the amalgam is like this: Let G be a group, p a prime. Let G be generated by two subgroups P_1 and P_2 and $B = P_1 \cap P_2$. We suppose that

(a) $B = N_{P_i}(O_p(B))$;

(b) No nontrivial normal subgroup of G is contained in B.

(c) $C_{P_i}(O_p(P_i)) \leq O_p(P_i)$.

If we know the isomorphism types of $P_i/O_p(P_i)$, $\cdot i = 1$, 2. We want to obtain the explicit information about the structure of P_i. Generally, here $O^{p'}(P_i)/O_p(P_i)$ is assumed to be a group of Lie type of rank 1 or a permutation group.

The amalgam in the above is said to be of characteristic p. If we assume $P_i/O_p(P_i) \cong M_i$, where M_i are given, we also use (M_1, M_2) to denote this amalgam.

To discuss this problem, one often define a graph Γ: The vertices of Γ are the cosets P_1g and P_2g, where $g \in G$. Two vertices are adjacent if and only if they have the forms P_1x and P_2y and $P_1x \cap P_2y \neq \emptyset$. Thus Γ is a bipartite graph and G is a group of automorphisms of Γ. G is edge-transitive but not vertex-transitive. We can easily see that the stabilizer in G of vertex P_ix is conjugate to P_i and the stablizer in G of each edge is conjugate to B. When G is the universal completion, Γ is a tree. Now that Γ is a graph, we can consider the graph properties of Γ such as degree, distance, etc, and these graph properties reflect the structure of G. The method above is due to Goldschmidt and is called the coset graph method. Stellmacher developed this method.

Recently Huang Jiahua obtained a series of results in the field of the amalgam.

In [10] he considered the amalgam $(L_3(2), \Sigma_5)$ of characteristic 2.

In [11] he considered the amalgam (Σ_5, Σ_3) of characteristic 2.

In [12] he considered the amalgam (Σ_3, X) of characteristic 3, where X is a group of Lie type of characteristic 3 and of rank 1.

In [22] he considered the amalgam $\Sigma_5, M_{11})$ of characteristic 3.

In [14] he considered the amalgam (Σ_6, M_{11}) of characteristic 3.

And in [15] he considered the amalgam in which some wreath products are involved.

Now we quote some results explicitly.

In [11] the following hypothesis was made:

Hypothesis A *Let G be a group generated by two subgroup P_1 and P_2, $B = P_1 \cap P_2$, $B = N_{P_i}(O_2(P_i))$ such that*

(i) $P_1/O_2(P_1) \cong \Sigma_5$, $P_2/O_2(P_2) \cong \Sigma_3$;

(ii) *No nontrivial normal subgroup of G is contained in B;*

(iii) $C_{P_i}(O_2(P_i)) \leq O_2(P_i)$, $i = 1, 2$.

Under this hypothesis the following theorem was proved.

Theorem *Suppose that G, P_1, P_2 and B satisfy hypothesis A. Then P_1 and P_2 are of type $Aut(U_4(2))$, M_{22}, $AutM_{22}$, HS, $Aut(HS)$, Ly or Ru.*

Here the pair (X_1, X_2) is of type X which means that X is a group containing $\tilde{X}_1$, $\tilde{X}_2$, such that $\tilde{X}_i \cong X_i$ and $\tilde{X}_1 \cap \tilde{X}_2 \cong X_1 \cap X_2$.

[11] is an important work in the amalgam theory and will be used in

the new proof of the classification of these groups.

In [14] set

$$\Delta = \{SL_2(3^n),\ L_2(3^n),\ SU(3^n),\ U_3(3^n),\ Ree(3^n),\ SL_2(5)\text{and } L_2(5)\}$$

and the following hypothesis was made.

Hypothesis B   *Let G be a group and P_1, P_2 be two finite subgroups. For $B = P_1 \cap P_2$, $S \in syl_3B$, the following hold:*

(i) $G = \langle P_1,\ P_2 \rangle$;

(ii) $O^{3'}(P_1/O_3(P_1)) \cong M_{11}$, $O^{3'}(P_2/O_3(P_2)) \cong X$, *where* $X \in \Delta$;

(iii) $B = N_{P_i}(S)$, $i = 1,\ 2$.

(iv) $C_{P_i}(O_3(P_i)) \leq O_3(P_i)$, $i = 1,\ 2$.

(v) *B contains no nontrivial normal subgroup of G.*

Then we have

Theorem   *Let G, P_1, P_2 and B satisfy the Hypothesis B, and let $L_i = O^{3'}(P_i)$, $i = 1,\ 2$. Then one of the assertions (I)-(III) holds.*

(I) *$L_1/O_3(L_1) \cong M_{11}$ and $O_3(L_1)$ is an elementary abelian group of 3^5; $L_2/O_3(L_2) \cong SL_2(3^2)$ and $O_3(L_2)$ is an extraspecial group of order 3^5;*

(II) *$L_1/O_3(L_1) \cong M_{11}$ and $O_3(L_1)$ is an elementary abelian group of order 3^5; $L_2/O_3(L_2) \cong SL_2(5)$ and $O_3(L_2)$ is an extraspecial group of order 3^6, where $|Z(L_2)| = |Z(O_3(L_2))| = 3^2$;*

(III) *$L_1/O_3(L_1) \cong M_{11}$ and $O_3(L_1)$ is an elementary abelian group of order 3^5; $L_2/O_3(L_2) \cong SL_2(3)$, $O_3(L_1)$ is an extraspecial group with $|Z(O_3(L_2))| = |Z(L_2)| = 3^2$.*

The above three possibilities occur in the sporadic simple groups Co_3, Ly and Suz respectively.

From these theorems we see that by using the amalgam method, we can determine the structure of $\{P_1,\ P_2\}$ (of $\{O_p(P_1),\ O_p(P_2)\}$, more exactly) if we know the isomorphism type of $\{P_1/O_p(P_1),\ P_2/O_p(P_2)\}$. If G possesses the parabolic system $\{P_1,\ P_2\}$ and a given simple group $\tilde{G}$ possesses a parabolic system $\{P_1^*,\ P_2^*\}$ such that $P_i/O_p(P_i) \cong P_i^*/O_p(P_i^*)$ and $O_p(P_i)$ resembles $O_p(P_i^*)$, that is as a normal subgroup of P_i and P_i^*, $O_p(P_i)$, $O_p(P_i^*)$ have the "same" principal factors of P_i and P_i^*, respectively. At this time if we know G is a simple group then we often can conclude that G is isomorphic to G^*. Hence by the aid of the amalgam we can identify a simple groups with a known simple group.

2. Some results on combinatorics

The following definition of a design is well-known. Let P be a finite set of size v, whose elements are called points, and let $\mathcal{B}$ be a collection of some subset of P called blocks with $|B| = k$ for every $B \in \mathcal{B}$. If there is an integer λ such that every pair of points of P is contained in exactly λ blocks, then we say that $\mathcal{D} = (P, \mathcal{B})$ is a $2 - (v, k, \lambda)$ design. Clearly if there are exactly b blocks in $\mathcal{B}$, then we have

$$\lambda v(v - 1) = bk(k - 1), \tag{1}$$

and every point occurs in exactly $r = \frac{bk}{v}$ blocks.

Now let g be a permutation on P. If g maps every block of $\mathcal{B}$ to a block, then we say that g is an automorphism of $\mathcal{D}$. All automorphism of $\mathcal{D}$ form a group which is called the automorphism group of $\mathcal{D}$ and is denoted by $\text{Aut}\mathcal{D}$.

Every projective plane is a $2 - (n^2 + n + 1, n + 1, 1)$-design, where n is the order of this plane. If D is a Desargusian plane of order q then $\text{Aut}\mathcal{D} = P\Gamma L(3, q)$.

The properties of the automorphism group of a design are strong restrictions on its construction. Thus we often assume that the automorphism groups satisfy some transitivity condition to study the designs. We say that $G \leq \text{Aut}\mathcal{D}$ is point-transitive (block-transitive) if G is transitive on the point set (on the block sets, respectively). A pair (p, B) is called a flag, if p is a point and B is a block and $p \in B$. If G is transitive on all flags of $\mathcal{D}$, then we say that G is flag-transitive. Obviously, if G is flag-transitive then G is both block-transitive and point-transitive. Also we can prove that if G is block-transitive then G is point-transtive (See [2], for example). Similarly we can define the point-primitivity and the block-primitivity of a group $G \leq \text{Aut}\mathcal{D}$. D.G.Higman and J.Machaughlin proved that if $G \leq \text{Aut}\mathcal{D}$ is flag-transitive on $\mathcal{D}$ then G is point-primitive (see [2]). Recently Buekeubout et al ([3]) determined all flag-transitive $2 - (v, k, 1)$ designs. Thus we need to consider the designs with block-transitive or block-primitive automorphism groups.

If $\mathcal{D}$ is a design and $\text{Aut}\mathcal{D}$ is block-transitive, then we say that $\mathcal{D}$ is a block-transitive design. Similarly we can speak of block-primitive designs, point-primitive designs and so on. About block-primitive designs there is a conjecture:

Conjecture *If $\mathcal{D}$ is a $2 - (v,\ k,\ 1)$-design and $G \leq AutD$ is block-primitive on $\mathcal{D}$, then G is point-primitive.*

This conjecture is true under each one of the following assumptions.

(1) $\mathcal{D}$ is a projective space of dimension $d \geq 2$;

(2) $k|v$;

(3) $v \geq \frac{1}{2}((\binom{k}{2}) - 1)^2$;

(4) $k < 30$;

(5) G has a subgroup acting regularly on the point set P of $\mathcal{D}$; and

(6) G has *rank* ≤ 7 when viewed as a permutation group on the block set $\mathcal{B}$.

In the last years several results have been obtained by my students and myself on this conjecture. We introduce four new parameters: Let

$$b_1 = (b,\ v), \qquad b_2 = (b,\ v-1),$$

$$k_1 = (k,\ v), \qquad k_2 = (k,\ v-1).$$

Then it is clear that

$$b = b_1 b_2, \qquad k = k_1 k_2$$

$$(b_1,\ b_2) = (b_1,\ k_2) = (k_1,\ b_2) = (k_1,\ k_2) = 1.$$

Now we suppose that G is block-primitive but point-imprimitive and

$$P = \Delta_1 \cup \Delta_2 \cup \cdots \cup \Delta_c \tag{2}$$

is a partition of P fixed by the action of G with $|\Delta_i| = s$. Then we can easily deduce that

$$s \equiv c \equiv 1 \pmod{b_2}. \tag{3}$$

From this we set

$$s = 1 + x b_2, \quad c = 1 + y b_2, \tag{4}$$

where x and y are integers. Then we have

$$xy < k_2^2 \tag{5}$$

and $k_1 < k_2(k_2 + 1)^2$.

Hence we have

Theorem *For a given k_2, there are at most finitely many counterexamples for this conjecture.*

We also proved

Theorem   ([7], [26])*If $k_2 \leq 10$, then the conjecture is true.*

Remark $k_2 = 1$ means that $k|v$, thus our results generalized the well-known Camina-Gagen theorem.

Also, as Delandtsheer proved if $(\mathcal{D}, G)$ is a counterexample, where $\mathcal{D}$ is a $2 - (v, k, 1)$ design and $G \leq \text{Aut}\mathcal{D}$ is block-primitive and point-imprimitive, then G is almost simple, that is there is a nonabelian simple group T such that $T \triangleleft G \leq$
$\text{Aut}T$. Thus we consider the case in which T is isomorphic to an alternating group. We have

Theorem ([24])   *If $(\mathcal{D}, G)$ is a counterexample and $T \leq G \leq \text{Aut}T$, then T is not isomorphic to any alternating group.*

Liu also considered the case in which T is a sporadic simple group.

In addition to this Liu proved that

Theorem ([25])   *The conjecture is true if $k \leq 40$.*

This theorem improved the result of above fact 4).

We also studied block-transitive $2 - (v, k, 1)$ designs for small value of k. The classification of block-transitive $2 - (v, 3, 1)$ designs was completed more than ten years ago. Recently in [4] Camina and Siemons studied the $2 - (v, 4, 1)$-designs admitting a solvable, block-transitive automorphism group. We consider the other case in which the block-transitive automorphism group is unsolvable, and we get

Theorem ([20])   *If $\mathcal{D}$ is a $2 - (v, 4, 1)$ design and $G \leq \text{Aut } \mathcal{D}$ is unsolvable, block-transitive, then G is flag-transitive.*

We also considered block-transitive $2 - (v, 5, 1)$ designs, We suppose that $G \leq \text{Aut}\mathcal{D}$, G is solvable and block-transitive and $\mathcal{D}$ is a $2 - (v, 5, 1)$ designs. Then $v = p^a$ for some prime p and an integer a and $G \leq A\Gamma L(1, v)$ with just one exceptionin which $\mathcal{D}$ is the Desarguesian projective plane of order 4. (Tong Wenwen [35], unpublished).

Remark   In the result of Camina and Siemons on $2 - (v, 4, 1)$ designs with solvable and block-transitive automorphism group and Tong's similar result about $2 - (v, 5, 1)$ designs they merely showed that $G \leq A\Gamma L(1, v)$ for some $v = p^a$. But they didn't tell if such a design exists.

Besides the study of block-transitive designs, I would like mention some results I got about other combinatorial structures.

Let Γ be a geometric lattice and $G \leq \text{Aut}\Gamma$ be a group of automorphisms of Γ. If $\text{Aut}\Gamma$ is transitive on the ordered bases of Γ, then we say that Γ is basis-transitive, and if $\text{Aut}\Gamma$ is transitive on the unordered bases of Γ, then we say that Γ is basis-homogeneous. All basis-transitive geometric lattices have been determined by W.Kantor. Thus we consider the basis-homogeneous geometric lattices. In [19] I determined all basis-homogeneous geometric lattices of dimension 2 and 3. Also I proved a reduction theorem:

Theorem   *Let Γ be a geometric lattice of dimension > 0 and let G be a group of automorphisms of Γ. If G is transitive on unordered bases of Γ, then Γ is the direct sum of sublattices Γ_1, $\Gamma_2, \ldots$, Γ_s, where every Γ_i has a basis-homogeneous and point-transitive automorphism group.*

Then A.Delandtsheer classified the basis-homogeneous geometric lattices of arbitrary dimension with at least one line of size greater than 2. Recently Delandtsheer and I proved another reduction theorem which enables us to reduce the study of basis-homogeneous and point-transitive geometric lattices to that of basis-homogeneous and point-primitive geometric lattices([5]).

3. Qiu Wen-shen's work on the Multipler Conjecture

We begin with the definition of difference sets in groups.

Let G be a finite group of order v and D a subset of G with $1 < |D| = k < |G|-1$. Let λ be an integer. If for any nontrivial element g of G, there are exactly λ pairs (c_i, d_i), where c_i, $d_i \in D$ and $i = 1, 2, \ldots, \lambda$, such that $c_i d_i^{-1} = g$, then D is called a (v, k, λ)-difference set in G. Clearly we have

$$\lambda(v - 1) = k(k - 1). \tag{1}$$

If we set $n = k - \lambda$, then we call n the order of D and then we have

$$\lambda v = k^2 - n. \tag{2}$$

There is a close connection between difference sets and symmetric designs. Let G be a group of order v and D a (v, k, λ)-difference set in D. Call the elements of G points and the subsets of form $Dg = \{dg \mid d \in D\}$, $g \in G$, blocks. Then we get a symmetric design, which admits G as a regular group of automorphisms. Conversely, if $\mathcal{B}$ is a symmetric design and G is a group of automorphisms of $\mathcal{B}$, which is regular on both point set and block set of $\mathcal{B}$. Then every block D of $\mathcal{B}$ is a (v, k, λ)-difference set in G.

Let D be a difference set in a finite group G. If an automorphism τ of G sends D to Dg for some $g \in G$, then τ is called a multiplier of D.

The first multiplier theorem was proved by M.Hall:

The first multiplier theorem   *Let G be an abelian group with a (v, k, λ)-difference set D and let p be a prime number dividing n but not v. If $p > \lambda$, then $\mu_p : g \to g^p$, $\forall g \in G$, is a multiplier of D.*

The condition "$p > \lambda$" in this theorem is crucial for the proof. But for every known difference set for a prime p, μ_p is multiplier if $p|n$ and $(p, v) = 1$. Thus we have the following conjecture:

Multiplier Conjecture   *The first multiplier theorem holds without the assumption that $p > \lambda$.*

Many efforts have been made by several authors to weaken the condition "$p > \lambda$". See [1], [17]. The following second multiplier theorem replaces the condition "$p > \lambda$" by "$n_1 > \lambda$":

Second Multiplier Theorem   *Let G be an abelian group with a (v, k, λ)-difference set D and let v_0 be the exponent of G. Let n_1 be a divisor of $n = k - \lambda$ such that $(n_1, v) = 1$ and $n_1 > \lambda$. Suppose t is an integer such that for every prime divisor p of n_1, there is a positive integer j such that $t \equiv p^j (\mod v_0)$. Then $\mu_t : g \to g^t$, $\forall g \in G$, is a multiplier of D.*

Recently Qiu Weisheng studied the Multiplier Conjecture using the representation theory of finite groups and the algebraic number theory and obtained a series of new results. Here we list some of them as follows:

(1) If $n = n_1$, then the second multiplier theorem holds without the assumption "$n_1 > \lambda$".

(2) If $\frac{n}{n_1} = d$ and every prime divisor q of d is a divisor of n_1, then the second multiplier theorem holds without the assumptiom "$n_1 > \lambda$".

(3) If $\frac{n}{n_1} = d$, d a prime and $(d, n_1) = 1$ and G is an elementary abelian group of order p^r ($r > 1$ and $p \neq d$), and the order ω of d modulo p satisfies $\omega > d^2/3$, then in the majority of cases, the assumption "$n_1 > \lambda$" may be removed.

(4) If $\frac{n}{n_1} = 2$, then the second multiplier theorem holds without the assumption "$n_1 > \lambda$", provided that one of the following conditions holds:

(i)$2|n_1$;

(ii) n_1 is odd and $(v, 7) = 1$;

(iii) n_1 is odd and $7^2|v$;

(iv) n_1 is odd, $7|v$ and t is a quadric residue $\mod 7$.

(5) If $\frac{n}{n_1} = 3$ and $(v, 3) = 1$, then in the majority of cases the assumption"$n_1 > \lambda$" can be removed.

(6) If $\frac{n}{n_1} = 4$ and $(v, 2 \cdot 7 \cdot 31) = 1$, then the assumption "$n_1 > \lambda$" can be removed.

To obtain these results Qiu Weisheng developed a method of studing the Multiplier Conjecture. Now we give a brief description of his method.

Let G be a finite abelian group. Then its complex irreducible characters form a group $\hat{G}$ isomorphic to G. Also the group ring KG over a field K is isomorphic to $K\hat{G}$. In particular the group ring $Z\hat{G}$ of $\hat{G}$ over the integer ring Z is isomophic to the ring of generalized characters of G. Now let $D = \{g_{r_1}, \ldots, g_{r_k}\}$ is a subset of G. Then D is a (v, k, λ)-difference set if and only if

$$\sum_{i=1}^{k} g_{r_i} \sum_{j=1}^{k} g_{r_j}^{-1} = ng_1 + \lambda \sum_{l=1}^{v} g_l \qquad (3)$$

holds in the group ring, where g_1 is the identity of G. Equivallently, D is a (v, k, λ)-difference set if and only if

$$\sum_{i=1}^{k} \chi_{r_i} \sum_{j=1}^{k} \bar{\chi}_{r_j} = n\chi_1 + \lambda \sum_{l=1}^{v} \chi_l \qquad (4)$$

holds in the group ring $Z\hat{G}$, where χ_l is the image of g_l under the isomorphism of G to $\hat{G}$.

Let m be an integer, for $A = \sum_{l=1}^{v} a_l g_l \in QG$. We set

$$\hat{A} = \sum_{l=1}^{v} a_l \chi_l, \quad A^{(m)} = \sum_{l=1}^{v} a_l g_l^m. \qquad (5)$$

and set

$$\hat{D} = \sum_{i=1}^{k} \chi_{r_i}. \qquad (6)$$

Using these notation we know that μ_t is a multiplier if and only if

$$\hat{D}^{(t)} = \chi_s \hat{D} \qquad (7)$$

for some χ_s. Equivalently, μ_t is a multiplier if and only if

$$\hat{D}^{(t)} \hat{D}^{(-1)} = n\chi_s + \lambda \bar{G} \qquad (8)$$

for some χ_s, where $\bar{G} = \sum_{l=1}^{v} \chi_l$.

Qiu's method consists of three parts:

(I) A necessary and sufficient condition for μ_t to be a multiplier.

Qiu proved a lemma by using the algebraic number theory.

Lemma   Let $G = \{g_1 = 1, \, g_2, \, \ldots, \, g_v\}$ be an abelian group of order v and let v_0 be its exponent. Let n be an integer and $n = dn_1$ $(n_1 > 1)$ with $(n_1, \, v) = 1$. Suppose t is an integer such that for every prime divisor p of n_1, there is a positive integer j such that $t \equiv p^j \pmod{n_0}$. Let $g_i \to \chi_i$ be the isomorphism of G to $\hat{G}$. Let $\hat{A} = \sum_{l=1}^{v} a_l \chi_l \in Z\hat{G}$ and $\sum_{l=1}^{v} a_l = k$. If

$$\hat{A}\hat{A}^{(-1)} - \lambda \bar{G} = n\chi_1, \tag{9}$$

where $n = k - \lambda$ and $k^2 - \lambda v = k$ for some λ, then there exists $\hat{C} = \sum_{l=1}^{v} c_l \chi_l \in Z\tilde{G}$, such that

$$\hat{A}^{(t)} \hat{A}^{(-1)} - \lambda \hat{G} = n_1 \hat{C} \tag{10}$$

and $\hat{C}$ satisfies

$$\sum_{l=1}^{v} C_l = d, \tag{11}$$

$$\hat{C}\hat{C}^{(-1)} = d^2 \chi_1. \tag{12}$$

We observe that for every $\chi_l \in \tilde{G}$, $\xi = d\chi_l$ satisfies (11) and (12). Now suppose $D = \{g_{i_1}, \, \ldots, \, g_{i_r}\}$ is a $(v, \, k, \, \lambda)$-difference set in G and μ_t is a multiplier of D. Then $\hat{D} = \sum_{l=1}^{k} \chi_{i_l}$ does satisfies (9) and (8) shows that there is a χ_s such that $\xi = d\chi_s$ satisfies (10) (11) and (12). We call equations (11) and (12) CH-equations of d for G and call $\xi = d\chi_l$ $(l = 1, \, 2, \, \ldots, \, v)$ a trivial solution of the CH-equations.

Then the following theorem is an easy consequence of the above lemma.

Theorem   Let D be a $(v, \, k, \, \lambda)$-difference set in an Abelian group G and n, n_1 and t satisfy the all conditions of the above lemma. Then μ_t is a multiplier of D if and only if no nontrivial solution ξ of the CH-equations d in G satisfies

$$\hat{D}^{(t)} \hat{D}^{(-1)} - \lambda \hat{G} = n_1 \xi. \tag{13.}$$

From this theorem we see that we need to consider the nontrivial solutions of the CH-equations.

(II) Qiu gave a method of finding all solutions of the CH-equations of d for G when $d = q^e$ is a prime power.

Indeed at this time Qiu shows that if $\hat{C} = \sum_{l=1}^{v} c_l \chi_l$ is a solution of the CH-equations, then there is a character χ_r such that

$$\hat{C}^{(q)} = \hat{C} \chi_r. \tag{14}$$

And so the set $\{c_1, \ldots, c_v\}$ of coefficients in the disjoint union of several subsets such that the coefficients is the same subset are equal.

(III) Qiu gave a method to check if a nontrivial solution of tha CH-equations satisfies the equation (10). Here the algebraic number theory is needed.

References

[1] T.Beth, D.Jungnicke and H.Lenz, *Design theory*, Cambridge University Press, Cambridge, (1986).

[2] F.Buekenhout, J.Doyen, A.Delandtsheer, *Finite linear spaces with flag-transitive groups*, J. of Comb Theory, Series A, **49** (1988), 268-293.

[3] F.Buekenhout, J.Doyen, and A.Delandtsheer, P.B.Kleidman, M.W.Liebeck and J.Sarl, *Linear spaces with flag-transitive automorphism groups*, Geom. Dedicata, **36** (1991), 89-94.

[4] A.Camina and J.Siemons, *Block-transitive automorphism groups of* $2 - (v, k, 1)$ *block designs*, J. Comb. Theory, Series A, **51** (1989), 268-276.

[5] A.Delandtsheer and Li Huiling, *Basis-transitive matroids*, J. Algebraic Combinatorics, 3 (1994), 285-290.

[6] A.Delegado, D.Goldschmidt, B.Stellmacher, *Groups and graphs: new results and new methods*, DMV seminar, Bd. 6, Bassel-Boston-Stuttgart, (1985).

[7] Fan Weidong and Li Huiling, *A generalization of Camina-Gagen theorem*, J. of Math., 13 (1993), 437-442.

[8] D.Goldschmidt, *Automorphism of trivalent graphs*, Ann. of Math., **111** (1980), 377-406.

[9] D.Gorenstein, *Finite simple groups*, Plenum Press, New York-London (1982).

[10] Huang Jianhua, *A characterization of the 3-local subgroups of HS and Ru*, Lecture Notes in Math., **1185**, 289-307.

[11] Huang, Stellmacher and Stroth, *Some parabolic systems of rank 2 related to sporadic groups*, J. Alg., **106** (1986), 78-118.

[12] Huang Jianhua, Ma Jimin and Stellmacher, *Amalgam of rank 2 in char. 3 involving* $L_2(5)$, Acta Math Sinica, **3** (1989), 263-270.

[13] D.Jungnickel, *Design theory: An Update*, Ars Combinatoria, **28** (1989), 129-199.

[14] Huang Jianhua and Li Huiling, *Amalgam of rank 2 involving* M_{11}, To appear in Science of China.

[15] Huang Jianhua, *Amalgam of rank 2 involving wreath products*, Unpublished.

[16] Huang Jianhua and Li Huiling, *Building, Amalgam and lucidence geometries, I, BN-pairs and Amalgam*, Chinese Adv. Math., , **18** (1989), 251-269.

[17] Li Huiling, *Two pairs of simpley connected geometries*, European J. of Comb., **6** (1985), 245-251.

[18] Li Huiling, *On 2-connectedness of geometries*, Lecture Notes in Math., **1185**, 308-314.

[19] Li Huiling, *On basis-transitive geometric lattices*, European J. of Combi., **10** (1989), 561-573.

[20] Li Huiling, *On block transitive* $2 - (v, 4, 1)$ *designs,* J. of Comb. theory, Series A, **69** (1995), 115 – 124.

[21] Li Huiling, *On block-transitive* $2 - (v, 5, 1)$ *designs*, In preparation.

[22] Li Huiling and Huang Jianhua, *A characterization of 3-local subgroups of Ly*, Act. Math. Sinica, **7** (1991), 66-74.

[23] Li Huiling and Huang Jianhua, *Buildings, Amalgam and Incidence geometries, II. Buildings and Incidence geometries*, Chinese Adv. Math, **18** (1989), 385 – 401.

[24] Li Huiling, *On the block-primitive designs, I*, In preparation.

[25] Liu Weijun, *Block-primitive* $2 - (v, k, 1)$-*designs for small value of k*, Submitted to Math. J. (in Chinese).

[26] Liu Weijun, *Block-primitive* $2 - (v, k, 1)$-*designs with* $(k, k-1)$ *small Unpublished*.

[27] Qiu Weisheng, *Proving the multiplier Theorem using the representation theory of groups.*, Northeastern Math. J., **9** (1993), 115-118.

[28] Qiu Weisheng, *A character approch to the multiplier conjecture and a new result on it*, submitted to Arc Comb.

[29] Qiu Weisheng, *A method of studing the multiplier conjecture and some partial solutions for it*, to appear in Arc Comb.

[30] Qiu Weisheng, *Further results on the multiplier conjecture for* $n = 2n_1$ *and* $n = 3n_1$.

[31] ——, *A new result on the Multiplier Conjecture for* $\frac{n}{n_1} = 4$, submitted to J. Comb. Designs.

[32] ——, *The Multiplier Conjecture for elementary abelian groups*, submitted to J. Comb. Designs.

[33] ——, *A necessary condition on the existence of Abelian Difference sets*, To appear in Discrete Math.

[34] ——, *On the Multiplier Conjecture*, To appear in Acta Math. Sinica.

[35] Tong Wenwen, *Solvable, block-transitive groups of automorphisms of* $2 - (v, 5, 1)$ *designs*.

On the Subgroup Structure
of Classical Groups

Shangzhi Li

Department of Mathematics, University of Science and Technology of China,
Hefei, Anhui 230026, China

It is an important topic in group theory to investigate the maximal subgroups of classical groups. The goal of this investigation is, of course, to give a complete classification of all these maximal subgroups. But this is very difficult. So, researchers try to make studies to approach this final goal step by step. They do classifications for certain types of maximal subgroups or for certain parts of classical groups. And they investigate the maximality of certain classes of subgroups. Among all the works on this topic, the most significant one is certainly the theorem of M.Aschbacher about the maximal subgroups of finite classical groups, published in [1]. Because of this theorem, and the works of P.Kleidman-M.Liebeck [2], G.Seitz [3], etc., the classification of maximal subgroups of finite classical groups is hopeful to complete. There are also many group theorists, such as R.H.Dye, O.King, N.A.Vavilov, doing works for the classical groups over infinite fields. However, so far there is no theorem for infinite classical groups like Aschbacher's for the finite case.

Chinese group theorists, including Mr. Zha Jian-guo, Mr. Ren Jin-jiang and the author, have also done much work on this topic. This paper is going to give an illustration of these studies.

§1. Definitions about classical groups

We need some notations about matrices. $\text{Mat}_{m \times n} R$ (resp. $\text{Mat}_n R$) denotes the set of all the $m \times n$ (resp. $n \times n$) matrices over a ring R. ^tA denotes the transpose of a matrix $A \in \text{Mat}_{m \times n} R$. E_{ij} denotes a matrix having a single 1 at the (i, j)th position, and all other entries zero. $T_{ij}(s) = I + sE_{ij}$ for any $i \neq j$ and $s \in R$.

Let K be an arbitrary division ring, $V = V(n, K)$ an n-dimensional left K-space. The *general linear group* $\mathrm{GL}(V)$ on V is by definition the multiplicative group of all the invertible K-linear transformations on V. The *special linear group* $\mathrm{SL}(V)$ is the subgroup of $\mathrm{GL}(V)$ generated by all the transvections $\tau_{u,\phi} : x \mapsto x + \phi(x)u$ on V, where $0 \neq u \in V$, and $0 \neq \phi \in V^*$ is a nonzero K-linear function on V satisfying $\phi(u) = 0$.

With respect to a basis $\{e_i \mid 1 \leq i \leq n\}$ we can write V as the space $\mathrm{Mat}_{1 \times n}K$ of n-dimensional K-rows. And then, $\mathrm{GL}(V)$ can be identified with the multiplicative group $\mathrm{GL}(n, K)$ of all the invertible matrices in $\mathrm{Mat}_n K$, each $A \in \mathrm{GL}(n, K) \subset \mathrm{Mat}_n K$ corresponds to the element $A_R : v \mapsto vA$ in $\mathrm{GL}(V)$. In this sense, $\mathrm{SL}(V)$ is just the matrix group $\mathrm{SL}(n, K)$ generated by all the $T_{ij}(s) \in \mathrm{GL}(n, K)$ $(i \neq j, 0 \neq s \in K)$.

If K admits an involutorial anti-automorphism $J : a \mapsto \bar{a}$, then we can consider the set $\mathrm{Sesq}(V)$ of all the J-*sesquilinear forms* $h(x, y) = xH{}^t\bar{y}$ on $V = \mathrm{Mat}_{1 \times n}K$, associated with matrices $H \in \mathrm{Mat}_n K$. For each $h \in \mathrm{Sesq}(V)$ we define $\bar{h} \in \mathrm{Sesq}(V)$ by $\bar{h}(x, y) = \overline{h(y, x)}$, $\forall x, y \in V$. For a given $\varepsilon = 1$ or -1 in K^*, an ε-*hermitian matrix* $H \in \mathrm{Mat}_n K$ (satisfying ${}^t\overline{H} = \varepsilon H$) corresponds to an ε-*hermitian form* $f(x, y) = xH{}^t\bar{y}$ which satisfies $\bar{f} = \varepsilon f$. If moreover, all the $f(x, x)$ $(x \in V)$ lie in the set $K_{J,\varepsilon} = \{a + \varepsilon\bar{a} \mid a \in K\}$, then f is said to be *trace-valued*, this is just to require all the diagonal entries of H lie in $K_{J,\varepsilon}$. A nondegenerate trace-valued ε-hermitian form $f(x, y) = xH{}^t\bar{y}$ on V determines a *unitary group* $\mathrm{U}(V, f) = \mathrm{U}(n, K, f) = \{g \in \mathrm{GL}(V) \mid f(xg, yg) = f(x, y), \forall x, y \in V\}$, which coincides with the matrix group $\mathrm{U}(n, K, H) = \{A \in \mathrm{Mat}_n K \mid AH{}^t\overline{A} = H\}$, and is a normal subgroup of the *generalized unitary group* $\mathrm{GU}(V, f) = \mathrm{GU}(n, K, f) = \{g \in \mathrm{GL}(V) \mid f(xg, yg) = \gamma_g f(x, y)$ for all $x, y \in V$ and a $\gamma_g \in K^*\}$. In the special case $J = 1$ (hence K is a field), $\mathrm{U}(n, K, f)$ is a *symplectic group* $\mathrm{Sp}(n, K, f)$ when $\varepsilon = -1$, or an *orthogonal group* $\mathrm{O}(n, K, f)$ when $\mathrm{char}\, K \neq 2$ and $\varepsilon = 1$, while $\mathrm{GU}(V, f)$ becomes $\mathrm{GSp}(v, f), \mathrm{GO}(V, f)$ respectively. $\mathrm{SU}(n, K, f) = \mathrm{U}(n, K, f) \cap \mathrm{SL}(n, K)$ is called the *special unitary group* on V. A unitary group $\mathrm{U}(n, K, f) \neq \mathrm{O}(n, K, f)$ with $\nu(f) \geq 1$ contains *unitary transvections* $\rho_{u,s} : x \mapsto x + f(x, u)su$ associated with isotropic $0 \neq u \in V$ (with $f(u, u) = 0$) and $s = -\varepsilon\bar{s} \in K^*$, all these unitary transvections generate a subgroup $\mathrm{TU}(n, K, f) \trianglelefteq \mathrm{U}(n, K, f)$. When $\mathrm{U}(n, K, f) = \mathrm{Sp}(n, K, f)$ we have $\mathrm{TU}(n, K, f) = \mathrm{Sp}(n, K, f)$. In case K is a field and $J \neq 1$ we have $\mathrm{TU}(n, K, f) = \mathrm{SU}(n, K, f)$.

A trace-valued ε-hermitian form f can be written in the form $f =$

$h + \varepsilon \overline{h}$ with $h \in \mathrm{Sesq}(V)$. Each such h determines a function $Q(x) = h(x,x) = xH {}^t\overline{x}$ on V, called a *pseudo-quadratic form*, which is a *quadratic form* when $J = 1$. Denote $K^{J,-\varepsilon} = \{a \in K \mid \overline{a} = -\varepsilon a\} \supseteq K_{J,-\varepsilon} = \{a - \varepsilon \overline{a} \mid a \in K\}$. Let L be an additive subgroup of K satisfying $K_{J,-\varepsilon} \subseteq L \subseteq K^{J,-\varepsilon}$ and $aL\overline{a} \subseteq L$ ($\forall a \in K$). (We may assume $\varepsilon = -1$ when $\mathrm{U}(V,f) \neq O(V,f)$, in such case $K_{J,-\varepsilon} = \mathrm{Tr}\, K$ consists of the traces $a + \overline{a}$ ($a \in K$), while $K^{J,-\varepsilon} = K_J$ consists of the symmetric elements.) We define $\mathrm{U}(V,h,L) = \mathrm{U}(n,K,h,L) = \{g \in \mathrm{U}(n,K,f) \mid Q(xg) - Q(x) \in L$, $\forall x \in V\}$, which we may consider as a *'pseudo-orthogonal group'*. And $\mathrm{GU}(V,h,L) = \{g \in \mathrm{U}(n,K,f) \mid Q(xg) - \gamma_g Q(x) \in L$ for all $x \in V$ and a $\gamma_g \in K^*\}$. The case $L \neq K$ holds when $\mathrm{U}(n,K,f) \neq \mathrm{Sp}(n,K,f)$, in such case $\mathrm{U}(n,K,h,L)$ coincides with $\mathrm{U}(V,Q,L) = \mathrm{U}(n,K,Q,L) = \{g \in \mathrm{GL}(n,K) \mid Q(xg) - Q(x) \in L, \forall x \in V\}$. In particular, $\mathrm{U}(n,K,Q,0)$ is just the *orthogonal group* $O(n,K,Q)$. We have $\mathrm{U}(n,K,h,L) \leq \mathrm{U}(n,K,f)$ by definition, the equality holds if and only if $L = K^{J,-\varepsilon}$. $\mathrm{U}(n,K,h,L) \lneq \mathrm{U}(n,K,f) \iff L \subsetneq K^{J,-\varepsilon} \implies K_{J,-\varepsilon} \subsetneq K^{J,-\varepsilon} \implies \mathrm{char}\, K = 2$. We require Q be L-regular, i.e., $Q(x) \notin L$ for all nonzero $x \in V^\perp$. When $\mathrm{char}\, K \neq 2$ this implies that f is nondegenerate. In case $\mathrm{char}\, K = 2$ and f is degenerate, the group $\mathrm{U}(V,h,L)$ is said to have *defect*. In such case we can write $V = W \oplus V^\perp$ with W nondegenerate, and identify $\mathrm{U}(V,h,L)$ with the $\mathrm{U}(W,h|_W,L_1)$ induced on $V/V^\perp \cong W$, for $L_1 = \{s + aQ(x)\overline{a} \mid s \in L, a \in K\} \supseteq L$, to reduce to the case in which f is nondegenerate. In the special case $K_{J,-\varepsilon} = 0$ (hence $J = 1$ and $\varepsilon = 1$), we have $O(n,K,Q) = \mathrm{U}(n,K,f) = O(n,K,f)$ when $\mathrm{char}\, K \neq 2$, while $O(n,K,Q) < \mathrm{U}(n,K,f) = \mathrm{Sp}(n,K,f)$ when $\mathrm{char}\, K = 2$. In the later case we have $K^{J,-\varepsilon} = K^{J,1} = K$, L is just a K^2-subspace of K, $\mathrm{U}(n,K,Q,L)$ with $L \neq 0$ is essentially an *orthogonal group with defect*, which will be written as $O(n,K,Q,L)$. We write $\Omega(n,K,Q), \Omega(n,K,Q,L)$ to denote the commutator group of an orthogonal group $O(n,K,Q)$ resp. $O(n,K,Q,L)$. A vector u with $Q(u) \in L$ is said to be *L-singular*. A subspace consisting of mutually orthogonal L-singular vectors is said to be *totally L-singular*. The *Witt index* $\nu_L(h)$ is defined to be the greatest dimension of totally L-singular subspaces of V, which is also written as $\nu_L(Q)$ when $L \neq K$, or as $\nu(Q)$ when $L = 0$. For a subspace W we write $\nu_L(W)$ to mean $\nu_L(h|_W)$. When $\nu_L(h) \geq 1$ and $L \neq 0$, all the $\rho_{u,s} : x + \phi(x,u)su$ with $0 \neq u \in V, Q(u) \in L$ and $0 \neq s \in L$ generate a subgroup $\mathrm{TU}(n,K,h,L) \trianglelefteq \mathrm{U}(n,K,h,L)$, also written as $\mathrm{TU}(n,K,Q,L)$ when $L \neq K$. Note that $\Omega(n,K,Q,L) = \mathrm{TU}(n,K,Q,L)$ when $L \neq 0$. For the convenience of notation, we denote $U'(n, K, h, L) = TU(n, K, h, L)$ when $L \neq 0$, while $U'(n, K, h, L) = \Omega(n, K, Q)$ when $L = 0$ (i.e., when $U'(n, K, h, L) = O(n, K, Q)$.) As usual, we write $O^+(2m, q), O^-(2m, q)$ to denote an orthogonal group $O(2m, F_q, Q)$ over a finite field F_q with $\nu(Q) = m, m - 1$ respectively, and write $\Omega^{\pm}(2m, q)$ to denote their commutator groups. Moreover, for an arbitrary field F we write $O^+(2m, F), \Omega^+(2m, F)$ to denote $O(2m, F, Q)$ resp. $\Omega(2m, F, Q)$ with $\nu(Q) = m$.

For more detailed discussion of the 'pseudo-orthogonal' groups, see [4].

§2. Maximal subgroups containing root subgroups

As groups of Lie types, classical groups have root subgroups. However, we can define root subgroups of the classical groups directly, without refering Lie theory.

A *root subgroup* of $\mathrm{SL}(n, K)$ is by definition a subgroup $T_{u,\phi} = \{\tau_{su,\phi} \mid s \in K\}$ for given u, ϕ, or in matrix language, a conjugate of any subgroup $\{T_{ij}(s) \mid s \in K\}$ with $i \neq j$.

For a $TU(n, K, h, L) \neq \Omega(n, K, Q, L)$, we define its *long-root subgroup* to be a $T_u = \{\rho_{u,s} \mid s \in L\}$ for any given L-singular line Ku. If $\nu_L(h) \geq 2$ moreover, then for each pair of u, w spanning a totally L-singular plane it associates with an element $\eta_{u,w} : x \mapsto x + f(x, w)u + f(x, u)w$ of $TU(n, K, h, L)$, called a *2-transvection*, and the subgroup $T_{u,w} = \{\eta_{su,w} \mid s \in K\}$ is called a *short-root subgroup* of $TU(n, K, h, L)$.

For an othogonal group $O(n, K, Q)$ with the Witt index $\nu(Q) \geq 1$, given an orthogonal pair of vectors u, w with $Q(u) = 0$ and $\dim\langle u, w\rangle = 2$, it associates an element $t_{u,w} : x \mapsto x + f(x, w)u - f(x, u)(w + Q(w)u)$ of $\Omega(n, K, Q)$, called a *root element* of $\Omega(n, K, Q)$, while the group $T_{u,w} = \{t_{su,w} \mid s \in K\}$ is called a *root subgroup* of $\Omega(n, K, Q)$. If $Q(w) = 0$ (resp. $Q(w) \neq 0$) moreover, we call $t_{u,w}$ a *long-root* (resp. *short-root*) *element* and call $T_{u,w}$ a *long-root* (resp. *short-root*) *subgroup*.

We hope to classify the maximal subgroups containing root subgroups for all the classical groups. This was done in a series of our papers [5–14] for the classical groups over fields. A topic related to this is to classify the irreducible subgroups generated by sets of root subgroups. J.McLaughlin did this in [15, 16] in 1969 for $\mathrm{SL}(n, F)$ over a field F. And I generalized this work in [17] in 1989 to $\mathrm{SL}(n, K)$ over an arbitrary division ring. B.S. Stark determined in [18] in 1974 the irreducible subgroups of an $O(n, F, Q)$ generated by sets of long-root subgroups, but she missed a class of example of such subgroups. In [19] in 1979, W. Kantor determined all the subgroups generated by sets of long-root elements for all the classical groups over finite fields. Mr. Ren Jin-Jiang, a graduate student directed by me, classified in his doctoral dissertation [20] all the irreducible subgroups generated by long-root subgroups and all the maximal subgroups containing long-root subgroups, for a $TU(n, K, Q, L)$ over a skewfield K. We also note that, in [21, 22] (F.G. Timmesfeld) the terminology of K-root subgroups is defined in view of abstract groups, and all the groups generated by K-root subgroups were classified up to isomorphism.

Our results about root subgroups are as follows.

1. The results for $SL(n, K)$ over an arbitrary division ring K:

Theorem 2.1 *Let X be an irreducible subgroup of $SL(n, K)$ over a division ring K, generated by a set of root subgroups of $SL(n, K)$. Then one of the following holds:*

1) $X = SL(n, K)$.

2) *K is a field, n is even, $X = Sp(n, K, f)$ for some f.*

3) *$K = F_2$, n is even and ≥ 4, X is a subgroup of $Sp(n, 2)$ isomorphic to orthogonal group $O^{\pm}(n, 2)$ ($n \neq 4$ in $O^+(n, 2)$ case) or symmetric group S_{n+1} or S_{n+2}.*

Corollary 2.1 *The subgroups in the following classes are maximal in $SL(n, K)$. Any maximal subgroup of $SL(n, K)$ containing root subgroups is necessarily in one of these classes.*

1) *The stabilizer of a nonzero proper subspace of V.*

2) *The stabilizer of a decomposition $V = V_1 \oplus \cdots \oplus V_k$ with all $\dim V_i = d = n/k \geq 2$ ($1 \leq i \leq k$), with exceptions: i) $F = F_2$ and $d = 2$; or ii) $F = F_2$, $n = 4$ and $d = 2$.*

3) *$SGSp(n, F, f) = GSp(n, F, f) \cap SL(n, F)$ for even n and some f.*

2. The maximal subgroups containing root subgroups, of $Sp(n, F, f)$, $SU(n, F, f)$ or $\Omega(n, F, Q)$ over a field F:

Theorem 2.2 *Let $G = Sp(n, F, f), SU(n, F, f)$ or $\Omega(n, F, Q)$ over a field F, acting on V, with the Witt index $\nu(f)$ or $\nu(Q)$ positive. Then the subgroups M listed in the following classes are maximal in G. Any maximal subgroup M of G containing root subgroups is necessarily in one of these classes.*

The classes for $G = Sp(n, F, f)$:

(1) *Stabilizer of a totally isotropic subspace of V.*

(2) *Stabilizer of a nondegenerate subspace of dimension $< n/2$.*

(3) *Stabilizer of an orthogonal decomposition $V = V_1 \perp \cdots \perp V_k$ into nondegernerate subspaces V_i $(1 \leq i \leq k)$ of the same dimension r, except in the case $F = F_2$, $r = 2$ and $n \geq 6$.*

(4) $M = O(2\nu, F_2, Q) < G = Sp(2\nu, F_2, f)$.

(5) *A symmetric group S_{4k+2} embedded in $G = Sp(4k, 2)$ $(k \geq 2)$.*

(6) *The Chevalley group $G_2(F)$ embedded naturally in $G = Sp(6, F, f) \cong \Omega(7, F, Q)$ $(\nu(Q) = 3)$ over a perfect field F of char $F = 2$.*

(7) *In case char $F \neq 2$, the stabilizer of a disjoint pair of maximal totally isotropic subspaces U_0, V_0, except in case $G = Sp(4, 3)$.*

(8) char $F \neq 2$, $M = Sp(2k, K, f_K) \rtimes \operatorname{Aut} K/F < G = Sp(4k, F, f)$ *for a quadratic extension field K of F.*

The classes for $G = SU(n, F, f)$ with $\nu(f) \geq 1$:

(1) *The stabilizer of a totally isotropic subspace.*

(2) *The stabilizer of a nondegenerate subspace $W \not\cong W^\perp$.*

(3) *The stabilizer of an orthogonal decomposition $V = W_1 \perp \cdots \perp W_k$ with all the W_i mutually isometric, all $\nu(W_i) \geq 1$ when $F \neq F_2$, and we require $r \geq 3$, or $r = 1$ in case $F = F_2$ but $n \neq 6$.*

(4) *The normalizer in $G = SU(6, 2^2)$ of $3 \cdot P\Omega^{-,\pi}(6, 3)$ (see [19]).*

(5) $M = G \cap GSp(2\nu, F, \begin{pmatrix} & I_{(\nu)} \\ -I_{(\nu)} & \end{pmatrix})$ *in* $G = SU(2\nu, F, \begin{pmatrix} & I_{(\nu)} \\ -I_{(\nu)} & \end{pmatrix})$.

(6) *In case $\nu(f) = \nu = n/2$, the stabilizer of a disjoint pair of maximal totally isotropic subspaces U_0, V_0, except in case $G = SU(4, 2^2), SU(4, 3^2)$.*

The classes for $G = \Omega(n, K, Q)$ with $\nu(Q) \geq 1$, of maximal subgroups containing a root subgroup $T_{u,w}$ with $\langle u, w \rangle \cap V^\perp = 0$:

(1) *The stabilizer of a totally singular subspace V_0 of V, with exception: $\nu(Q) = n/2$ and $\dim V_0 = n/2 - 1$.*

(2) *The stabilizer of one of the following regular subspace W which is not isometric to $W^\perp$:*

i) $W = W_1 \oplus V^\perp$, *with W_1 nondegenerate, $\nu(W^\perp) \geq 1$ and $\dim W_1 \geq 3$, and require $\nu(W) = 0$ and $W^\perp \cong V(3,3)$ or $V^+(4,2)$ when $F = F_2, F_3$ and $\dim W = 2$.*

ii) char $F = 2$ and $\dim(V/V^\perp) \geq 4$, W *is a nonsingular line not lying in $V^\perp$, $W^\perp$ is regular, with $\nu(W^\perp) \geq 1$.*

iii) char $F = 2$ and $V^\perp \neq 0$, W *is a maximal subspace of V not containing $V^\perp$, when $\dim(V/V^\perp) = 2$ we require that F coincide with the field generated by $L = \{Q(x) \mid x \in V^\perp\}$.*

(3) *The stabilizer of a set of mutually orthogonal and isometric subspaces W_i $(1 \leq i \leq k)$, all containing $V^\perp$, with $\sum_i W_i = V$, and satisfying*

the conditions: (a) $\nu(W_i) \geq 1$ and $\dim(W_i/V^\perp) \geq 3$ when $F \neq F_2, F_3$; (b) when $F = F_3$, $\dim W_i \geq 4$, or $\dim W_i = 1$ but $n \neq 8$; (c) when $F = F_2$, $\dim W_i \geq 2$, but $W_i \not\cong V^+(2,2), V^+(4,2), V(3,2)$.

(4) A symmetric group S_m or an alternating group A_m embedded in $G = \Omega(V,Q)$ over F_2 or F_3 in the following way:

i) $M = S_m$ (when $m = 4k+2 \geq 10$) or A_m (when $m = 7$, or $m \geq 9$ and $m \equiv 0$ or $1 (\mathrm{mod}4)$) acting on the basis $\{h_i \mid 1 \leq i \leq m\}$ of $V(m, F_2, Q)$ with all $Q(h_i) = 0$ and $f(h_i, h_j) = 1$ $(i \neq j)$, $V = V_{m-1}/\langle \sum_i h_i \rangle$ (when $m = 4k$) or $V = V_{m-1}$ (otherwise) for the subspace $V_{m-1} = \{\sum_i c_i h_i \mid \sum_i c_i = 0\}$ of $V(m, F_2, Q)$.

ii) $M = A_m$ ($m = 5$, or $m \equiv 0$ or $1 (\mathrm{mod}3)$ and $m \geq 7$) acting on the basis $\{h_i \mid 1 \leq i \leq m\}$ of $V(m, F_3, Q) = \langle h_1 \rangle \perp \cdots \perp \langle h_m \rangle$ with all $Q(h_i) = 1$, $V = V_{m-1}/(V_{m-1}^\perp)$ for the subspace $V_{m-1} = \{\sum_i c_i h_i \mid \sum_i c_i = 0\}$ of $V(m, F_3, Q)$.

(5) $V = V_8$ or V_7 for $V_8 = V(8, F_3, Q) = \langle h_1 \rangle \perp \cdots \perp \langle h_8 \rangle$ with all $Q(h_i) = 1$ and $V_7 = \{\sum_i c_i h_i \mid \sum_i c_i = 0\}$, M is the stabilizer of the set of isotropic lines $\langle c_i h_i + c_j h_j \rangle$ $(i \neq j, c_i c_j \neq 0)$ and $\langle \sum_i c_i h_i \rangle$ $(\prod_i c_i = 1)$.

(6) The stabilizer of a disjoint pair of maximal totally singular subspaces U_0, V_0 in $G = \Omega(4k, F, Q)$ with $\nu(Q) = 2k$, but $G \neq \Omega^+(8,2), \Omega^+(8,3)$.

(7) An irreducible image Ω_7 in $G = \Omega(8, F, Q)$ with $\nu(Q) = 4$ of a reducible $\Omega(7, F, Q) < G$ under $\mathrm{Aut}\, G$.

(8) The stabilizer $M = G \cap (\mathrm{GSp}(2k, F, f_1) \otimes \mathrm{GSp}(2, F, f_2))$ $(k \geq 2)$ in $G = \Omega(V,Q) = \Omega(4k, F, Q)$ of the tensor product structure $V = V(2k, F, f_1) \otimes V(2, F, f_2)$, with exceptions: (i) $F = F_2$; (ii) $\mathrm{char}\, F = 2$ and $4k = 8$.

(9) $M = \mathrm{SU}(m, K, f) \rtimes \mathrm{Aut}\, K/F$ (m even) or $M = \mathrm{SU}(m, K, f)$ (m odd) for an hermitian K-form f, embedded in $G = \Omega(2m, F, Q)$ with $F = \{a \in K \mid \bar{a} = a\}$ and $Q(x) = f(x, x)$, except the case $m = 4$.

(10) The Chevalley group $G_2(F)$ embedded in $G = \Omega(7, F, Q)$ $(\nu(Q) = 3)$ naturally.

(11) $M = O(m, F_4, Q)$ embedded in $G = \Omega(2m, F_2, Q_0)$ with $Q_0(x) = \mathrm{Tr}_{F_4/F_2} Q(x)$.

We note that the maximality of some of the subgroups in above classes, is also proved in other authors' works. For example, O.H.King proved in [23–29] the maximality of the reducible and the imprimitive maximal subgroups, R.H.Dye proved in [30–32] the maximality of $\mathrm{SGSp}(2\nu, F)$ in $\mathrm{SL}(2\nu, F)$, of $O(2\nu, F, Q)$ in $\mathrm{Sp}(2\nu, F, f)$ (F a perfect field with $\mathrm{char}\, F = 2$), and of symmetric groups in an $O^\pm(n, 2)$ or $\mathrm{Sp}(n, 2)$.

3. The results for $TU(n, K, h, L)$ over a skewfield K:

Theorem 2.3 *Let X be an irreducible subgroup of $G = TU(V, Q, L) = TU(n, K, Q, L) \leq TU(n, K, f)$ over a skewfield K with center K_c, generated by a set of long-root subgroups of G. Then the equality $X = G$ holds in all cases except the following:*

K is a generalized quaternion skewfield, $L = \text{Tr} K = K_c$, $V = K \otimes_F V_F$ for an F-space V_F over a subfield F of K with $\overline{F} = F$, such that the restrictions Q_F, f_F of Q, f on V_F are F-forms, and one of the following holds:

i) $\dim V = n = 2m$ is even, $\nu_L(h) = m \geq 2$, $F = L$, $X = Sp(V_F, f_F)$.

ii) $F = L[\delta]$ for a $\delta \in K$ with $\overline{\delta} \neq \delta$, $X = SU(V_F, f_F)$.

iii) char $K = 2$, $F = L[\delta]$ for a $\delta \in K_J \setminus L$, $X = TU(V_F, h_F, L)$. iv) $n = 4$, $\nu_L(h) = 2$. Write $V = \langle u_1, v_1 \rangle \perp \langle u_2, v_2 \rangle$ with u_i, v_i $(i = 1, 2)$ L-singular and $f(u_i, v_i) = 1$. Let $P = L \oplus L\delta_1 \oplus L\delta_2$ for two non-symmetric elements δ_1, δ_2 of K with $\delta_1 \notin L$ and $\delta_2 \notin L[\delta_1]$; and let Σ be the set of the L-singular lines $K v_2 \cup \{K(u_2 + bv_2) \mid b \in L\} \cup \{K(v_1 + a_2 u_2 + b_2 v_2) \mid a_2, b_2 \in P$ and $a_2 \overline{b_2} \in L\} \cup \{K(u_1 + b_1 v_1 + a_2 u_2 + b_2 v_2) \mid b_1 \in K, a_2, b_2 \in P, b_1 a_2 \in P, b_1 b_2 \in P, b_2 \overline{a_2} - \overline{b_1} \in L\}$. Then X is generated by all the long-root subgroups T_w with $Kw \in \Sigma$.

Theorem 2.4 *Let $G = TU(V, Q, L) = TU(n, K, h, L) \leq TU(n, K, f)$ over a skewfield K with center K_c. Then the following subgroups M are maximal in G. Any maximal subgroup of G containing long-root subgroups is necessarily in one of the following class.*

(1) The stabilizer of a totally L-singular subspace of V.

(2) The stabilizer of a decomposition $V = V_1 \oplus \cdots \oplus V_k$ of V into mutually L-isometric and orthogonal subspaces V_i $(1 \leq i \leq k)$ with $\nu_L(V_i) \geq 1$.

(3) The stabilizer of a nondegenerate subspace W not L-isometric to $W^\perp$ with either $\nu_L(W) \geq 1$ or $\nu(W^\perp) \geq 1$.

(4) In case char $K = 2$ and $L \subsetneq K_J$, the stabilizer of an f-isotropic but L-nonsingular vector x with $\nu_L(x^\perp) \geq 1$.

(5) K is a generalized quaternion skewfield, $L = \text{Tr} K = K_c$, n is even, $V = K \otimes_L V_L$, the restriction f_L of f on V_L is a nondegenerate alternating F-form, and we require $n \geq 6$ when char $K = 2$, M is the normalizer of $Sp(V_L, f_L)$ in G.

(6) K is a generalized quaternion skewfield, $L = \text{Tr} K = K_c$, $n \geq 3$, $F = L[\delta]$ for a $\delta \neq \overline{\delta}$ with $\delta + \overline{\delta} = 1$, $V = K \otimes_F V_F$ for an F-space V_F spanned by some L-singular vectors, with $f_F = f|_{V_F}$ being an anti-hermitian F-form relative to $J|_{V_F}$, M is the normalizer of $SU(V_F, f_F)$ in

G.

(7) K is a generalized quaternion skewfield with char $K = 2$, $L = \mathrm{Tr}\, K = K_c$, $n = 2m$ is even and $\nu(f) = m \geq 2$, $F = L[\delta]$ for some $\delta \in K_J \setminus L$, $V = K \otimes_F V_F$ for an F-space V_F spanned by L-singular vectors, with $Q_F = Q|_{V_F}$ being a pseudo-quadratic F-form, and we require $m \geq 3$ when $\nu_L(Q_F) = 2$, M is the normalizer of $\mathrm{TU}(V_F, Q_F, L)$ in G.

(8) K is a generalized quaternion skewfield, $L = \mathrm{Tr}\, K = K_c$, $n = 4$, $\nu_L(Q) = 2$. Let $V = \langle u_1, v_1 \rangle \perp \langle u_2, v_2 \rangle$, with u_i, v_i $(i = 1,2)$ L-singular and $f(u_i, v_i) = 1$; let $P = L \oplus L\delta_1 \oplus L\delta_2$ for two non-symmetric elements δ_1, δ_2 of K with $\delta_1 \notin L$ and $\delta_2 \notin L[\delta_1]$; and let $\Sigma = Kv_2 \cup \{K(u_2 + bv_2) \mid b \in L\} \cup \{K(v_1 + a_2 u_2 + b_2 v_2) \mid a_2, b_2 \in P \text{ and } a_2\overline{b_2} \in L\} \cup \{K(u_1 + b_1 v_1 + a_2 u_2 + b_2 v_2) \mid b_1 \in K, a_2, b_2 \in P, b_1 a_2 \in P, b_1 b_2 \in P, b_2\overline{a_2} - \overline{b_1} \in L\}$. Then M is the stabilizer in G of the set Σ of L-singular lines.

§3. Aschbacher's reduction theorem for finite classical groups

Let G be a classical group over a finite field $F = F_q$, $V = V(n, F)$ the underlying space of G and M a maximal subgroup of G. According to the main theorem in Aschbacher's paper [1], either of the following holds:

1. M is a member of one of the 8 classes $C_1 - C_8$ of subgroups described in [1].

2. M is a nonabelian almost simple group, with its action on V absolutely irreducible, tensor indecomposable, defined over no proper subfield of F.

The definition of the classes $C_1 - C_8$ is as follows:

C_1 consists of reducible maximal subgroups, i.e., the stabilizers of nonzero proper subspaces U of V such that U is nondegenerate (not isometric to $U^\perp$) or totally singular, or U is a nonsingular line when $G = O(n, 2^m, Q)$.

C_2 consists of imprimitive maximal subgroups, i.e., the stabilizers of certain sets S of subspaces of V such that $V = \oplus_{U \in S} U$.

C_3 consists of the stabilizers of vector space structures $V = V(n/r, K)$ over extension fields K of F, with prime indices $r = [K : F]$ dividing n.

C_4 consists of the stabilizers of tensor product structures $V = U \otimes_F W$.

C_5 consists of the stabilizers of the n-dimensional K-subspaces U of V over maximal subfields K of F, such that $V = F \otimes_K U$.

C_6 consists of the normalizers of certain extra-special r-groups, where an r is a prime not equal to $p = \operatorname{char} F$, and $n = r^m$ for an m.

C_7 consists of the stabilizers of tensor product structures $V = V_1 \otimes \cdots \otimes V_m$ with $m \geq 3$ and all the V_i ($1 \leq i \leq m$) are mutually similar.

C_8 consists of the sub-classical groups having the same underlying space $V = V(n, F)$ as G.

Aschbacher's theorem reduces the classification of the maximal subgroups of finite classical groups to the following two steps: 1. to determine which of the groups in classes $C_1 - C_8$ is actually maximal; 2. to determine all the almost simple maximal subgroups not in the classes $C_1 - C_8$. The first step was done by P. Kleidman and M. Liebeck in [2], the analysis depends heavily on the classification of finite simple groups. The second step is much more difficult, and G. Seitz has achieved a lot in this direction in [3].

For the infinite classical groups (over infinite fields or division rings), so far there is not a similar reduction theorem. But still we can define the similar classes of subgroups and investigate the maximality of the subgroups in these classes. Since the method in Kleidman-Liebeck [2] depends heavily on the classification of finite simple groups, it can hardly apply to the infinite classical groups. We need some other method to treat it. For the reducible or imprimitive subgroups in the classes C_1 and C_2, the maximality has been proved in O.King [23–29] as well as in our paper [5–14] about the maximal subgroups containing root subgroups, since most of the reducible or imprimitive maximal subgroups contain root subgroups. We also prove in [33] the maximality of the monomial groups in linear groups over division rings, which contains no root subgroup. (For finite fields, this was done in J.D.Key [34, 35], R.H.Dye [36], by using classifications of groups generated by reflections or homologies, whereas I just perform elementary matrix calculations in treating monomial subgroups over arbitrary division rings.) All the subgroups in the classes $C_3 - C_8$ are irreducible and primitive. In the following sections §4–§7 of this paper we shall give an illustration of our work on the maximality of these subgroups. The results were partially declaired in [37, 38].

§4. Embedding of one classical group in another

Aschbacher's class C_8 consists of such subgroups M of G, that M and G are classical groups acting on the same space $V = V(n, K)$. For example, M is a unitary (or symplectic, orthogonal) group in a linear group G; or $M = U'(n, K, h, L) < G = GU(n, K, f)$ for an $L \subsetneq K_{J,-\varepsilon}$ in case char $K = 2$. To investigate the maximality of M in G, we need only to classify the overgroups of the smallest $M = U'(n, K, h, K_{J,-\varepsilon})$ in the greatest $G = GL(n, K)$. This is done in [39]-[42], with results as follows.

Theorem 4.1 *Let $N = U'(n, K, h, L_0)$ for $L_0 = K_{J,-\varepsilon}$, with $n \geq 3$ and $\nu_{L_0}(h) \geq 1$, $N \leq X \leq GL(n, K)$. Then either of the following holds:*

(i) $X \trianglerighteq SL(n, K)$.

(ii) $X \trianglerighteq U'(n, K, h, L)$ *for some $L \supseteq L_0$.*

Theorem 4.2 *Let $N = TU(2, K, Q, L_0)$ with $L_0 \neq 0$, and $N \leq X \leq GL(2, K)$. Then one of the following holds:*

(i) $X \trianglerighteq SL(2, K)$.

(ii) $X \trianglerighteq TU(2, K, Q, L)$ *for some $L \supseteq L_0$.*

(iii) $N = \Omega(2, K, Q, L_0) = \Omega(2, K_0, Q, L_0)$ *over a proper subfield K_0 of K generated by L_0, $X \trianglerighteq \Omega(2, E, Q, L)$ for some $L \supseteq L_0$ and some subfield E of K containing L.*

(iv) K *is a generalized quaternion ring over its center F, $N = SL(2, F)$, $X \trianglelefteq SL(2, F(\alpha))$ for some $\alpha \in K \setminus F$; or $X \trianglelefteq \Omega(2, F(\alpha), Q_1, L_1)$ for some Q_1, L_1; or $X \trianglelefteq TU(2, K, f_1)$ for some f_1 respect to an involution $J_1 : a \mapsto e^{-1}\bar{a}e$ of K (where $0 \neq e \in K^*$, $a \mapsto \bar{a}$ is the standard involution of K as a generalized quaternion ring).*

(v) $N = SU(2, 3^2) = SL(2, 3)$, $X/\{\pm 1\} \cong A_5$.

From these theorems we obtain results about the maximality of one classical group in another.

Corollary (1) *The normalizer of $SU(n, K, f)$ with $\nu(f) \geq 1$ is maximal in $SL(n, K)$, except in the case when $n = 2$, K is a generalized quaternion ring of characteristic 2 and $SU(2, K, f) = SL(2, F)$ for the center F of K.*

(2) *Let char $K = 2$, and $n \geq 3$ when $L_1 = 0$. Then the normalizer of $U'(n, K, h, L_1)$ with $\nu_{L_1}(h) \geq 1$ is maximal in $U'(n, K, h, L)$, provided that L_1 is maximal in L up to admission. In particular, if L_1 is a maximal K^2-subspace of L, and assume $n \geq 3$ when $L_1 = 0$, then the normalizer of $\Omega(n, K, Q, L_1)$ is maximal in $\Omega(n, K, Q, L)$; if L_1 is a maximal K^2-subspace of K, then $O(n, K, Q, L_1)$ is maximal in $Sp(n, K, f)$.*

In case K is a field there are some other papers to refer to. R.Dye determined in [31] the overgroups of $Sp(n, K)$ in $SL(n, K)$, and he proved in [32] the maximality of $O(n, K, Q)$ in $Sp(n, K)$ for perfect fields K of characteristic 2. In [10] the author proved the maximality in $SL(n, K)$

of $SGU(n, K, f)$ $(n \geq 3, \nu(f) \geq 1)$ and the maximality in $SL(n, K)$ of $SGO(n, K, f)$ (char $K \neq 2, n \geq 5$, and $\nu(f) \geq 2$). In [11] the author proved the maximality of $\Omega(n, K, Q, L_1)$ in $\Omega(n, K, Q, L)$ (char $K = 2, \nu(Q) \geq 1$), provided that L_1 is a maximal K^2-subspace L. In [43] O.King determined the overgroups of $SU(n, K, f)$ $(\nu(f) \geq 1)$ in $SL(n, K)$, and in [44] he determined the overgroups of $SO(n, K, f)$ (char $K \neq 2, \nu(f) \geq 1$) in $SL(n, K)$.

§5. Extension of the division base rings

Let $K \supset F$ be two division rings, with $\dim_F K = r < \infty$. An n-dimensional left K-space $V = V(n, K)$ can also be viewed as an nr-dimensional left F-space $V(nr, F)$. Hence the $GL(n, K)$ on $V(n, K)$ is embedded in the $GL(nr, F)$ on $V(nr, F)$ as a subgroup. Let $G(K) = SL(n, K)$ or $TU(n, K, h_K, L_K)$ acting on $V(n, K)$. We hope to determine all the overgroups of $G(K)$ in $GL(nr, F)$. For a division ring E between F and K, viewing $V(n, K)$ as an E-space $V(nd, E)$ we obtain $GL(nd, E)$ acting on $V(nd, E)$, as an overgroup of $GL(n, K)$ in $GL(nr, F)$. Moreover, certain classical groups on $V(nd, E)$ can be obtained as overgroups of $G(K)$ in $GL(n, K)$. The problem is: is there any group between $G(K)$ and $GL(nr, F)$ other than those corresponding to intermediate division rings E? We solve this problem in [45],[46] for most of the cases. Our answer is 'Yes!' for almost all the cases, with results as follows. When F is a maximal division subring of K, we obtain as corollaries the maximality of the subgroups in classes C_3, i.e., of the stabilizers of the K-structure $V(n, K)$ in the classical groups on $V(nr, F)$. For the case when $K = F_{q^r}, F = F_q$ are finite fields, there are some other authors' work to refer to: W. Kantor determines in [47] the overgroups of $F_{q^r}^*$ (i.e., $GL(1, q^r)$) in $GL(r, q)$, R.H.Dye determines in [48]-[50] the overgroups of $Sp(n, q^r)$ in $Sp(nr, q)$ for the smallest $r = 2, 3$.

Theorem 5.1 *Let $n \geq 2$, $N = SL(n, K) \leq X \leq G = GL(nr, F)$. Then one of the following holds:*

(i) $SL(nd, E) \trianglelefteq X \leq GL(nd, E) \rtimes \operatorname{Aut} E/F$ for a division ring E between F and K, where $d = \dim_E K$.

(ii) $n = 2$, K is commutative, $N = Sp(2, K) = Sp(2, K, f)$ for any non-degenerate alternating K-form f, $X \trianglerighteq Sp(2d, E, f_E)$ for a field E between F and K (where $d = \dim_E K$) and an alternating E-form $f_E = \phi_E f$ obtained from a nonzero $\phi \in Hom_E(K, E)$.

(iii) $N = \mathrm{SL}(2,4) \cong A_5$, $G = \mathrm{SL}(4,2)$, while $X = (\mathrm{Sp}(4,2))' \cong A_6$ or $X \cong A_7$.

Theorem 5.2 Let $N = \mathrm{Sp}(2\nu, K, f) \leq X \leq G = \mathrm{GL}(2\nu r, F)$. Then one of the following holds:

(i) $\mathrm{SL}(2\nu d, E) \trianglelefteq X \leq \mathrm{GL}(2\nu d, E) \rtimes \mathrm{Aut}\, E/F$ for a field E between F and K, where $d = \dim_E K$.

(ii) $\mathrm{Sp}(2\nu d, E, f_E) \trianglelefteq X \leq \mathrm{GSp}(2\nu d, E, f_E) \rtimes \mathrm{Aut}\, E/F$ for a field E between F and K (where $d = \dim_E K$) and an alternating E-form $f_E = \phi_E f$ obtained from a nonzero $\phi \in \mathrm{Hom}_E(K, E)$.

(iii) $N = \mathrm{Sp}(2, F_4, f) = \mathrm{SL}(2,4) \cong A_5$, $G = \mathrm{SL}(4,2)$, $X = (\mathrm{Sp}(4,2))' \cong A_6$ or $X \cong A_7$.

Theorem 5.3 Let $N = \mathrm{U}'(n, K, h, L)$ with $\nu_L(h) \geq 2$, $N \leq X \leq \mathrm{GL}(nr, F)$. Then there exists a division ring E with $F \subseteq E \subseteq K$, such that one of the following holds (where $d = \dim_E K$):

(i) $X \trianglerighteq \mathrm{SL}(nd, E)$.

(ii) $X \trianglerighteq \mathrm{U}'(nd, E, h_E, L_E) > N$ for some $h_E = \phi_E h$ $(0 \neq \phi_E \in \mathrm{Hom}_E(K, E))$ and some L_E.

(iii) K is a field, $J \neq 1$, $N = \mathrm{SU}(4, K, f)$, $E = K_J = \{a \in K \mid \bar{a} = a\}$, and X normalizes an irreducible embedding Ω_7 of an $\Omega(7, E, Q_E)$ $(\nu(Q_E) = 3)$ in $\mathrm{SL}(8, E)$ obtained by spinor representations.

Corollary Let F be a maximal division subring of a division ring K, with $\dim_F K = r < \infty$.

(i) Let $\mathrm{GL}(nr, F) \geq G \geq \mathrm{SL}(nr, F)$. Then the normalizer M of $\mathrm{SL}(n, K)$ in G is maximal in G, provided $G = M \cdot \mathrm{SL}(nr, F)$, with an exception $\mathrm{SL}(2, K)$ and $\mathrm{Norm}_{K/F}^{-1}(\det G) \subseteq F^*$ (where $\mathrm{Norm}_{K/F} : K \to F$ is the norm mapping). In the exceptional case we have $M \lneq (\mathrm{GSp}(2r, F) \rtimes \mathrm{Aut}\, K/F) \cap G \lneq G$.

(ii) In case K is a field (and F is a maximal subfield of K), let $\mathrm{GSp}(2\nu r, F) \geq G \geq \mathrm{Sp}(2\nu r, F)$, then the normalizer M is maximal in G, provided $G = M \cdot \mathrm{Sp}(2\nu r, F)$, with an exception $G = \mathrm{Sp}(4, 2)$ and $M = \mathrm{Sp}(2, 4)$.

(iii) The normalizer of $\mathrm{U}'(n, K, h, L)$ with $\nu_L(h) \geq 2$ is maximal in $\mathrm{U}'(nr, F, \phi_F h, \phi_F L) > \mathrm{U}'(n, K, h, L)$, $(0 \neq \phi_F \in \mathrm{Hom}_E(K, E))$, except in the case when $\mathrm{U}'(n, K, h, L) = \mathrm{SU}(4, K, f)$ over a field K and $\mathrm{U}'(nr, F, \phi_F h, \phi_F L) = \Omega(8, F, Q_F)$ with $Q_F(x) = h(x, x)$. In particular, when $\mathrm{char}\, K \neq 2$ the normalizer of $\mathrm{U}'(n, K, f)$ with $\nu(f) \geq 2$ is maximal in $\mathrm{U}'(nr, F, \phi_F f)$.

(iv) In case K is a field (and F is a maximal subfield of K), let $Q_F = \phi_F Q$ for a nonzero $\phi_F \in \mathrm{Hom}_F(K, F)$. Then the normalizer of $\Omega(n, K, Q)$

is maximal in $\Omega(nr, F, Q_F)$.

§6. Tensor product structures

Let $V = V_1 \otimes_F \cdots \otimes_F V_m$ be the tensor product of vector spaces V_i $(1 \le i \le m)$ over a field F. Then any set of $g_i \in \mathrm{GL}(V_i)$ $(1 \le i \le m)$ associates an element $g_1 \otimes \cdots g_m : v_1 \otimes \cdots \otimes v_m \mapsto v_1 g_1 \otimes \cdots v_m g_m$ in $\mathrm{GL}(V)$. All these $g_1 \otimes \cdots \otimes g_m$ form a subgroup $\Gamma_0 = \mathrm{GL}(V_1) \otimes \cdots \mathrm{GL}(V_m)$ of $\mathrm{GL}(V)$. If all $\dim V_i$ $(1 \le i \le m)$ are equal, then we can view all $V_i = V_1$ $(2 \le i \le m)$ by means of F-linear isomorphisms, and have an embedding π of the symmetric group S_m into $\mathrm{GL}(V)$, with $\pi(\sigma) : v_1 \otimes \cdots \otimes v_i \otimes \cdots v_m \mapsto v_{\sigma(1)} \otimes \cdots \otimes v_{\sigma(i)} \otimes \cdots v_{\sigma(m)}$ for each $\sigma \in S_m$. The group $\Gamma = \Gamma_0 \rtimes \pi(S_m)$ is just the stabilizer of the tensor product structure in $\mathrm{GL}(V)$.

If on each V_i above it defines an ε_i-hermitian form f_i with respect to a same involution $J \in \mathrm{Aut}\, F$, then it induces an ε-hermitian form $f = f_1 \otimes \cdots f_m$ on V for the $\varepsilon = \varepsilon_1 \cdots \varepsilon_m$, defined by $(f_1 \otimes \cdots f_m)(u_1 \otimes \cdots \otimes u_m, v_1 \otimes \cdots \otimes v_m) = f_1(u_1, v_1) \cdots f_m(u_m, v_m)$. The group $\Gamma_0(f) = \{g_1 \otimes \cdots g_m \mid g_i \in \mathrm{GU}(V_i, f_i)\ \forall 1 \le i \le m\}$ is embedded in $\mathrm{GU}(V, f)$, and it is embedded in $GO(V, Q)$ if f is symmetric, for the quadratic form Q satisfying $Q(x+y) = Q(x) + Q(y) + f(x, y)\ \forall x, y \in V$ and $Q(u_1 \otimes \cdots \otimes u_m) = 0, \forall u_i \in V_i, 1 \le i \le m$. If moreover, all the V_i $(1 \le i \le m)$ can be identified with V_1 by means of similarities, then the embedding $\pi : S_m \to \mathrm{GL}(V)$ defined above sends S_m into $\mathrm{GU}(V, f)$ (or $GO(V, Q)$). The group $\Gamma(f) = \Gamma_0(f) \rtimes \pi(S_m)$ is the stabilizer in $\mathrm{GU}(V, f)$ resp. $GO(V, Q)$ of the tensor product structures.

Aschbacher's classes C_4, C_7 for a classical group on V consist of the stabilizers of the tensor product structures $V = V_1 \otimes \cdots \otimes V_m$, for the case $m = 2$ or $m \ge 3$ respectively.

We hope to prove the maximality of the subgroups in these two classes. The class C_7 seems rather difficult to treat, so far nothing has been done for the case when the ground field is infinite. Whereas for the class C_4 we have done a lot in [51, 52]. Instead of just investigating the maximality of the subgroups in C_4, we determine all the overgroups in $\mathrm{GL}(V_1 \otimes V_2)$ of an $\mathrm{SL}(V_1) \otimes \mathrm{SL}(V_2)$ or an $\mathrm{U}'(V_1, f_1) \otimes \mathrm{U}'(V_2, f_2)$. This has been done perfectly [51] for the overgroups of $\mathrm{SL}(V_1) \otimes \mathrm{SL}(V_2)$ or $\mathrm{Sp}(V_1, f_1) \otimes \mathrm{SL}(V_2, f_2)$, with results as follows. We denote $\Gamma_0 = \mathrm{GL}(V_1) \otimes \mathrm{GL}(V_2)$. And we denote by Γ the normalizer of Γ_0 in $\mathrm{GL}(V)$, namely, $\Gamma = \Gamma_0$ when $\dim V_1 \neq \dim V_2$, or $\Gamma = \Gamma_0 \rtimes \pi(S_2) = \Gamma_0 \cup \Gamma_0 \tau$ when $\dim V_1 = \dim V_2$, where $\tau = \pi(\sigma)$ for the non-identity $\sigma \in S_2$.

Theorem 6.1 *Let* $N = \mathrm{SL}(n, F) \otimes \mathrm{SL}(r, F) \le X \le \mathrm{GL}(nr, F)$, *say* $n \ge r \ge 2$. *Then one of the following holds:*

(i) $N \trianglelefteq X \le \Gamma$.

(ii) $X \trianglerighteq \mathrm{SL}(nr, F)$.

(iii) $N = \mathrm{SL}(n, 2) \otimes \mathrm{SL}(2, 2)$, $X = \Gamma L(n, 4)$.

(iv) $\mathrm{char}\, F = 2$, $N = \mathrm{SL}(2, F) \otimes \mathrm{SL}(2, F) = \Omega(4, F, Q)$ *with* $\nu(Q) = 2$, $X \trianglerighteq \Omega(2, F, Q, L)$ *for an* F^2-*subspace* L *of* F.

(v) $N = \mathrm{SL}(2, 2) \otimes \mathrm{SL}(2, 2) = \Omega^+(4, 2) \cong (S_3 \times S_3 \times S_2) \cap A_8$, $\mathrm{GL}(nr, F) = \mathrm{SL}(4, 2) \cong A_8$, $X \cong (S_3 \times S_5) \cap A_8$.

Theorem 6.2 *Let* $N = \mathrm{Sp}(n, F, f_1) \otimes \mathrm{Sp}(r, F, f_2) \le X \le \mathrm{GL}(V)$, *say* $n \ge r$ *and* $n \ge 4$, *and when* $n = r = 4$ *and* $F = F_2$ *we require* $X \le O(V, Q)$ *moreover. Then one of the following holds.*

(i) $X \le \Gamma$.

(ii) $X \trianglerighteq \mathrm{SL}(V)$.

(iii) $X \trianglerighteq \Omega(V, Q)$, *or* $X \trianglerighteq \Omega(V, Q, L)$ *for an* F^2-*subspace* L *of* F *when* $\mathrm{char}\, F = 2$.

(iv) $\mathrm{Sp}(V_2, f_2) = \mathrm{SL}(2, 2)$, X *normalizes* $\mathrm{Sp}(n, 4)$ *or* $\mathrm{SU}(n, 2^2)$ *or* $\mathrm{SL}(n, 4)$.

(v) $N = \mathrm{Sp}(4, F) \otimes \mathrm{Sp}(2, F)$ *with* $\mathrm{char}\, F = 2$, X *normalizes an irreducible image* Ω_7 *of a reducible* $\Omega(7, F) < \Omega^+(8, F)$ *under certain* $\sigma \in \mathrm{Aut}\,(\Omega^+(8, F))$ *obtained by spinor representation; or* $N = \mathrm{Sp}(4, F) \otimes \mathrm{Sp}(4, F)$, $X \trianglerighteq \Omega_{10}$, *or* $X \trianglerighteq \Omega_9$ *when* $\mathrm{char}\, F = 2$, *where* Ω_{10} *and* Ω_9 *are the images of* $\Omega^+(10, F)$ *and* $\Omega(9, F) < \Omega^+(10, F)$, *respectively, under an embedding* $\sigma : \Omega^+(10, F) \to \mathrm{GL}(16, F)$ *obtained by spinor representation.*

From these theorems we can deduce the maximality of certain subgroups in class C_4.

Corollary 6.1 *Let* $\mathrm{SL}(nr, F) \le G \le \mathrm{GL}(nr, F)$. *Then the normalizer* $\Gamma \cap G$ *of* $\mathrm{SL}(n, F) \otimes \mathrm{SL}(r, F)$ *in* G *is maximal in* G, *provided* $\{\det g \mid g \in \Gamma \cap G\} = \{\det g \mid g \in G\}$, *with exceptions:* (i) $r = 2$ *and* $F = F_2$; (ii) $n = r = 2$ *and* $\mathrm{char}\, F = 2$.

Corollary 6.2 *The normalizer of* $\mathrm{Sp}(n, F, f_1) \otimes \mathrm{Sp}(r, F, f_2)$ *is maximal in* $\Omega(nr, F, Q)$, *with exceptions:* (i) $r = 2$ *and* $F = F_2$; (ii) $\mathrm{char}\, F = 2$, $n = 4$, $r = 2$ *or* 4.

The overgroups in $\mathrm{GL}(V)$ of other $\mathrm{U}'(V_1, f_1) \otimes \mathrm{U}'(V_2, f_2)$ are investigated in [52], with the results as follows.

Theorem 6.3 *Put $n = \dim V_1$, $r = \dim V_2$, $\nu = \nu(f_1)$ and $\mu = \nu(f_2)$. Let $N = N_1 \otimes N_2 \leq X \leq \mathrm{GL}(V)$, with one of the following holding:*

(a) *char $F \neq 2$, $N_1 = \mathrm{Sp}(V_1, f_1)$ and $N_2 = O(V_2, f_2)$, with the case $n = r = 2$ or 4 remaining to be treated.*

(b) *$N_1 = \mathrm{SU}(V_1, f_1)$ and $N_2 = \mathrm{SU}(V_2, f_2)$, with $\nu^2 > 2r$.*

(c) *char $F \neq 2$, $N_1 = \Omega(V_1, f_1)$ and $N_2 = \Omega(V_2, f_2)$, with $\nu(\nu - 1) > 4r$.*

Then either (i) or (ii) in the following holds, with exceptions stated in the following (iii)–(vi) for small r or F.

(i) *$X \leq \Gamma$.*

(ii) *$X \trianglerighteq U'(V, f)$ $(= \mathrm{Sp}(V, f), \mathrm{SU}(V, f), \Omega(V, f)$ resp. in the cases (a),(b),(c)).*

(iii) *X is reducible or imprimitive when N is.*

(iv) *$N_2 \leq U(V_2, f_2) \leq \mathrm{GU}(V_2, \tilde{f}_2)$ for another $\tilde{f}_2$ on V_2, $X \trianglerighteq U'(V, f_1 \otimes \tilde{f}_2)$.*

(v) *$N_2 \leq O(2, F, f_2)$ with $\nu(f_2) = 0$, thus N_2 normalizes a quadratic extension field $K \subset \mathrm{GL}(2, F) \cup \{0\}$ of F, $X \trianglerighteq \mathrm{Sp}(n, K)$ (in the case (b)), or $X \trianglerighteq \mathrm{SU}(n, K)$ (in case (b) or (d)), or $X \trianglerighteq \Omega(n, K)$ (in the case (d)); or $N_2 = \Omega(4, F, f_2)$ with $\nu(f_2) = 0$ in the case (d), and $\prod_{i=1}^{4} f_2(w_i, w_i) \in F^{*2}$ for an f_2-orthogonal basis $\{w_i \mid 1 \leq i \leq 4\}$ of V_2, N_2 normalizes a quaternion skewfield $D \subset \mathrm{GL}(4, F) \cup \{0\}$ over F, $X \trianglerighteq \mathrm{SU}(n, D, f_D) > N_1 \otimes 1_{V_2}$ for an hermitian D-form f_D on $V = V(n, D)$*

(vi) *$N_2 = \Omega^+(4, F) = \mathrm{Sp}(W_2, f_{W_2}) \otimes \mathrm{Sp}(W_3, f_{W_3})$ in the case (d), with $\dim W_2 = \dim W_3 = 2$, $W_2 \otimes W_3 = V_2$ and $f_2 = f_{W_2} \otimes f_{W_3}$, $X \trianglerighteq \mathrm{Sp}(V_1 \otimes W_2, f_1 \otimes f_{W_2}) \otimes \mathrm{Sp}(W_2, f_{W_2})$ or $X \trianglerighteq \mathrm{SL}(V_1 \otimes W_2) \otimes \mathrm{SL}(W_2)$.*

Corollary 6.3 (a) *Let char $F \neq 2$, and exclude the case $n = r = 2$ or 4 (which remains to be settled). Then the normalizer of $\mathrm{Sp}(n, F, f_1) \otimes O(r, F, f_2)$ is maximal in $\mathrm{Sp}(nr, F, f)$, with exceptions (i) $r = 2$, or (ii) $r = 3$ and $F = F_3$.*

(b) *Let $\nu(f_1)^2 > 2r$. Then the normalizer of $\mathrm{SU}(n, F, f_1) \otimes \mathrm{SU}(r, F, f_2)$ is maximal in $\mathrm{SU}(nr, f)$, except in the case $r = 2$ and $F = F_4$.*

(c) *Let char $F \neq 2$, and $\nu(f_1)(\nu(f_1) - 1) > 4r$. Then the normalizer of $\Omega(n, F, f_1) \otimes \Omega(r, F, f_2)$ is maximal in $\Omega(nr, F, f)$, with exceptions (i) $r=2$, (ii) $r = 3$ and $F = F_3$.*

As a generalization of Theorem 6.1 to vector spaces over division rings, we have the following result [51], which would be improved if the requirement $n > r$ could be removed.

Theorem 6.4 *Let K be a division ring, F its center, F_0 a subfield of F. Let $V_1 = V(n, K)$ and $V_2 = V(r, F)$ be n-dimensional K-space and r-dimensional F-space respectively, $V = V_1 \otimes_F V_2$. Suppose $n > r$, and $\mathrm{SL}(n, K) \otimes \mathrm{SL}(r, F_0) \leq \mathrm{GL}(nr, K)$. Then one of the following holds: (i) $X \leq \mathrm{GL}(n, K) \otimes \mathrm{GL}(r, F)$; (ii) $X \trianglerighteq \mathrm{SL}(nr, K)$; (iii) $\mathrm{SL}(r, F_0) = \mathrm{SL}(2, 2)$, $X \trianglerighteq \mathrm{SL}(n, K_1)$ over the ring $K_1 = \{ \begin{pmatrix} a & b \\ b & a+b \end{pmatrix} \mid a, b \in K \}$.*

In [53] we generalize Theorem 6.1 to division ring in another way as follows. Observe that the tensor product $V = V(\acute{n}, F) \otimes_F V(r, F)$ over a field F can be written as the F-space $\mathrm{Mat}_{n \times r} F$ of all the $n \times r$ matrices over F, while $\mathrm{GL}(V_1) \otimes \mathrm{GL}(V_2)$ becomes the group of all the mappings $A_L \otimes B_R : X \mapsto AXB$ with $A \in \mathrm{GL}(n, F), B \in \mathrm{GL}(r, F)$. In case $n = r$, the $\tau : u_1 \otimes u_2 \mapsto u_2 \otimes u_1$ is just the transpose $X \mapsto {}^t X$ of the matrices $X \in V$. For a skewfield K we can also take $V = \mathrm{Mat}_{n \times r} K$ and consider the group $\Gamma_0 = \mathrm{GL}(n, K)_L \otimes \mathrm{GL}(r, K)_R = \{ A_L \otimes B_R \mid A \in \mathrm{GL}(n, K), B \in \mathrm{GL}(r, K) \}$, where $A_L \otimes B_R$ sends $X \in V$ to AXB. In general, Γ_0 does not lie in the $\mathrm{GL}(nr, K)$ acting on $V = V(nr, K)$. However, Γ_0 lies in the $\mathrm{GL}(V, F)$ acting on V as a vector space over the center F of K. If $\dim_F K = m < \infty$, then $\Gamma_0 < \mathrm{GL}(nrm, F)$. Moreover, if $n = r$ and K admits an anti-automorphism J fixing all the elements in F, then τ_J sending each $X = (x_{ij})_{n \times n} \in V$ to the transpose ${}^t X^J = (x_{ji}^J)_{n \times n}$ of $X^J = (x_{ij}^J)_{n \times n}$ also lies in $\mathrm{GL}(nrm, F)$. Denote by Γ the normalizer of Γ_0 in $\mathrm{GL}(nrm, F)$. Then $\Gamma = \Gamma_0 \cup \Gamma_0 \tau_J$ when τ_J exists, or $\Gamma = \Gamma_0$ otherwise. Our result is as follows.

Theorem 6.5 *Let K be a skewfield with center F, $\dim_F K = m < \infty$, $n \geq r \geq 2$. Let $N = \mathrm{SL}(n, K)_L \otimes \mathrm{SL}(r, K)_R$ defined as above, $N \leq X \leq \mathrm{GL}(nrm, F)$. Then one of the following holds: (i) $N \trianglelefteq X \leq \Gamma$; (ii) $X \trianglerighteq \mathrm{SL}(nrm, F)$.*

As a generalization of Theorem 6.3, a reasonable thing to be done is to classify the overgroups in $\mathrm{GL}(nrm, F)$ of $\mathrm{U}'(n, K, f_1)_L \otimes \mathrm{U}'(r, K, f_2)_R$.

§7. The restriction to a subring of the division base ring

Let $G(K)$ be a classical group acting on $V = V(n, K)$ over a division ring K. With respect to a K-basis of V we may write the elements of $G(K)$ into the $n \times n$ matrices over K. For a subring F of K, we can define the group $G(F) = G(K) \cap \mathrm{GL}(n, F)$, which consists of all the elements in $G(K)$ with entries lying in F. Then the class C_5 for $G(K)$ consists of the normalizers of the $G(F)$'s. Instead of investigating the maximality of these normalizers, we hope to classify all the overgroups of $G(F)$ in $G(K)$. Obviousely, for a ring E between K and F, we have a group $G(E)$ between $G(K)$ and $G(F)$. The question is: are there any groups between $G(F)$ and $G(K)$ other than these $G(E)$ corresponding to intermediate ring E?

When K is finite, F is a finite subfield of K, the overgroups were classified by N.Burgoyne, R.Griess and R.Lyons in [54].

For the case that K is infinite and $[K:F] = r < \infty$, we have the following results in [55],[56]:

Theorem 7.1 *Let F be a subfield of a field K, with $[K:F] = r < \infty$. Let $N \leq X \leq \mathrm{GL}(n,K)$, with one of the following holding:*

(a) $N = \mathrm{SL}(n,F)$, *and* $n > r$ *or* $n^2 > 4r$.

(b) $N = \mathrm{Sp}(n,F,f)$, *and* $n > r$ *or* $n(n+2) > 8r$.

(c) $N = \mathrm{SU}(n,F,f)$, *and* $\nu^2 > 2r$ *for* $\nu = \nu(f)$.

(d) $N = \Omega(n,F,Q)$, *and* $\nu(\nu - 1) > 4r$ *for* $\nu = \nu(Q)$.

Then we have a field E between K and F, such that one of the following holds:

(i) $\mathrm{SL}(n,E) \trianglelefteq X \leq \mathrm{GL}(n,E) \cdot K^*$.

(ii) $\mathrm{U}'(n,E,f_E) \trianglelefteq X \leq \mathrm{GU}(n,E,f_E) \cdot K^*$, *where f_E is obtained from f by E-linear extension.*

(iii) *char $K = 2$, and the case (b) or (d) holds, $X \trianglerighteq \Omega(n,E,Q_E,L) \geq N$ for a suitable Q_E and an E^2-subspace L of E, ($L \supseteq K$ in case (b)).*

When F is a maximal subfield one can easily obtain a corollay about the maximality of the groups over F in the groups over K. An infinite field K might have a maximal subring R which is not a field. In [55] we prove the following result about the maximality of the linear group over R in that over K.

Theorem 7.2 *Let K be an infinite field, R a maximal subring of K but not a field. Then the normalizer of $\mathrm{SL}(n,R)$ is maximal in $\mathrm{SL}(n,K)$.*

References

[1] M. Aschbacher, On the maximal subgroups of finite classical groups, *Invent. Math.* **76** (1984), 469-514.

[2] P. Kleidman and M. Liebeck, The subgroup structure of the finite classical groups, Cambridge Univ. Press, Cambridge, 1989.

[3] G. Seitz, The maximal subgroups of classical algebraic groups, *Mem. AMS* **67**, No.365 (1987).

[4] A. Hahn and O. T. O'meara, The classical groups and K-theory, Springer-Verlag, Berlin, 1989.

[5] S. Li and J. Zha, Some results about maximal subgroups in classical groups, *Journal of China University of Science and Technology*, **11** (1981) No.2, 124-125.

[6] S. Li and J. Zha, On certain classes of maximal subgroups in projective special unitary groups over finite fields, *Scientia Sinica* (Series A), **2** (1982), 125-131 (Chinese Edition); **25** (1982) No.8, 808-815 (English Edition).

[7] S. Li and J. Zha, On certain classes of maximal subgroups in PSp($2n, F$), *Scientia Sinica* (Series A), **6** (1982), 481-486 (Chinese Edition); **25** (1982) No.12, 1250-1257 (English Edition).

[8] S. Li, Maximal subgroups containing root subgroups infinite classical groups, *Chinese Science Bulletin*, **28** (1983) No.5, 257-260 (Chinese Edition); **29** (1984) No.1, 14-18 (English Edition).

[9] S. Li, On the maximality of certain orthogonal groups embedded in symplectic and unitary groups resp., *Journal of Mathematical Research and Exposition*, **3** (1983) No.1, 101-103.

[10] S. Li, Certain classes of maximal subgroups in PSL(n, F), *Acta Math. Sinica*, **26** (1983) No.5, 613-621 (in Chinese).

[11] S. Li, Maximal subgroups in P$\Omega(n, F, Q)$ with root subgroups, *Scientia Sinica* (Series A), **5** (1985), 193-205 (Chinese Edition); **28** (1985) No.8, 826-838 (English Edition).

[12] S. Li and J. Zha, Maximal subgroups containing root subgroups in Sp($2n, F_2$), *Journal of Mathematical Research and Exposition*, **5** (1985) No.2, 45-48 (in Chinese).

[13] S. Li, Maximal subgroups containing root subgroups in PSU(n, K, f) ($\nu(f) \geq 1$), *Acta Math. Sinica*, **29** (1986) No.5, 632-641 (in Chinese).

[14] S. Li, Maximal subgroups containing short-root subgroups in PSp($2n, F$), *Acta Math. Sinica* (New Series), **3** (1987) No.1, 82-91.

[15] J. Mclaughlin, Some groups generated by transvections, *Arch. Math.* **18** (1967), 364-368.

[16] J. McLaughlin, Some subgroups of SL$_n(F_2)$, *Ill. J. Math.* **13** (1969), 108-115.

[17] S. Li, Irreducible subgroups of SL(n, K) generated by root subgroups, *Geom. Dedicata* **31** (1989), 41-44.

[18] B. Stark, Irreducible subgroups of orthogonal groups generated by groups of root type 1, *Pacific J. Math.* **53** (1974), 611-625.

[19] W. Kantor, Subgroups of classical groups generated by long root elements, *Trans. Amer. Math. Soc.* **248** (1979), 347-379.

[20] J. Ren, Maximal subgroups containing long-root subgroups in the unitary transvection groups over skewfield, Doctor thesis, Univ. of Sci. and Tech. of China, (preprint).

[21] F. G. Timmesfeld, Groups generated by k-transvections, *Invent. Math.* **100** (1990), 167-206.

[22] F. G. Timmesfeld, Groups generated by K-root subgroups, *Invent. Math.* **106** (1991), 575-666.

[23] O. H. King, Some maximal subgroups of the classical groups, *J. Algebra* **68** (1981), 109-120.

[24] O. H. King, Maximal subgroups of the classical groups associated with non-isotropic subspaces of a vector space, *J. Algebra* **73** (1981), 350-375.

[25] O. H. King, Maximal subgroups of the orthogonal group over a field of characteristic two, *J. Algebra* **76** (1982), 540-548.

[26] O. H. King, Imprimitive maximal subgroups of the general linear, special linear, symplectic and general symplectic groups, *J. London Math. Soc.* (2) **25** (1982), 416-424.

[27] O. H. King, Imprimitive maximal subgroups of the orthogonal, special orthogonal, unitary and special unitary groups, *Math. Z.* **182** (1983), 193-203.

[28] O. H. King, Imprimitive maximal subgroups of the symplectic, orthogonal and unitary groups, *Geom. Dedicata* **15** (1984), 339-353.

[29] O. H. King, Imprimitive maximal subgroups of the general linear, special linear, symplectic and general symplectic groups, *J. London Math. Soc.* (2) **25** (1982), 416-424.

[30] R. H. Dye, Symmetric groups as maximal subgroups of orthogonal and symplectic groups over the field of two elements, *J. London Math. Soc.* (2) **20** (1979), 227-237.

[31] R. H. Dye, Maximal subgroups of $GL_{2n}(K), SL_{2n}(K), PGL_{2n}(K)$ and $PSL_{2n}(K)$ associated with symplectic polarities, *J. Algebra* **66** (1980), 1-11.

[32] R. H. Dye, On the maximality of the orthogonal groups in the symplectic group in characteristic two, *Math. Z.* **172** (1980), 203-212.

[33] S. Li, The maximality of monomial subgroups of linear groups over division rings, *J. Algebra* **127** (1989), No.1, 22-39.

[34] J. D. Key, Some maximal subgroups of $PSL(n,q), n \geq 3, q = 2^r$, *Geom. Dedicata* **4** (1975), 377-386; erratum, **6** (1977), 389.

[35] J. D. Key, Some maximal subgroups of certain projective unimodule groups, *J. London Math. Soc.* (2) **19**, No.2 (1979), 219-300.

[36] R. H. Dye, Maximal subgroups of finite orthogonal and unitary groups, *Math. Z.* **189** (1985), 111-129.

[37] S. Li, Maximal subgroups in classical groups over arbitrary fields, *Proc. Sympos. Pure Math.* **47** (1987), Part II, 487-493.

[38] S. Li, Overgroups of certain subgroups in the classical groups over division rings, *Contemp. Math.* **82** (1989), 53-57.

[39] S. Li, The maximality of unitary groups in linear groups over division rings, *Chinese Science Bulletin* **39** (1988), No.21, 1608-1610.

[40] S. Li, Overgroups of $SU(n, K, f)$ or $\Omega(n, K, Q)$ in $GL(n, K)$, *Geom. Dedicata* **33** (1990), 241-250.

[41] S. Li, A new type of classical groups over division rings of characteristic 2, *J. Algebra* **138** (1991), 399-419.

[42] S. Li, Overgroups of a unitary group in $GL(2, K)$, *J. Algebra* **149** (1992), 275-286.

[43] O. H. King, On subgroups of the special linear group containing the special unitary group, *Geom. Dedicata* **19** (1985), 297-310.

[44] O. H. King, On subgroups of the special linear group containing the special orthogonal group, *J. Algebra* **96** (1985), 178-193.

[45] S. Li, Overgroups in $GL(nr, F)$ of certain subgroups of $SL(n, K)$,I, *J. Algebra* **125** (1989), 215-235.

[46] S. Li, Overgroups in $GL(nr, F)$ of certain subgroups of $SL(n, K)$,II, preprint.

[47] W. Kantor, Linear groups containing a singer cycle, *J. Algebra* **62** (1980), 232-234.

[48] R. H. Dye, A maximal subgroup of $\mathrm{PSp}_6(2^m)$ related to a spread, *J. Algebra* **84** (1983), 128-135.

[49] R. H. Dye, Maximal subgroups of symplectic groups stabilizing spreads, *J. Algebra* **87** (1984), 493-509.

[50] R. H. Dye, Maximal subgroups of $\mathrm{PSp}_{6n}(q)$ stabilizing spreads of totally isotropic planes, *J. Algebra* **99** (1986), 191-209.

[51] S. Li, Overgroups in $\mathrm{GL}(U \otimes W)$ of certain subgroups of $\mathrm{GL}(U) \otimes \mathrm{GL}(W)$,I, *J. Algrbra* **137** (1991), 338-368.

[52] S. Li, Overgroups in $\mathrm{GL}(U \otimes W)$ of certain subgroups of $\mathrm{GL}(U) \otimes \mathrm{GL}(W)$,II, preprint.

[53] S. Li, $\mathrm{SL}(n, K)_L \otimes \mathrm{SL}(m, K)_R$ over a skewfield K, preprint.

[54] N. Burgoyne, R. Griess & R. Lyons, Maximal subgroups and automorphisms of Chevalley groups, *Pacific J. Math.* **71** (1977), 365-403.

[55] S. Li, Overgroups of $\mathrm{SL}(n, K)$ in $\mathrm{GL}(n, F)$ $(K \subset F)$, *Acta Math. Sinica* **33** (1990) No.6, 774-778 (in Chinese).

[56] S. Li, Overgroups of the subclassical groups defined over a subfield, preprint.

Complex Lie Algebras in China

Lu Caihui

Department of Mathematics, Capital Normal University

Zhang Hechun

Department of Applied Mathematics, Tsinghua University

Zhao Kaiming

Institute of Systems Science, Acadamia Sinica

Hsio-Fu Tuan is the earliest Chinese mathematician who worked on Lie algebras of characteristic 0 and related topics. In the mid-forties of this century together with C. Chevalley he studied the correspondence between algebraic Lie groups and algebraic Lie algebras. Their results ([60–62]) are the beginning of the modern theory of algebraic groups. In the present paper we review only the work on complex Lie algebras in China in recent years.

Chapter 1. Infinite dimensional Lie algebras

§1. Kac-Moody algebras

An integral matrix $A = (a_{ij})_{ij=1}^{n}$ is called a generalized Cartan matrix (GCM) if

(1) $a_{ii} = 2$, $i = 1, 2, \cdots, n$,

(2) $a_{ij} \leq 0$, if $i \neq j$,

(3) $a_{ij} = 0 \Longrightarrow a_{ji} = 0$.

If all principal minors of A are positive, the GCM A is a Cartan matrix.

A GCM A is called decomposable if it can be transformded into the following type by the same permutation of rows and columns: $\begin{pmatrix} A_1 & 0 \\ 0 & A_2 \end{pmatrix}$. Otherwise A is called indecomposable.

A realization of A is a triple $(H, \Pi, \Pi^\vee)$, where H is a complex vector space, $\Pi = \{\alpha_1, \cdots, \alpha_n\} \subset H^*$, and $\Pi^\vee = \{\alpha_1^\vee, \cdots, \alpha_n^\vee\} \subset H$ are indexed subset in H^* and H respectively, satisfying the following three conditions

(a) both sets H^* and H are linearly independent;

(b) $\langle \alpha_i^\vee, \alpha_j \rangle = a_{ij}$ $(i, j = 1, \cdots, n)$;

(c) $n - rank\, A = \dim H - n$.

Two realization $(H, \Pi, \Pi^\vee)$ and $(H_1, \Pi_1, \Pi_1^\vee)$ are called isomorphic if there exists a vector space isomorphism $\phi H \to H_1$ such that $\phi(\Pi^\vee) = \Pi_1^\vee$ and $\phi^*(\Pi) = \Pi_1$.

For every A there exists a unique up to isomorphism realization.

The Lie algebra $\tilde{g}(A)$ is the Lie algebra with generators e_i, $f_i(i = 1, \cdots, n)$ and $\mathcal{H}$, and the following relations:

$$\begin{cases} [e_i, f_j] = \delta_{ij}\alpha_i^\vee & (i, j = 1, \cdots, n) \\ [h, h'] = 0 \;\; (h, h' \in \mathcal{H}) \\ [h, e_i] = \langle \alpha_i, h \rangle e_i \\ [h, f_i] = -\langle \alpha_i, h \rangle f_i & (i = 1, \cdots, n; h \in \mathcal{H}). \end{cases}$$

Let $\mathcal{R}$ be the maximal ideal in $\tilde{g}(A)$, which intersects $\mathcal{H}$ trivially. We call

$$g(A) = \tilde{g}(A)/\mathcal{R}$$

the Kac-Moody algebra associated with A.

Remark   For every $n \times n$ complex matrix A we can similiarly define its realization and the Lie algebra $g(A)$.

In 1968, V. Kac [16] and R. Moody [50] independently constructed a Lie algebra $g(A)$ now called the Kac-Moody algebra associated with A. If A is an indecomposable Cartan matrix, then $g(A)$ is a simple finite dimensional Lie algebra. So $g(A)$ can be regarded as generalizations of simple finite dimensional Lie algebras. Although $g(A)$ has many similar properties to simple finite dimensional Lie algebras, they have essential differences.

In the recent twenty years, the theory of Kac-Moody algebras has been found to have various applications and profound impacts in many fields of mathematics, for instance, combinatorics, number theory, finite groups, topology, differential equations, in particular it has deep relations with mechanics and soliton theory.

1.1 Classification of generalized Cartan matrices

The first stimulant problem is the classification of indecomposable GCM's.

Let $u = \begin{pmatrix} u_1 \\ \vdots \\ u_n \end{pmatrix} \in \mathbf{R}^n$. If $u_i > 0$ for all i we write $u > 0$. If $u_i \leq 0$ for all i we write $u \leq 0$.

Theorem (Vinberg [63]) *Let A be an indecomposable GCM. Then A and $^t A$ satisfy only one of the following conditions:*

(Fin) *$\det A \neq 0$, and there exists $u > 0$ such that $Au > 0$, and $Av \geq 0 \Longrightarrow v > 0$ or $v = 0$.*

(Aff) *Corank $A = 1$, and there exists $u > 0$ such that $Au = 0$, and $Av \geq 0 \Longrightarrow Av = 0$.*

(Ind) *There exists $u > 0$ such that $Au < 0$, and $v \geq 0$, $Av \geq 0 \Longrightarrow v = 0$.*

If A is of type Fin, Aff, Ind we call the corresponding Lie algebra $g(A)$ Kac-moody algebra of type Fin, Aff, Ind.

A classification of affine GCM's was given by Kac [16] and Moody [50]. The classification of the indefinite GCM's is unsolved. But the classification of indefinite GCM's called of hyperbolic type has been given by Li Wanglai [26] in 1988. (An indecomposable GCM A is called a GCM of hyperbolic type if any proper connected subdiagram of $S(A)$ is of finite type or affine type.) Zhang Hechun [71] gave a new description of the GCM of hyperbolic type.

Theorem *Let A be an $n \times n$ GCM. Then A is of hyperbolic type if and only if for $u \in \mathbf{R}^n$, $Au \leq 0$ implies $u \geq 0$, hence $A^{-1} \leq 0$. Furthermore, A is of strictly hyperbolic type if and only if $A^{-1} < 0$, namely, all of the entries in A^{-1} are negative.*

1.2 Structures of Kac-moody algebras

Let $g(A)$ be the Kac-Moody algebra associated with a GCM A, H a Cartan subalgebra (CSA). Denote a basis of simple roots and coroots of $g(A)$ with respect to H by $\Pi = \{\alpha_1, \alpha_2, \cdots, \alpha_n\}$ and $\Pi^\vee = \{\alpha_1^\vee, \alpha_2^\vee, \cdots, \alpha_n^\vee\}$ respectively. Denote the lattice of roots and coroots respectively by $Q = \sum_{i=1}^{n} \mathbf{Z}\alpha_i$ and $Q^\vee = \sum_{i=1}^{n} \mathbf{Z}\alpha_i^\vee$. $Q_+ = \sum_{i=1}^{n} \mathbf{Z}_+\alpha_i$. Denote the Weyl group of $g(A)$ by W. Therefore we have a partial order in H^* as follows:

$$\lambda \geq \mu \iff \lambda - \mu \in Q_+, \ \lambda, \mu \in H^*.$$

Denote the set of roots of $g(A)$ by Δ, the set of positive roots of $g(A)$ by Δ_+, the set of negtive roots of $g(A)$ by Δ_-. Then

$$\Delta = \Delta_+ \cup \Delta_-, \ \Delta_+ \cap \Delta_- = \emptyset.$$

When A is a GCM of non-finite type, Kac [16] proved the following statements in 1968.

(1) There exists $\alpha \in \Delta$, such that $\alpha \notin \cup_{i=1}^{n} W\alpha_i$.

(2) $W\Delta_+^{im} \subseteq \Delta_+^{im}$.

(3) If $\alpha \in \Delta_+^{im}$ and $k \in \mathbf{Z}\backslash\{0\}$, then $k\alpha \in \Delta_+^{im}$.

The properties of roots are determined by the type of the GCM A. We know that if A is of non-finite type there are two different kinds of roots for $g(A)$. To distinguish these the concepts of real and imaginary roots are introduced. We call $\alpha \in \Delta$ a real root, if there is $w \in W$ such that $w(\alpha) = \alpha_i$ for some $i = 1, \cdots, n$. Otherwise α is called imaginary.

It is familiar to us that if α is a real root then $k\alpha \in \Delta$ implies $k = 1$ or -1. Denote the set of real roots of $g(A)$ by Δ^{re}, the set of imaginary roots of $g(A)$ by Δ^{im}. Then $\Delta = \Delta^{re} \cup \Delta^{im}$.

How to discribe the set of imaginary roots is a very important fundamental problem.

Lu shirong [42] gave a necessary and sufficient condition for Δ_+^{im} to be a semi-group.

Theorem Δ_+^{im} *is a semi-group if and only if any two subdiagram of Dynkin diagram $S(A)$ of affine or indefinite type are connected.*

Using the highest weight integrable representations, Zhang Hechun studies the imaginary roots of Kac-Moody algebras. In particular, he [71] gave a description of minimal imaginary roots ($\alpha \in \Delta_+^{im}$ is called minimal imaginary root if α is minimal in Δ_+^{im} with respect to the partial ordering in H^*).

Theorem (a) *The number of minimal imaginary roots of $g(A)$ is finite.*

(b) *Let $\alpha, \beta \in \Delta_+^{im} \cap (-C^v)$, α be not a null root, $\beta \geq \alpha$. Then $-\beta \in P(-\alpha)$.*

(c) *If $\alpha \in \Delta_+^{im} \cap (-C^v)$, then $P(-\alpha) \subset \Delta_-^{im}$.*

(d) *Let $\alpha \in \Delta_+^{im} \cap (-C^v)$. α is a minimal imaginary root if and only if $\alpha - \alpha_i \notin \Delta_+^{im}$, for all i.*

(e) *Let $\alpha, \beta \in \Delta_+^{im} \cap (-C^v)$, $\beta > \alpha$. Then there exists a series of positive roots $\beta_1, \beta_2, \cdots, \beta_s$, such that $\beta = \alpha + \beta_1 + \cdots + \beta_s$, and $\alpha + \beta_1 + \cdots + \beta_i \in \Delta_+^{im} \cap (-C^v)$, $i = 1, 2, \cdots, s$.*

For GCMs of strictly hyperbolic type, Lu Caihui [35] proved there is a unique minimal imaginary root and he calculated the minimal imaginary roots explicitly.

$\alpha \in \Delta_+^{im}$ is called strictly imaginary if for every $\gamma \in \Delta^{re}$, either $\alpha + \gamma$ or $\alpha - \gamma$ is a root. Zhang Hechun [71] proved

Theorem *If $\alpha \in \Delta_+^{im}$ and Suppα is connected with every point of $S(A)$, then there exists a positive integer k such that $k\alpha$ is strictly imaginary.*

The groups of automorphisms of simple finite dimensional Lie algebras are well known. Liu Lishi [34] discussed the structure of group of $g(A)$ preserving a CSA.

Using the results on affine Lie algebras, Zhao Kaiming [79] studied the quasirational automorphisms of finite dimensional simple Lie algebras explicitely.

An automorphism σ of order m of a finite dimensional Lie algabra g is called quasirational if for the associated $\mathbf{Z}/m\mathbf{Z}$-gradation $g = \oplus g_j$ one has $dim \ g_i = dim \ g_j$ if $(i, m) = (j, m)$. An automorphism σ of g is called rational if σ^k is conjugate to σ for every k such that k and m are relatively prime.

Theorem (a) *An automorphism σ of g is quasirational if and only if for $i \in \mathbf{Z}$ such that $(i, m) = 1$, the characteristic polynomials of σ and σ^i are the same.*

(b) *The power of a quasirational (resp. rational) is also quasirational (resp. rational).*

Let σ be an automorphism of g of $(s_o, \cdots, s_l; 1)$ type (for details refer to §8.6 in [18]). Then the order of σ is $m = \sum_i a_i s_i$. Set

$$\phi : \Delta \to \mathbf{Z}/m\mathbf{Z},$$

$$\sum_i k_i \alpha_i' \mapsto \sum_i k_i s_i.$$

Set $T_\sigma = \phi(\Delta')$, and $i \in T_\sigma$ repeats $\phi^{-1}(i)$ times.

$T \subset \mathbf{Z}/m\mathbf{Z}$ is called quasirational if for $i, j \in \mathbf{Z}/m\mathbf{Z}$, i and j appear the same times in T whenever $(i, m) = (j, m)$.

Theorem *Let σ be an automorphism of g of $(S; 1)$ type. Then σ is quasirational if and only if T_σ is quasirational.*

Similarly, Zhao Kaiming studied the quasirational automorphisms of type (S;2) and (S;3). And he calculated the quasirational automorphisms of A_1, A_2, G_2, B_2 in detail.

Lu Caihui and Wan Zhexian [41] solved completely the problem on the minimal number of generators of $g(A)$ associate with an arbitrary complex matrix A.

Jia Yuting [13,14] determined the Q-graded ω_0-invariant subalgebras of nontwisted affine Lie algebras.

Li Haisheng [23] proved

Theorem *Let A be a complex $n \times n$ matrix. Then the Chevalley generators $e_i, f_i (i = 1, \cdots, n)$ are locally nilpotent if and only if A is equivalent to a matrix of the form $\begin{pmatrix} A_1 & A_2 \\ 0 & 0 \end{pmatrix}$, where A_1 is a $r \otimes r$ GCM and A_2 is a $r \otimes (n - r)$ matrix of nonpositive integers. Furthermore, $\dim g(A) < \infty$ if and only if A_1 is a direct sum of some GCM of finite type.*

Let g be a Lie algebra. $x \in g$ is called locally finite if for any $y \in g$, $\dim \sum_{i \geq 0} \mathbf{C}(adx)^i y < \infty$. $x \in g$ is called locally nilpotent if for any $y \in g$, there exists a positive integer N such that $(adx)^N y = 0$. We use F_g to denote the set of all locally finite elements in g, g_{fin} the subalgebra of g generated by F_g. A Lie algebra g is called integrable if $g = g_{fin}$.

Su Yucai [57] proved

Theorem (a) *$gl(V)_{fin} = gl(V)$, i.e., $gl(V)$ is integrable.*

(b) *Let G be the group associated with $gl(V)$. Then G is isomorphic to $GL(V)$.*

1.3 Representations of $g(A)$

Let A is an arbitrary complex $n \times n$ matrix. A module V over $g(A)$ is called belonging to the category $\mathcal{O}$, if

(1) V is an H-diagonalizable Harish-Chandra module, i.e., $V = \oplus_{\lambda \in H^*} V_\lambda$, where $V_\lambda = \{v \in V | hv = <\lambda, h > v, h \in H\}$. and $\dim V_\lambda < \infty$, $\forall \lambda \in H^*$.

(2) The weight set $P(V)$ of V satisfies: $P(V) \subseteq \cup_{i=1}^s D(\lambda_i)$, $\lambda_i \in H^*$, and $D(\lambda_i) = \{\mu \in H^* | \mu \leq \lambda_i\}$, $i = 1, \cdots, s$.

Obviously the highest weight modules over $g(A)$ belong to $\mathcal{O}$. We use $L(\lambda)$ to denote the irreducible highest weight module with the highest weight λ. And denote the weight set of $L(\lambda)$ by $P(\lambda)$.

Lu Caihui [36] gave a criterion of completely reducibility.

Theorem *Let V be a module from category $\mathcal{O}$. Then V is completely reducible if and only if:*

(1) *If $x \in V$ is a primitive weight vector, and $n_+ x = 0$, then $U(g(A))x$ is irreducible.*

(2) *if $x \in V$ is a primitive weight vector, U is a submodule of V, and $x \notin U$, $n_+ x \subset U$, then there exists $a \in U$ such that $n_+(x - a) = 0$.*

Let A be a GCM. An H-diagonalizable module V over $g(A)$ is called integrable if all e_i and $f_i (i = 1, \cdots, n.)$ are localy nilpotent on V.

Wang Xiaodong [65,66] constructed the level 1 standard modules for C_n^1 and the level 2 standard modules for $A_3^{(1)}$ and $A_5^{(1)}$. Jiang Cuipo

[15] constructed some level 2 standard module for $B_3^{(1)}$, by using vertex operators.

Remark In [49], Misra constructed the level 2 standard modules for $A_{2l+1}^{(1)}$ completely.

Lepowsky [22] defined a class of generalized vertex operators by using the central binomial coefficients. Zhou Shanyou proved the closedness of the generalized vertex operators under Lie bracket [86]. Xu Yichao and Jiang Cuipo [68,69] constructed level-one vertex representations for affine Lie algebras with the first kind.

Xie Chuanfu [67] calculated the maximal weights of the standard fundamental modules for $C_2^{(1)}$ and $B_2^{(1)}$.

The above modules are all in $\mathcal{O}$. Obviously integral modules not belonging to $\mathcal{O}$ are another kind of significant modules. Lu Caihui [37] constructed irreducible modules V over $g(A_1^{(1)})$ satisfying the following conditions:

(1) V has a weight space decomposition

$$V = \oplus V_\lambda, \; dim V_\lambda < \infty,$$

for all $\lambda \in H^*$.

(2) For $0 \neq v \in V$, there is an e_i such that $e_i v \neq 0$.

(3) $cv = 0$, for all $v \in V$. c is the center element of $g(A)$.

(4) For $v \in V_\lambda$ if $e_0 v = 0(resp. \; f_0 v = 0)$, then $e_1 v$ (resp. $f_1 v$) is contained in an irreducible $g_0 = \mathbf{C}e_0 + \mathbf{C}\alpha_0^v + \mathbf{C}f_0$ module.

Using Lu's method, Su Yucai did the similar work for affine Kac-Moody algebra of $A_2^{(2)}$ type [55].

Here we only exhibit the construction of $A_2^{(2)}$ irreducible modules as follows [55].

Let $a \in \mathbf{C}, m \in 2\mathbf{Z}_+, V = \oplus_{n \in \mathbf{Z}} V^{(n)}$ be a vector space, $V^{(n)}$ be a subspace of dimensional $m+1$ with basis $\{v_i^{(n)} | 1 \leq i \leq m+1\}$. Define

$$cV = 0, d|_{V^{(n)}} = (a+n)I_{V^{(n)}},$$

$$f_0 v_i^{(n)} = v_{i+1}^{(n)}, e_0 v_i^{(n)} = i(m+1-i)v_{i-1}^{(n)},$$

$$\alpha_0^v v_i^{(n)} = (m-2i)v_i^{(n)},$$

$$f_1 v_{2i+1}^{(n)} = i(m/2-i)v_{2i-1}^{(n-1)}, \; f_1 v_{2i}^{(n)} = i(m/2+1-i)v_{2i-2}^{(n-1)},$$

$$e_1 v_i^{(n)} = v_{i+2}^{(n+1)}, \; \alpha_1^v v_i^{(n)} = (-m/2+i)v_i^{(n)}.$$

It is easy to verify that we can construct an irreducible integrable $A_2^{(2)}$-module which satisfies the above conditions. We use $V(a, m)$ to denote this module. Lu Caihui and Su Yucai [37, 55] proved the following theorem.

Theorem   (a) *Let V be an integrable $A_1^{(1)}$ or $A_2^{(2)}$ module which satisfies above conditions. Then V is completely reducible.*

(b) *Every irreducible integrable $A_2^{(2)}$-module is isomorphic to $V(a, m)$ for some $a \in C(\mathrm{mod}\, m), m \in 2\mathbf{Z}_+$.*

Remark   More general, Chari and Pressley [2,3,4] classified the irreducible integrable Harish-Chandra modules over affine Lie algebras completely. Lu's work is in the same time with Chari's work, and Lu's method is different from Chari's. But Lu's work is more concrete.

A GCM A is called symmetrizable if it can be written as $A = DB$, where D is an invertible and diagonal matrix and B is a symmetric matrix. It is known that [18]

(1) GCM's of finite and affine types are symmetrizable.

(2) A is symmetrizable if and only if $g(A)$ has a nondegenerate symmetric invariant bilinear form $(\cdot | \cdot)$.

In the following of this subsection, we always assume that A is symmetrizable.

V. Chari and A. Pressley [3] studied a class of integrable modules, i.e., hightest$\otimes$lowest: $U(\lambda, \mu) = L(\lambda) \otimes L(\mu)^*$, where $\lambda, \mu \in P_+$, $L(\mu)^*$ is the lowest weight module with the lowest weight $-\mu$. Chari and Pressley proved that:

Theorem   *Let $\lambda, \mu \in P_+$. Then $U(\lambda, \mu)$ is generated by $v_\lambda \otimes v_\mu^*$ of the highest and lowest weight vectors in $L(\lambda)$ and $L(\mu)^*$, with relations*

$$e_i^{<\mu, \alpha_i^v>+1}(v_\lambda \otimes v_\mu^*) = 0,$$

$$f_i^{<\lambda, \alpha_i^v>+1}(v_\lambda \otimes v_\mu^*) = 0,$$

$$h(v_\lambda \otimes v_\mu^*) = \langle \lambda - \mu, h \rangle (v_\lambda \otimes v_\mu^*)$$

for all simple coroots α_i^v and $h \in H$.

Chari and Pressley proposed an open problem: Whether $U(\lambda, \mu)$ is reducible or not? Zhang [72] answered this problem completely.

Theorem   (a) *Let V be an arbitrary irreducible integrable $g(A)$ module. Then there exist $\lambda, \mu \in P_+$ such that V is a quotient of $U(\lambda, \mu)$. So we call $U(\lambda, \mu)$ universal integrable module.*

(b) *Let A be a GCM of affine or indefinite type, $\lambda, \mu \in P_+$. Then $End_{g(A)}U(\lambda, \mu) \cong \mathbf{C}$. So $U(\lambda, \mu)$ is indecomposable.*

(c) *Let A be a GCM of affine or indefinite type, $\lambda, \mu \in P_+$, both λ and μ be nontrivial. Then $U(\lambda, \mu)$ is reducible. So it is not completely reducible.*

(d) *Let A be a GCM of finite type. Then $U(\lambda, \mu)$ is irreducible if and only if $\lambda = 0$ or $\mu = 0$.*

(e) *Let A be an indefinite type GCM, V be an irreducible integrable $g(A)$-module which is neither the highest nor the lowest. Then $P^{re}(V) = \{\lambda \in P(V) | W\lambda$ has neither the maximal nor the minimal element$\} \neq \phi$.*

(f) *Let A be a strictly hyperbolic type GCM, V be an irreducible integrable $g(A)$-module which is neither the highest nor the lowest and $P(V) \subset Q$. Then $P_+^{im} = \{\lambda \in P(V) | W\lambda$ has minimal elements$\} = \Delta_+^{im} \cup \{0\}$. $P_-^{im}(V) = \{\lambda \in P(V) | W\lambda$ has maximal elements$\} = \Delta_-^{im} \cup \{0\}$.*

Liu Lishi [33] got the Kostant's formula for symmetrizable Kac-Moody algebras. Li Zhenheng [28] gave a basis and the corresponding bracket of compact real form $k(A)$ of the Kac-Moody algebra $g(A)$. He also proved that the eigenvalues of the adjoint representation of $k(A)$ are imaginaries if A is of affine type.

Zhang Zhixeu [77] discussed some properties of weight sets of integrable modules over Kac-Moody algebras $g(A)$. He proved that the weight sets of integrable modules with the highest weight Λ over Kac-Moody algebras $g(A)$ is the smallest saturated weight set containing Λ, and there exists only one integrable module $M(\Lambda)$ with the highest weight Λ such that every integrable highest weight module with the highest weight Λ is a homomorphic image of $M(\Lambda)$.

Zhang Zhixeu [78] discussed some properties of weight sets of integrable modules over generalized Kac-Moody algebras $g(A)$. He constructed the weight sets of integrable modules with the highest weight Λ (strong dorminant weight) over generalized Kac-Moody algebras $g(A)$. These results can be regard as generalizations of corresponding results of Kac-Moody algebras.

Li Zhenheng [31] investigated some similar properties of modules over generalized Kac-Moody algebras.

1.4 The homorlogy of $g(A)$

Let g be a finite dimensional semisimple Lie algebra over an algebric closed field k.

Qiu Sen [52] gave the structure of the homology of a Kac-Moody algebra $g(A)$ with coefficients in an arbitrary generalized Verma module $V^{m(\lambda)}$.

Let A be a symmetrizable GCM, d the space spanned by some derivations on $g(A)$. Set $g^e = d \times g(A), H^e = d \oplus H, \rho \in (H^e)^*, < \rho, \alpha_i^v >= 1$, for all i. Fix a subdiagram S of finite type of $S(A)$. Let $r^e = g(A_S) + H$, where $g(A_S)$ is the subalgebra of g(A) generated by e_i, f_i, $i \in S$. He proved

Theorem   (a) *Let $\lambda, \mu \in H^*$. If $\mu \in P_+$ and μ is not greater than λ, then*

$$Ext_g^n(L(\mu), M(\lambda)) = 0,$$

where $M(\lambda)$ is the Verma module with highest the weight λ.

(b) *The homology of g^e is given as follows:*

(1) *if $\lambda \neq w\rho - \rho$ for all $w \in W_S^1 = \{w \in W | w^{-1}\Delta_+^S \subset \Delta_+\}$, then*

$$H_j(g^e, (V^{M(\lambda)})^t) = 0, \forall j \geq 0.$$

(2) *if $\lambda = w_0\rho - \rho$, for some $w_0 \in W_S^1$, then*

$$H_j(g^e, (V^{M(\lambda)})^t) = (H^{j-l(w_0)}(r^e))^* = H_{j-l(w_0)}(r^e), \forall j \geq 0$$

and $H_j(g^e, (V^{M(\lambda)})^t) = 0$, unless $l(w_0) \leq j \leq dim r^e + l(w_0)$.

Su Yucai [56] computed the low dimensional cohomology of $g(A)$, for a complex matrix A. Let H' be a complementary subspace of $\sum_i \alpha_i^v$ in H the CSA of $g(A)$.

Theorem   *Let A be an indecomposable $n \times n$ complex matrix of rank r. $r' = n - r$. Then*

(1) $H^2(g(A)) \cong Int^2(g(A)) \cong C^2(H'), \ dim H^2(g(A)) = \frac{1}{2}r'(r - 1)$;

$$\frac{1}{6}r'(r' - 1)(r' - 2) + n \leq dim H^3(g(A)) \leq 1/6 r'(r' - 1)(r' - 2) + \frac{1}{2}n(n + 1).$$

(2) *If A is a GCM, then*

$$H^2(g'(A)) = 0, \ H^3(g(A)) \cong H^3(g'(A)) \oplus C^3(H'), \ dim H^3(g'(A)) = n,$$

$$C^3(H') \subset Int^3 g(A), \ dim C^3(H') = \frac{1}{6}r'(r' - 1)(r' - 2).$$

§2. The Virasoro Algebra

The Virasoro algebra Vir is an infinite dimensional Lie algebra over **C**, with basis $\{c, d_i, | i \in \mathbf{Z}\}$ and commutation relations

$$[c, d_i] = 0,$$

$$[d_i, d_j] = (i - j)d_{i+j} + \delta_{i,j}\frac{i^3 - i}{12}c.$$

The Virasoro algebra Vir is another important infinite dimensional Lie algebra over $\mathbf{C}$, which has many applications in physics and mathematics. In 1992, O. Mathieu [43] proved the conjecture proposed by Kac in 1982, i.e., an irreducible Harish-Chandra module V over the Virasoro algebra is isomorphic to one of the following modules:

(i) irreducible highest weight module $V(c, h)$,

(ii) irreducible lowest weight module $V^*(c, h)$,

(iii) irreducible module $V_{\alpha,\beta}$ with all weight spaces being 1-dimensional.

The module $V(c, h)$ and $V_{\alpha,\beta}$ are discussed in detail in [21]. It is natural to ask what properties of the module $V_{\alpha,\beta} \otimes M(c, h)$ has. Zhang Hechun [75] proved that $V_{\alpha,\beta} \otimes M(c, h)$ is an indecomposable module, but reducible. He also got

Theorem *If the Verma module $M(c, h)$ is reducible and has a non-trivial maximal weight vector of degree m, and the degree of α on $\mathbf{Q}(\beta, c, (c-1)(c-25)^{\frac{1}{2}})$ is greater than m, then $V'_{\alpha,\beta} \otimes M(c, h)$ is irreducible.*

Zhao Kaiming also gave a class of indecomposable modules for Vir [81].

Theorem (a) *For any α, $\beta \in \mathbf{C}$ we have*

$$V'_{\alpha,\beta} \otimes V'_{\alpha,\beta} = V^+ \oplus V^-,$$

where V^+ and V^- are indecomposable submodule, and

$$End_{Vir}V^+ \cong \mathbf{C}, End_{Vir}V^- \cong \mathbf{C}, \quad End_{Vir}V'_{\alpha,\beta} \otimes V'_{\alpha,\beta} \cong \mathbf{C} \oplus \mathbf{C}.$$

(b) *If $V'_{\alpha,\beta}$ is not isomorphic with $V'_{\alpha',\beta'}$, then*

$$End_{Vir}V'_{\alpha,\beta} \otimes V'_{\alpha',\beta'} \cong \mathbf{C}.$$

So $V'_{\alpha,\beta} \otimes V'_{\alpha',\beta'}$ *is indecomposible.*

Zhao Kaiming [80] determined the automorphisms and endomorphisms of the Virasoro algebra.

In [84] Zhao Kaiming proved that

Theorem (a) *$V'_{\alpha,\beta}$ has a nondegenerate contravariant Hermitian form if and only if one of the following conditions holds:*

(1) $\beta + \bar{\beta} = 1$, $\alpha + \beta = \bar{\alpha} + \bar{\beta}$,
(2) $\alpha \in \mathbf{R}$, $\beta = 0$,
(3) $\alpha \in \mathbf{R}$, $\beta = 1$.

(b) $V'_{\alpha,\beta}$ has a nondegenerate contravariant bilinear form if and only if one of the following conditions holds: $\beta = 0$, 1 or $\frac{1}{2}$.

§3. The Lie algebra of differential operators

The algebra $\mathcal{L} = \mathbf{C}[t, \ t^{-1}, \ d/dt]$ of differential operators is an important algebra. It has many applications in mathematical physics.

Zhao Kaiming [83] proved that the extreior derivative algebra is one dimensional, and the unique exterior derivative(up to a constant) is σ:

$$\sigma(t) = \sigma(t^{-1}) = 0, \ \sigma(td/dt) = 1.$$

If L is a Lie algebra, it is well known that all 1-dimensional central extensions of L are determined by the 2-cocycle group of L. A 2-cocycle is a biliear $\mathbf{C}$-form ψ satisfying the following conditions:

(1) $\psi(a, b) = -\psi(b, a)$, (2) $\psi([a, b], c]) + \psi([b, c], a]) + \psi([c, a], b]) = 0$, for $a, b, c \in L$.

If a 2-cocycle is induced by a linear function on L, this 2-cocycle is called trivial. A 2-cocycle ϕ is equivalent to a 2-cocycle ψ, if $\phi - \psi$ is trivial.

Let $D = d/dt$, $\mathcal{A} = \mathbf{C}[t, \ t^{-1}, \ D]$, $U = \mathcal{A} \otimes gl_n(\mathbf{C})$. Define a bilinear form α on $\mathcal{A}$

$$\alpha(t^{l+m}D^l, \ t^{l'+m'}D^{l'}) = \delta_{m,-m'}(-1)^l l! l'! \binom{m+l}{l+l'+1},$$

for $l, l' \geq 0, m, m' \in \mathbf{Z}$. α is a 2-cocycle of $\mathcal{A}$.

Define a bilinear form ψ on U :

$$\psi(t^{l+m}D^l \otimes A, \ t^{l'+m'}D^{l'} \otimes A') = \delta_{m,-m'}(-1)^l l! l'! \binom{m+l}{l+l'+1} tr AA',$$

for $l, l' \geq 0, m, m' \in \mathbf{Z}, A, A' \in gl_n(\mathbf{C})$. It is easy to check that ψ is a 2-cocycle on U.

V.G. Kac proposed the following problems: Is every 2-cocycle on $\mathcal{A}$ equivalent to a multiple of α? Is every 2-cocycle on U equivalent to a multiple of ψ ?

Li Wanglai [25] gave these problems a positive answer.

Chapter 2. Finite dimensional Lie Algebras

§1. Complete, solvable and nilpotent Lie algebras

The Lie algebra G is called a complete Lie algebra, if the center of G is 0 and all the derivations are inner. G is called a simple complete Lie algebra, if any proper ideals of a complete Lie algebra G are not complete.

A maximal solvable subalgebra of a Lie algebra G is called a Borel subalgebra, and a subalgebra containing a Borel subalgebra is called a parabolic subalgebra. Let $\delta(G)$ be the Lie algebra of derivations of the Lie algebra G. We have the classical extension with derivations: $\eta(G) = G + \delta(G)$. $\eta(G)$ is called the holomorph of G.

Meng Daoji [44] studied the completeness of parabolic subalgebras of semisimple Lie algebras over **C**. He obtained

Theorem   (a) *Let G be a complete Lie algebra. Then*

(1) *G is simple complete iff G cannot be decomposed into a direct sum of nontrivial ideals,*

(2) *G can be decomposed into a direct sum of simple complete ideals.*

(b) *Parabolic subalgebras of a semisimple Lie algebra are complete Lie algebras.*
Parabolic subalgebras of a simple Lie algebra are simple complete Lie algebras.

(c) *G is complete iff $\eta(G)$ has a direct sum of ideals:*

$$\eta(G) = G \oplus C_{\eta(G)}(G),$$

where $C_{\eta(G)}(G)$ is the centralizer of G in $\eta(G)$.

In recent years, Meng Daoji [45,46,47] studied the uniqueness of decompositions of complete Lie algebras and some special complete Lie algebras. Lu Caihui also got some results on solvable Lie algebras [38].

The classification of solvable finite-dimensional Lie algebras over a field of characteristic 0 has not solved still. Malcev turned the studying of the classification and the structure of a solvable Lie algebra to the studying of a nilpotent Lie algebra. But the nilpotent Lie algebras are still not clear. A. G. Micael [48] classified metabelian Lie algebras in 1973. L.J. Santharoubane [53] classified the nilpotent Lie algebras of a maximal rank in 1982.

Let R be a finite dimensional Lie algebra over a field of characteristic 0. The intersection F_R of all maximal subalgebras of R is called the Frattini subalgebra of R. The intersection J_R of all maximal ideals of R is called the Jacobson radical of R. It is well known that F_R and J_R are nilpotent characteristic ideals (i.e. stable under all derivations).

A Lie algebra R is called complemented, if for any subalgebra A of R there exists a subalgebra B such that: $A \cap B = 0$, and R can be generated by A and B.

Lu Caihui [38] classified a kind of solvable Lie algebras in 1985.

Theorem (a) *Let R be a solvable Lie algebra with 1-dimensional Jacobson radical over a field of characteristic 0, $J_R = \mathbf{F}$ and $\dim R = n+1$. Then R is one of the following two types:*

(1) *R is a nonnilpotent solvable Lie algebra, with a basis: $e, c_1, c_2, \cdots, c_n$, and the following commutative relations*

$$[e, c_1] = e; \ [e, c_i] = 0, 2 \le i \le n; \ [c_i, c_j] = 0, \ i, j = 1, 2, \cdots, n.$$

(2) *R is metabelian, with a basis: $c_1, c_2, \ldots, c_{2l}, e, z_1, z_2, \cdots, z_k$, and the following commutative relations*

$$[c_{2i-1}, c_{2i}] = e, \ i = 1, 2, \cdots, l,$$

and other brackets are 0. And there are $\left[\frac{n}{2}\right]$ isomorphic classes of R.

(b) *Let R be a solvable Lie algebra over an algebraically closed field Φ of characteristic 0, $\dim R = n + 2(n \ge 1)$, $\dim J_R = 2$. Then R is one of the following ten types.*

(1) *R has a basis: $x_1, x_2, c_1, \ldots, c_n$, with the following commutative relations*

$$(A_1): \begin{cases} [x_1, c_1] = \lambda_1 x_1, & 0 \ne \lambda_1 \in \Phi, \\ [x_2, c_1] = x_2, \\ \text{other brackets are 0.} \end{cases}$$

(2) *R has a basis: $x_1, x_2, c_1, \ldots, c_n$, with the following commutative relations*

$$(A_2): \begin{cases} [x_1, c_1] = x_1, \\ [x_2, c_2] = x_2, \\ \text{other brackets are 0.} \end{cases}$$

In the above two cases $\dim F_R = 0$.

(3) R *has a basis:* $x, f, c_1, \ldots, c_n$, *with the following commutative relations*

$$(B_1): \begin{cases} [x, c_1] = x + \rho f, & 0 \neq \rho \in \Phi, \\ [f, c_1] = f, \\ \text{other brackets are } 0. \end{cases}$$

(4) R *has a basis:* $x, f, c_1, \ldots, c_n$, *with the following commutative relations*

$$(B_2): \begin{cases} [x, c_1] = x, \\ [x, c_2] = f, \\ [f, c_1] = f, \\ \text{other brackets are } 0. \end{cases}$$

(5) R *has a basis:* $x, f, z, c_1, \ldots, c_{2l}$, $2l + 1 = n$, *with the following commutative relations*

$$(B_3): \begin{cases} [x, z] = x \\ [c_{2j-1}, c_{2j}] = f, & j = 1, 2, \ldots, l. \\ \text{other brackets are } 0. \end{cases}$$

(6) R *has a basis:* $x, f, c_1, \ldots, c_{2l}$, *where* $2l = n$. *For this basis there is an integer* s $(1 \leq s \leq l)$, *such that* R *has the following commutative relations*

$$(B_4): \begin{cases} [x, c_{2j-1}] = x, & 1 \leq j \leq s \\ [c_{2j-1}, c_{2j}] = f, & 1 \leq j \leq l, \\ \text{other brackets. are } 0. \end{cases}$$

In $(B_1), (B_2), (B_3), (B_4)$, $F_R = \mathbf{C}f$.

(7) R *has a basis:* x_1, x_2, c_1, c_2, $(\dim R = 4)$, *with the following commutative relations*

$$(C_1): \begin{cases} [x_2, c_1] = x_1, \\ [c_1, c_2] = x_2, \\ \text{other brackets are } 0. \end{cases}$$

(8) R *has a basis:* $x_1, x_2, c_1, \cdots, c_{2l+2}$, *where* $n = 2l + 2, l \geq 1$ *with the following commutative relations*

$$(C_2): \begin{cases} [x_1, c_1] = x_1, \\ [c_{2j+1}, c_{2j+2}] = x_1, & j = 1, 2, \ldots, l, \\ [c_1, c_2] = x_2, \\ \text{other brackets are } 0. \end{cases}$$

(9) R has a basis: $x_1, x_2, c_1, \ldots, c_{2l+1}$, where $n = 2l + 1$, with the following commutative relations

$$(C_3) : \begin{cases} [x_2, c_1] = x_1, \\ [c_{2j}, c_{2j+1}] = x_1, & j = 1, 2, \ldots, l, \\ [c_1, c_{2j}] = x_2, & j = 1, 2, \ldots, s, 1 \leq s \leq l \\ \text{other brackets are 0.} \end{cases}$$

In C_1, C_2, C_3, R is nilpotent, and $\dim F_R = 2$.

(10) R is metabelian.

Cui Yimin studied the case that $\dim F_R = 1, \dim J_R = 3$, and J_R is commutative. She also gives the commutation relations [6].

Recently Lu Caihui [39] studied a kind of solvable finite dimensional Lie algebras which have nondegenerate, symmetric, invariant bilinear forms.

Let g be a finite dimensional Lie algebra over $\mathbf{C}$. If $\forall x \in g$, the semi-simple and nilpotent parts in the Jordan decomposition of adx are inner derivations, then g is called splitable. A subalgebra H is called toral if every nonzero element in H is toral.

At first, a noncommutative solvable finite dimensional Lie algebra g over $\mathbf{C}$ with a nondegenerate, symmetric, invariant bilinear form $(\cdot|\cdot)$ can be decomposed into a direct sum of ideals

$$g = A \oplus Z_0 \oplus g_1.$$

And

(1) A is commutative, $(A|Z_0 \oplus g_1) = 0$.

(2) Z_0 is commutative, $(Z_0|Z_0) = 0$.

(3) g_1 is a nonzero solvable ideal, $(\cdot|\cdot)$ is nondegenerate on g_1. Let Z_1 be the center of g_1. Then $Z_1 \subset [g_1, g_1], (Z_1|Z_1) = 0$.

(4) $[gg] = [g_1 g_1]$.

So the study of g is turned to the study of a Lie algebra satisfying

[**Condition 1**] g is a noncommutative solvable finite dimensional Lie algebra over $\mathbf{C}$ with a nondegenerate, symmetric, invaraint bilinear form $(\cdot|\cdot)$ which is 0 on the center Z of g, and $Z \subset [g, g]$.

The Lie algebra g satisfying condition 1 is called indecomposable if g does not contain a nondegenerate proper ideal. Otherwise g is called decomposable.

Theorem *Any Lie algebra g satisfying condition 1 can be decomposed into an orthogonal direct sum of indecomposable solvable ideals*

$$g = \oplus_{i=1}^{s} g_i$$

such that

(1) *Each g_i satisfies condition 1,*

(2) *Each g_i is nondegenerate nilpotent or nondegenerate nonnilpotent solvable.*

Let g be a splitable Lie algebra. A nonzero element h of g is called toral if $ad_g h$ is diagonalizable and has a nonzero eigenvalue. For a splitable indecomposable algebra g satisfying condition 1 we have a root space decomposation of g with respect to a maximal toral subalgebra H:

$$g = g_0 \oplus \oplus_{\alpha \in \Delta} g_\alpha,$$

where $g_\alpha = \{\, x \in g | [h, x] = \alpha(x)x, \forall\, h \in H; \alpha \in H^* \},$ and

(1) H is abelian,

 If $\alpha, \beta \in \Delta \cup \{0\}$, then $[\, g_\alpha,\ g_\beta] \subset\ g_{\alpha+\beta}.$

(2) $\forall \alpha, \beta \in \Delta \cup \{0\}$, if $\alpha + \beta \neq 0$, then $(\,g_\alpha|\ g_\beta) = 0.$

(3) $(\,|\,)$ is nondegenerate on g_0.

(4) If $\alpha \in \Delta$ then $-\alpha \in \Delta$. $(\,g_\alpha|\ g_\beta) \neq 0, \dim\ g_\alpha = \dim\ g_{-\alpha}.$

Li Zhenheng gave the structures of a class of Lie algebras with chain conditions discussed in [27, 29].

Wang Shuqin [64], Zhou Ruqi [85] also undertook some studies on finite dimensional nonsemisimple Lie algebras.

§2. Finite-dimensional semisimple Lie algebras

Kac classified the symmetric spaces using the structure theory of affine Lie algebras in 1968. And then he applied the result to the classification of automorphisms of finite order of finite-dimensional simple Lie algebras, see [19,20].

In 1987 Zhang Zhixue [76] gave a simple proof of the classification of outer automorphisms of finite order of finite-dimensional simple Lie algebras using representation theory, extended Cartan groups and Gantmacher theorem.

G.B. Seligman [54] defined finite dimensional characreristic semisimple Lie algebras, characteristic simple Lie algebras and complete characteristic simple Lie algebras.

An ideal A of a finite dimensional Lie algebra L over a field is called characteristic if for any derivation D of L we have $D(A) \subset A$.

For a finite dimensional Lie algebra L there exists a unique maximal characteristic solvable ideal R. R is called the characteristic radical, or simply called C-radical. L is called a characteristic semisimple Lie algebra (c.s.s.) if the C-radical of L is 0. L is called a characteristic simple Lie algebra if L and $\{0\}$ are the only characteristic ideals and $[L, L] = L$. L is called a complete semisimple Lie algebra if L can be decomposed into a sum of characteristic simple ideals: $L = \sum_{i=1}^{s} L_i$.

Hochshild [12] got the following result: the Lie algebra of derivations $\mathcal{D}(L)$ of a finite dimensional Lie algebra L over a field of characteristic 0 is semisimple iff L is semisimple. Lu Caihui extended the result for Lie algebras with 0 center to Lie algebras over fields of any characteristic [40].

Theorem *Let L be a finite dimensional Lie algebra over a field F.*

(a) *If the center of L is 0, then $\mathcal{D}(L)$ is semisimple iff L is characteristic semisimple.*

(b) *If L is semisimple finite-dimensional, then L can be decomposed into a direct sum of simple ideals iff L is complete semisimple.*

(c) *If L is complete semisimple, then any nonzero ideals are semisimple.*

We denote the 3-dimensional simple Lie algebra $sl(2, \mathbf{C})$ by $\bar{g}$. Let x, y, h be a standard basis:

$$[x, y] = h, \quad [h, x] = 2x, \quad [h, y] = -2y.$$

Zhao Kaiming and Su Yucai constructed all indecomposable Harish-Chandra modules over $\bar{g}$. They [82] proved the following classification theorem

Theorem *An indecomposable Harish-Chandra module over $\bar{g}$ is isomorphic to one of the following ten types, where $r \in \mathbf{N}$, $\lambda, \alpha \in \mathbf{C}$, $V = \sum_{i \in \mathbf{Z}} V_i$, $V_i = \oplus_{s=1}^{r} \mathbf{C} v_i^{(s)}$:*

(1) $hv_i^s = (\lambda - 2i)v_i^{(s)}$, $yv_i^{(s)} = v_{i+1}^{(s)}$,

$$x\begin{pmatrix} v_i^{(1)} \\ \vdots \\ v_i^{(r)} \end{pmatrix} = \begin{pmatrix} i(\lambda - i + 1) + \alpha & 1 & \cdots & 0 \\ 0 & i(\lambda - i + 1) + \alpha & \cdots & 0 \\ \cdots & \cdots & \cdots & \cdots \\ 0 & 0 & \cdots & i(\lambda - i + 1) + \alpha \end{pmatrix} \begin{pmatrix} v_{i-1}^{(1)} \\ \vdots \\ v_{i-1}^{(r)} \end{pmatrix};$$

(2) $hv_i^s = (\lambda - 2i)v_i^{(s)}$, $xv_i^{(s)} = v_{i-1}^{(s)}$,

$$y\begin{pmatrix} v_i^{(1)} \\ \vdots \\ v_i^{(r)} \end{pmatrix} = \begin{pmatrix} (i+1)(\lambda - i) + \alpha & 1 & \cdots & 0 \\ 0 & (i+1)(\lambda - i) + \alpha & \cdots & 0 \\ \cdots & \cdots & \cdots & \cdots \\ 0 & 0 & \cdots & (i+1)(\lambda - i) + \alpha \end{pmatrix} \begin{pmatrix} v_{i+1}^{(1)} \\ \vdots \\ v_{i+1}^{(r)} \end{pmatrix};$$

(3) $hv_i^s = (\lambda - 2i)v_i^{(s)}$,

$$x\begin{pmatrix} v_i^{(1)} \\ \vdots \\ v_i^{(r)} \end{pmatrix} = \begin{pmatrix} k_1 - i & 1 & \cdots & 0 \\ 0 & k_1 - i & \cdots & 0 \\ \cdots & \cdots & \cdots & \cdots \\ 0 & 0 & \cdots & k_1 - i \end{pmatrix} \begin{pmatrix} v_{i-1}^{(1)} \\ \vdots \\ v_{i-1}^{(r)} \end{pmatrix};$$

$$y\begin{pmatrix} v_i^{(1)} \\ \vdots \\ v_i^{(r)} \end{pmatrix} = \begin{pmatrix} k_2 + i & 1 & \cdots & 0 \\ 0 & k_2 + i & \cdots & 0 \\ \cdots & \cdots & \cdots & \cdots \\ 0 & 0 & \cdots & k_2 + i \end{pmatrix} \begin{pmatrix} v_{i+1}^{(1)} \\ \vdots \\ v_{i+1}^{(r)} \end{pmatrix};$$

where $k_1, k_2 \in \mathbf{Z}$, $\lambda = k_1 - k_2, k_1 + k_2 \neq 1$;

(4) *In case 1, let* $i(\lambda - i + 1) + \alpha = -(i - k_1)(i - k_2)$, $k_1 \in \mathbf{Z}$, *and* $V^{(i)} = \oplus_{j \in \mathbf{Z}} \mathbf{C} v_j^{(i)}$. *Then* V *has the following indecomposable submodules:*
$\oplus_{s=2}^{r} V^{(s)} \oplus_{i \geq k_1} \mathbf{C} v_i^{(1)}$;

(5) *In case 2, let* $i(\lambda - i + 1) + \alpha = -(i - k_1)(i - k_2)$, $k_1 \in \mathbf{Z}$, *and* $V^{(i)} = \oplus_{j \in \mathbf{Z}} \mathbf{C} v_i^{(i)}$. *Then* V *has the following indecomposable submodules:*
$\oplus_{s=2}^{r} V^{(s)} \oplus_{i < k_1} \mathbf{C} v_i^{(1)}$;

(6) *In case 1, let* $i(\lambda - i + 1) + \alpha = -(i - k_1)(i - k_2)$, $k_1, k_2 \in \mathbf{Z}, k_1 < k_2$, *and* $V^{(i)} = \oplus_{j \in \mathbf{Z}} \mathbf{C} v_i^{(i)}$. *Then* V *has the following indecomposable submodules:*
$$\oplus_{s=3}^{r} V^{(s)} \oplus_{i \geq k_2} \mathbf{C} v_i^{(1)} \oplus_{j \geq k_2} \mathbf{C} v_j^{(2)};$$

(7) *In case 2, let* $i(\lambda - i + 1) + \alpha = -(i - k_1)(i - k_2)$, $k_1, k_2 \in \mathbf{Z}$, *and* $k_1 < k_2$, *and* $V^{(i)} = \oplus_{j \in \mathbf{Z}} \mathbf{C} v_j^{(i)}$. *Then* V *has the following indecomposable submodules:*
$$\oplus_{s=2}^{r} V^{(s)} \oplus_{i < k_1} \mathbf{C} v_i^{(1)} \oplus_{j < k_2} \mathbf{C} v_j^{(2)};$$

(8) *In case 3, V has indecomposable submodules:*

$$\oplus_{s=2}^{r} V^{(s)} \oplus_{i \geq k_1} \mathbf{C} v_i^{(1)}, \quad \oplus_{s=2}^{r} V^{(s)} \oplus_{i \leq -k_2} \mathbf{C} v_i^{(1)};$$

(9) *In case 3, if $k_1 + k_2 > 1$, V has indecomposable submodules:*

$$\oplus_{s=2}^{r} V^{(s)} \oplus_{i \geq k_1} \mathbf{C} v_i^{(1)} \oplus_{j \leq -k_2} \mathbf{C} v_j^{(2)};]$$

(10) *In case 3, if $k_1 + k_2 \leq 0$, V has indecomposable submodules:*

$$\oplus_{s=2}^{r} V^{(s)} \oplus_{-k_2 \geq i \geq k_1} \mathbf{C} v_i^{(1)}.$$

Chapter 3. Lie superalgebras

Chen Yongqing [5] considered graded operations of Bosen-Fermion operators. Using two pairs of Bosen operators and a pair of Fermion operators he gave the expressions of generators of the Lie superalgebra $OSP(1,2)$. So a new form of Bosen-fermion representation was given.

Dai Guisheng [7] discussed the category of modules over basic classical Lie superagebras. He proved that the indecomposable projective objects in this category have the index set $H \times \mathbf{Z}_2$. And the daulity theorem was given.

Dai Guisheng [8] studied the homomorphisms between Verma modules over the basic classical Lie superalgebras. He discussed the dimensions of the homomorphic spaces and the kernels of the homomorphisms.

Dai Guisheng [9] gave the following result. If G is a basic classical Lie superalgebra of type I or $G = B(0,n)$, every module over G has a unique minimal submodule. It is not true for $G = F(A)$.

In [10] Dai Guisheng discussed the following objects. Let $g = g_{-1} \oplus g_0 \oplus g_1$ be the Lie superalgebra of linear mappings of the graded space $V = V_0 \oplus V_1$. By induced comodules he defined the g-module structure of the outer algebra $\Lambda(g_0)$, where $g_1 = V_1 \oplus V_0^*$. And all $(g_0 + g_1), (g_0 + g_{-1})$, g modules are determined. Li Zhenheng [30] introduced the concept of complete Lie superalgebras and discussed some of its properties. He obtained the uniqueness of the decomposations of finite dimensioal complete Lie superalgebras.

H. Garland constructes an integrable form U_z in the universal enveloping algebra $U(\mathbf{g}(\tilde{A}))$ of affine Lie algebra $\mathbf{g}(\tilde{A})$, and defines, for any field

K, $U_k = U_z \otimes_z K$ to be the affine hyperalgebra of $\mathbf{g}(\tilde{A})$. Certain irreducible U_k module $L(\lambda)$ is defined by Liu Changkun in [32]. A one-one correspondence between $L(\lambda)$ and dorminant weights λ is proved, analogous to that by J. E. Humphreys. Besides, for each $n \in \mathbf{Z}^+$, J. E. Humphreys defines a subalgebra u_n of U_k. Liu Changkun obtained a basis of u_n.

Su Yucai [58] gave a classification of indecomposable Harish-Chandra modules over $B(0,1)$. In [59] he not only gave a complete classification of finite dimensional indecomposable modules over the Lie superalgebra $sl(2/1)$ with diagonal Cartan subalgebra, but also completely classified those modules with nondiagonal Cartan subalgebra.

References

[1] A. Borel and J. P. Serre, *Sur certains sous-groups des groups de Lie compacts*, Comment. Math. Helv., **27** (1953), 128–283.

[2] V. Chari, *Integrable representations of affine Lie algebras*, Invent. Math., **85** (1986), 317–355.

[3] V. Chari and A. N. Pressly, *A new family of irreducible integrable modules for affine Lie algebras*, Math. Ann., **277** (1987), 543–562.

[4] V. Chari and A. N. Pressly, *Integrable representations of twisted affine Lie algebras*, J. Algebra, 113 (1988), 438–464.

[5] Chen Yongqing, *A new form of Bosen-Fermion representations of OSP(1,2)*, J. of Hunan teachers University, **12** 1 (1987) 25–27.

[6] Cui Yimin, *The determination of a kind of solvable Lie algebras*, J. of Beijing Teacher's College, **12** 1 (1991), 1–9.

[7] Dai Guisheng, *Category of modules over basic classical Lie superalgebras and duality theorem*, J. of Suzhou University, **6** 1 (1990), 10–15.

[8] Dai Guisheng, *Homomorphisms between the Verma modules over Lie superalgebras*, J. of Suzhou University, **10** 3 (1990), 141–145.

[9] Dai Guisheng, *The uniqueness of minimal submoduls of Verma modules over Lie superalgebras*, J. of Suzhou University, **7** 2 (1991), 125–127.

[10] Dai Guisheng, *The gl(m,n)-module structures of $\Lambda gl(m,n)$*, Chinese Annual of Math., **11A** 5 (1990), 600–606.

[11] Gao Yongcun, *On the Structure of Some Modules for $g(A_1^{(1)})$*, Chinese Adv.in Math. **16** 1 (1990), 105–106.

[12] G. Holchschild, *Semisimple Lie algebras and generalized derivations*, Amer. J. Math., **64** (1942), 677–694.

[13] Jia Yuting, *Subalgebras of nonskiew affine Lie algebras*, Acta Chinese Math., **33** 2 (1990), 197–204.

[14] Jia Yuting, *Q-grading ω_0-invariant subalgebras of $g(B_l^{(1)})$, $g(C_l^{(1)})$, $g(D_l^{(1)})$*, Acta Chinese Math., **33** 4 (1990), 433–444.

[15] Jiang Cuibo, *A Class of Level 2 Standard Modules for $B_3^{(1)}$*, Journal of Yantai Teacher's College, **2** (1991).

[16] V. G. Kac, *Simple graded Lie algebras of finite growth,* Funkt. Analis i ego prilozh, **1** 4 (1967), 82–83; English translation: Funct. Anal. Appl., **1** (1967), 328–329

[17] V. G. Kac, *Infinite dimensional algebras, Dedekind's η-function, classical Mobious function and very strange formul,* Adv. in Math., **30** (1978), 85–136

[18] V. G. Kac, "Infinite Dimensional Lie Algebras" Second edition, Cambridge Press, 1985.

[19] V. G. Kac, *Graded Lie algebras and symmetric spaces,* Funkt. Anklis. ieg Prilozh., **2** (1968), 93–94.

[20] V. G. Kac, *Automorphisms of finite order of semisimple Lie algebras.,* Funkt. Anklis. ieg Prilozh., **3** (1969), 94–96.

[21] V. G. Kac and K. A. Raina, *Bombay lectures on highest weight representations of finite dimensional Lie algebras,* Adv. Series in Math. Phys., **2** (1986).

[22] J. Lepowsky, *Calculus of twist vertex operators,* Proc. Natl. Acad. Sci. USA, 1985.

[23] Li Haisheng, *Structure of a Class of Integrable Lie Algebras,* Chinese Adv. in Math. **16** 1 (1990), 106–108.

[24] Li Wanglai, *The Cohomology of the Algebra of Differential Operators with Coefficients in Laurent Polynomial Ring,* Chinese Science Bulletin, **35** 11 (1990), 807–809.

[25] Li Wanglai, *2-Cocycles on the Algebras of Differential Operators,* J. Alg. **122** (1989), 64–80.

[26] Li Wanglai, *Classification of Generalized Cartan Matrices of Hyperbolic Type,* Chinese Annal of Math., **9B** (1988), 68–77.

[27] Li Zhenheng, *The structures of Lie algebras satisfying chain conditions,* J. of Hebei University, **2** (1989), 1–5.

[28] Li Zhenheng, *The structures and the eigenvalues of the adjoint representation of $k(A)$,* J. of Hebei University, **5** (1989).

[29] Li Zhenheng, *The uniqueness of decomposation of complete Lie superalgebras,* Northeast Math. J., **9** 3 (1993), 403–405.

[30] Li Zhenheng, *Some properties of generalized Kac-Moody algebras,* Chinese Science Bulletin, **38** 15 (1993).

[31] Li Zhenheng, *The structures of a class of Lie algebras with chain conditions,* J. of Hebei University, **3** (1993).

[32] Liu Changkun, *On the Affine Hyperalgebre U_k,* Journal of East China Normal University, **1** (1985), 14–22.

[33] Liu Lishi, *Kostant's formular for Kac-moody algebras,* J. Alg., **149** (1992), 155–178.

[34] Liu Lishi, *Some Automorphism Groups of Affine Lie Algebra $g(A)$,* Chinese journal of Math. Research and Exposition, **10** 3 (1990), 315–326.

[35] Lu Caihui, *The minimal imaginary roots of Kac-Moody algebras of indefinite type,* Journal of Beijing Teacher's College, 1991.

[36] Lu Caihui, *A necessary and sufficient condition for the completely reducibility of a $g(A)$-module,* Chinese Science Bulletin, **31** 1 (1986).

[37] Lu Caihui, *The classification and construction of a class of irreducible integrable $A_1^{(1)}$-modules,* Acta. Chinese Math., **30** 5 (1987), 626–640.

[38] Lu caihui, *The basis structure conditions and classifications of solvable Lie algebras with 1 or 2-dimensional Jacobson ridicals,* Acta Math. Sinica, **28** 2 (1985), 251–262.

[39] Lu caihui, *Finite dimensional solvable Lie algebras with nondegenerate, symmetric, invariant bilinear forms,* Acta Math. Sinica, **35** 1 (1992), 121–132.

[40] Lu caihui , *Characteristic semisimple Lie algebras and complete semisimple Lie algebras over any field* , J. of Math. Reseach and exposition, **3** 1 (1983).

[41] Lu Caihui and Wan Zhexian, *On the minimal number of generators of Lie Algebra $g(A)$*, J. Alg., **101** (1986), 470–472.

[42] Lu Shirong, *A Necessary and Sufficient Condition for Δ_+^{im} to be a Semi-Group*, Chinese Adv. in Math., **16** 2 (1988), 294–298.

[43] O. Mathieu, *Classification of Harish-Chandra Modules over the Virasoro algebra*, Invt Math., **107** Fasc 2 (1992), 225–234.

[44] Meng Daoji, *On complete Lie algebras*, Acta Scientiarum Naturalium Universitatis NanKaiensis, **2** (1985), 11–19.

[45] Meng Daoji, *The Uniqueness of the Decomposations of Complete Lie Algebras*, Acta Scientiarum Naturalium Universitatis NanKaiensis, **3** (1990).

[46] Meng Daoji, *Complete Lie algebras with abelian nilpotent radicals*, Acta Scientiarum Naturalium Universitatis NanKaiensis, **34** (1990), 191–202.

[47] Meng Daoji, *The Lie algebras of derivations of Lie algebras with center 0*, Acta Scientiarum Naturalium Universitatis NanKaiensis, **4** (1990), 84–87.

[48] Michael A. Gauger, *On the classification of metabelian Lie algebras*, Trans. Amer. Math. Soc., **179** (1973), 293–329.

[49] K. C. Misra, *Relization of the level two standar modules over $sl(2k+1, \mathbf{C})^\vee$-Modules*, Trans. Amer. Math. Soc., **316** 1 (1989).

[50] R. V. Moody , *Lie algebras associated with generalized Cartan matrices*, Bull. Amer. Math. Soc., **73** (1967), 217–221.

[51] R. V. Moody, *Root system of hyperbolic type*, Adv. in Math., **33** (1979), 144–160.

[52] Qiu Sen, *The Cohomology of Kac-Moody Lie Algebras with Coefficients in a General Verma Module*, J. Alg., **90** (1984), 10–17.

[53] L. J. Santharoubane, *Kac-Moody Lie Algebras and the classification of nilpotent Lie algebras of maximal rank*, Canadian Journal of Math., **34** 6 (1982).

[54] G. B. Seligman, *Characteristic ideals and the structure of Lie algebras*, Proc. Amer. Math. Soc., **8** (1957), 159–164.

[55] Su Yucai , *A Complete Classification of Some Integrable Modules of Affine Lie Algebra $A_2^{(2)}$*, Acta. Chinese Math., **32** 3 (1989), 331–337.

[56] Su Yucai, *On the Low Dimensional Cohomology of Kac-Moody Algebras with Coefficients in Complex Field*, Chinese Adv. in Math., **18** 3 (1989), 346–351.

[57] Su Yucai , *General Integrable Lie Algebra $gl(V)_{fin}$ and Associated Group*, Chinese Science Bulletin, **17** (1989), 1287–1290.

[58] Su Yucai, *A Complete Classification of Indecomposable Harish-Chandra Modules of the Lie Superalgebra $B(0,1)$*, Comm. Alg. (1993) (to appear).

[59] Su Yucai, *Classification of Finite Dimensional Modules of the Li Superalgebra $sl(2/1)$*, Comm. Alg., **20** (1992), 3259–3278.

[60] Tuan Shiofu, *A note on the replicas of nolpotent matrices*, Bull. Amer. Math. Soc., **51** (1945), 305–313.

[61] Tuan Shiofu and C. Chevalley, *On algebraic Lie algebras*, Proc. Nat. Sci. U.S.A., **31** (1945), 195–196.

[62] Tuan Shiofu and C. Chevalley , *Algebraic Lie algebras and their invariants*, Acta. Math. Sinica, (1951), 215–242.

[63] Vinberg, *Discrete linear groups generated by reflections*, Izvestija AN USSR, **35** , 1072–1112.

[64] Wang Shuqin, *The generators and defininy relations of a class of nonsemisimple Lie algebras – on constructions of complementary Lie algebras*, J. Harbin Teacher's University.

[65] Wang Xiaodong, *The Structure of Level One Standard $C_n^{(1)}$-Modules*, Chinese Science Bulletin **37** 10 (1992).

[66] Wang Xiaodong, *The Structure of Level Two Standard A_{n-1}-Modules*, Chinese Science Bulletin, (1987)**32** 1 (1971), 11–13.

[67] Xie Chuanfu, *The Maximal Weights of Fundamental Modules over Affine Algebras*, Chinese Adv.in Math., **18** 2 (1989), 226–231.

[68] Xu Yichao and Jiang Cuipo, *Vertex operators of $G_2^{(1)}$ and $B_n^{(1)}$* , J. Phys. A. Math. Gen., **23** (1990), 3105–3121.

[69] Xu Yichao, *Vertex operators of affine Lie algebras with first kind*, Proc. SEAMS Conference on Ordered Structures and Algebra of Computer Languages, June (1991), 280–299.

[70] Zhang Hechun, *Some Remarks on Irreducible $g(A)$-Modules*, Chinese Science Bulletin, **32** 17 (1987), 1288–1290.

[71] Zhang Hechun , *The Imaginary Root Systems of Kac-Moody Algebras*, Acta. Chinese Math., **33** 1 (1990), 1–6.

[72] Zhang Hechun, *The Universal Integrable Representations of Kac-Moody Algebras*, Chinese Science Bulletin, **35** 7 (1990), 489–491.

[73] Zhang Hechun, *The Reducibility of Universal Integrable Modules*, Chinese Science Bulletin, **24** (1991), 1844–1846.

[74] Zhang Hechun, *Some Remarks on Inmaginary Root System*, Journal of He-bei Teacher's College, **2** (1991).

[75] Zhang Hechun, "The Representation Theory of Infinite Dimensional Lie Algebras", Doctor Thesis 1990.

[76] Zhang Zhixue, *Outer automorphisms of finite order of complex semisimple Lie algebras*, Chinese annual of mathematics, 1987.

[77] Zhang Zhixue, *Note on the sets of weights of integrable modules over $g(A)$*, Chinese Science Bulletin, **18** (1992).

[78] Zhang Zhixue, *Some properties of modules over generalized Kac-Moody algebras*, Chinese Science Bulletin, **9** (1993).

[79] Zhao Kaiming, *Qusirational Automorphisms of Simple Finite Dimensional Lie Algebras*, Chinese Science Bulletin, **1** (1990).

[80] Zhao Kaiming, *The Automorphosms and Endmorphisms of the Virasoro Algebra*, Sys. Sci. Math., **1** (1992).

[81] Zhao Kaiming, "The Virasoro Algebra and its Representations", Doctor Thesis, 1991.

[82] Zhao Kaiming and Su Yucai, *Indecomposable Harish-Chandra modules*, Chinese Science Bulletin, **36** 7 (1991), 486–489.

[83] Zhao Kaiming, *The Lie Algebra of Derivations of the Algebra of Differential Operators*, Chinese Science Bulletin, **38** 2 (1993), 100–103.

[84] Zhao Kaiming, *Irreducible representatioñs $V'_{\alpha,\beta}$ of the Virasoro algebra*, Chinese Science Bulletin, **36** 18 (1991).

[85] Zhou Ruqi, *DI-Lie algebras with ad-nilpotent elements*, J. Harbin Teachers University, **2** (1985), 1–4.

[86] Zhou Shanyou , *Closedness of General Vertex Operators in Central Binomial Coefficients*, Acta. Chinese Math., **3** (1992).

Lie Algebras of Prime Characteristic

Guang-Yu Shen

East China Normal University

Historically, the concept of a Lie algebra was introduced as a tool to analyze the local structure of a Lie group, and accordingly was at first only considered over the field of complex numbers. It became apparent later that Lie algebras are purely algebraic objects that can be defined and discussed over an arbitrary field and the theory can be developed independently of Lie groups. However, examples show that most of the powerful methods and results of the classical theory are not transferable to the case when the ground field F is of characteristic $p > 0$. To develop the modular theory, special concepts and techniques are needed. For instance, N. Jacobson has introduced the notion of a restricted Lie algebra which, roughly, requires that the p-th power of an inner derivation is also inner (cf. [2]). The derivation algebra of a (not necessarily associative) algebra and the Lie algebra of an algebraic group are examples of restricted Lie algebras.

The present article is a brief and informal sketch of some recent advances in the theory of modular Lie algebras, in particular those obtained by Chinese research workers in this field.

1. Classicfication of simple Lie algebras

A central problem in the theory of modular Lie algebras is the classification of simple Lie algebras. It was known early that there exist simple modular Lie algebras which are not classical Lie algebras (i.e., analogues of simple complex Lie algebras). The first example of a nonclassical Lie algebra is the Witt algebra introduced in 1937, of which a thorough investigation was given by H.-J. Chang in [3]. Subsequently, many more of these nonclassical simple Lie algebras were found and constructed by different authors and by diversified methods. The interrelations among

them were not quite clear at first. It was pointed out by A. I. Kostrikin
and I. R. Šafarevič in their 1966 paper [26] that if the characteristic is
not very low, every then known nonclassical restricted simple Lie algebra is isomorphic to one of *the restricted Lie algebras of Cartan type* W_n,
S_n, H_n and K_n (or $W(n,1)$, $S(n,1)$, $H(n,1)$ and $K(n,1)$ in more recent
notations) which are finite-dimensional analogues of the four classes of
infinite-dimensional complex Lie algebras obtained by E. Cartan in 1909
when investigating infinite-dimensional Lie pseudogroups [2]. They posed
the following

Kostrikin-Šafarevič Conjecture (KŠC). A restricted simple Lie algebra
over an algebraically closed field of characteristic $p > 7$ is either classical
or of Cartan type.

W_n, S_n, H_n and K_n were generalized to *the graded Lie algebras of
Cartan type* $W(n,\mathbf{m})$, $S(n,\mathbf{m})$, $H(n,\mathbf{m})$ and $K(n,\mathbf{m})$ by Kostrikin and
Šafarevič [27], and further to *the Lie algebras of generalized Cartan type*
$W(n,\mathbf{m},\Phi)$, $S(n,\mathbf{m},\Phi)$, $H(n,\mathbf{m},\Phi)$ and $K(n,\mathbf{m},\Phi)$ independently by
V. G. Kac [24] and R. L. Wilson [49]. Roughly, $W(n,\mathbf{m})$ is the Lie alge-
bra of special derivations of the divided power algebra $\mathfrak{A}(n,\mathbf{m})$, $X(n,\mathbf{m})$,
$X = S$, H or K, is a subalgebra of $W(n,\mathbf{m})$ preserving a certain differential form ω_X and $X(n,\mathbf{m},\Phi)$ is the twist of $X(n,\mathbf{m})$ by an automorphism
Φ of the infinite-dimensional Lie algebra of Cartan type $W(n)$ (for details,
cf. [27], [24] and [49]). The central point of the classification problem is
to prove KŠC and the following

Generalized Kostrikin-Šafarevič Conjecture (GKŠC). A simple Lie algebra over an algebraically closed field of characteristic $p > 7$ is either
classical or of Cartan type.

Many people has made contributions to the solution of the problem.
KŠC was proved by R. E. Block and R. L. Wilson in 1984 [1] and GKŠC
was finally verified by H. Strade and R. L. Wilson in 1991 [46].

G.-Y. Shen in [33] (which, though the main results are obtained in the
late 60's, was not published until 1983) constructed a class of Lie algebras
$\Sigma = \Sigma(n,m,\mathbf{r},G)$ where n, m are nonnegative numbers, $\mathbf{r}$ is a $(2n+1)$-
tuple of positive numbers and G a subgroup of the additive group of the
ground field F. If $1 \notin G$ and $n + 2 \not\equiv 0 \pmod{p}$, $\Sigma^* := \Sigma$ is simple. If
$1 \notin G$ and $n + 2 \equiv 0 \pmod{p}$, $\widetilde{\Sigma} := [\Sigma, \Sigma]$ is simple. When $1 \in G$, Σ
contains a one-dimensional center $\mathfrak{C} \subseteq [\Sigma, \Sigma]$ and $\overline{\Sigma} := [\Sigma, \Sigma]/\mathfrak{C}$ is simple.
The dimensions of Σ^*, $\widetilde{\Sigma}$ and $\overline{\Sigma}$ are p^N, p^{N-1} and $p^N - 2$ respectively, where
N is the sum of m and the components of $\mathbf{r}$. The derivation algebras of

Σ^*, $\widetilde{\Sigma}$ and $\overline{\Sigma}$ are determined. In particular, the dimensions of the outer derivation algebras are $N-(2n+1)$, $N-2n$ and $N+2$ respectively. When $m=0$ (then $G=\{0\}$), Σ^* and $\widetilde{\Sigma}$ are the Lie algebras of Cartan type K. If $m>0$, then Σ^*, $\widetilde{\Sigma}$ and $\overline{\Sigma}$ are explicit realizations of Lie algebras of generalized Cartan type H. It was shown in [33] that Σ^* is a new class of simple Lie algebras (that are explicitly constructed). By determining the associated graded Lie algebra, Q.-Y. Fei [15] showed that $\widetilde{\Sigma}$ is also new. It was shown in [37] that $\overline{\Sigma}$ is isomorphic to a simple Lie algebra associated with a noncommutative Jordan algebra. By the explicit form of these algebras, detailed investigations are possible. For instance, using the K-like gradation structure (cf. [33]), Y.-Z. Zhang [53] showed that when $n+3 \equiv 0 \pmod{p}$, Σ^* admits a nonsingular associate form (cf. also [45, p.178]). To the author's knowledge, up to the present time, not all the Lie algebras of generalized Cartan type, especially those of type H, are explicitly realized.

2. Forms and invariant filtrations of Lie algebras of generalized Cartan type

Let L' be a simple Lie algebra over a (not necessarily algebraically closed) field F'. By [21, Thm.10.1.3], L' may be regarded as a central simple Lie algebra over an extension field of F'. Hence to classify simple Lie algebras over an arbitrary field, not too much is lost if we consider only the central simple ones. If F is an extension field of F', L a Lie algebra over F and $L' \otimes_{F'} F \cong L$, then L' is said to be an F'-form of L. Let F be algebraically closed and Char $F > 7$. For a classical Lie algebra L over F, except $L = D_4$, E_6, E_7 or E_8, all F'-forms are known. By GKŠC (which is a theorem now), the determination of all simple Lie algebras over F' (Char $F' > 7$) will essentially follow from the determination of F'-forms of Lie algebras of generalized Cartan type. The determination of forms of restricted Lie algebras of Cartan type was completed by S. Serconek and R. L. Wilson [31]. Their approach depends substantially on the field extension theory and the higher derivations and seems not applicable to general cases. Later, W. C. Waterhouse [47] determined the forms of $W(n, \mathbf{m})$ by showing that all automorphisms of $W(n, \mathbf{m})$ over a commutative ring R of characteristic p are standard, i.e., are induced by automorphisms of the associated divided power algebra $\mathfrak{A}(n, \mathbf{m})$ over R. The forms of $W(n, \mathbf{m})$ are the Witt-Ree algebras which were given by R. Ree in 1956 [30]. In

[43], S. M. Skryabin gave a general discussion of the forms of (generalized) Cartan type Lie algebras. He defined the concept of the standard forms. A standard F'-form L' of a Lie algebra L of generalized Cartan type is, roughly, the corresponding Lie algebra of derivations of an L'-invariant form of $\mathfrak{A}(n, \mathbf{m})$ and thus can be completely described. A main result of [43] asserts that if all automorphisms of $L_R := L \otimes_F R$ over a commutative ring R of characteristic p are standard, then all forms of L are standard.

Thus the problem was reduced to whether the automorphisms of L_R are all standard. When $R = F$ is a field, the answer is well-known to be affirmative. It is a consequence of the existence of an invariant filtration of L. Recall that $L = X(n, \mathbf{m}, \Phi)$, $X = W$, S, H or K, is a Lie algebra of special derivations of $\mathfrak{A}(n, \mathbf{m})$ which is spanned by $x^{(\alpha)}$, $\alpha = (\alpha_1, \dots, \alpha_n) \in Z_+^{(n)}$. The gradation of $\mathfrak{A}(n, \mathbf{m})$ defined by letting $\deg x^{(\alpha)} = \sum_{i=1}^{n-1} \alpha_i + (1 + \delta_{XK})\alpha_n)$ induces a gradation, and so a filtration, of type $1 + \delta_{XK}\varepsilon_n$ of L which is invariant under the automorphismsof L. Invariant filtrations also exist in infinite-dimensional Cartan type Lie algebras. The invariance of filtrations in different situations has been investigated by various authors and by different approaches. In [22], N. Jin gave a more unifying treatment of this problem. He discussed various types of nilpotency of infinite-dimensional Cartan type Lie algebras, determined the sets N of nilpotent elements, and showed that the zeroth filtering space L_0 in the filtration of type $1 + \delta_{XK}\varepsilon_n$ is completely determined by N and so is invariant. Since the filtration of type $1 + \delta_{XK}\varepsilon_n$ is a Cartan-Weisfeiler filtration which is determined by L_0, the filtration is invariant. The invariance of the filtrations of the (finite-dimensional) Lie algebras of generalized Cartan type can be proved by a similar approach. Jin's result was more general that it covered some cases which had not been treated before. His approach has also the advantage that it can be used to discuss the invariance of filtrations of Cartan type Lie algebras over a commutative ring. In [23], N. Jin and G.-Y. Shen proved the following

Theorem 2.1 *Let R be a commutative ring of prime characteristic, N the nilpotent radical of R, L_R a Lie algebra of generalized Cartan type X over R, $X = W$, S, H or K, and $\{RL_i\}$ the filtration of type $1 + \delta_{XK}\varepsilon_n$ of L_R. Then the filtration $\{RL_i + NL\}$ is invariant under the automorphisms of L_R.*

Let Ψ be an R-algebra automorphism of L_R. In [43], Skryabin showed that if (a) the natural map of the p-envelop $p(L)$ to $\mathrm{Der}(L_0, L/L_0)$ is surjective, and (b) $\Psi(RL_0 + aRL) = RL_0 + aRL$ for every $a \in R$ satistying $a^p = 0$, then there exists a standard automorphism Ψ' of L_R such that $\Psi\Psi'RL_0 \subseteq RL_0$. In [42], B. Shu, by a case by case computation, showed that the condition (a) is satisfied for an arbitrary Lie algebra of generalized Cartan type. Combining this to Theorem 2.1. and applying Skryabin's result, he is able to prove that every automorphism of L_R is standard. This, as mentioned above, implies that the forms of L are all standard and thereby are completely determined.

3. Universal graded Lie algebras

In this section, F is an arbitrary field of any characteristic.

Gradation plays an impotent role in Lie algebra theory. The most important Lie algebras usaually admit natural graded Lie algebra structures. Let $K^- = \oplus_{i<0} K_i$ be a negatively graded Lie algebra. We consider the class of graded Lie algebras of type K^- which consists of graded Lie algebras with negative parts isomorphic to K^-. The algebra $S := K^-$ is the "smallest" one of the class which can be characterized as the graded Lie algebra such that if L is a graded Lie algebra of type K^- and ψ is a graded isomorphism of S^- onto L^-, then there exists a unique graded homomorphism φ of S into L satisfying $\pi_L \varphi = \psi \pi_S$ (where π_L and π_S are canonical projections), i.e., the diagram

$$
\begin{array}{ccc}
L^- & \xleftarrow{\ \psi\ } & S^- \\
\uparrow \pi_L & & \uparrow \pi_S \\
L & \xleftarrow{\ \varphi\ } & S
\end{array}
$$

is commutative. To obtain the opposite concept, we have

Definition 3.1. A graded Lie algebra $U = U(K^-)$ of type K^- is called *universal of type* K^- if for every Lie algebra L of type K^- and a graded isomorphism ψ of L^- onto U^-, there exists a unique graded homomorphism φ of L into U satisfying $\varphi \iota_L = \iota_U \psi$ (where ι_L and ι_U are canonical imbeddings), i.e., the diagram

$$
\begin{array}{ccc}
L^- & \xrightarrow{\ \psi\ } & U^- \\
\downarrow \iota_L & & \downarrow \iota_U \\
L & \xrightarrow{\ \varphi\ } & U
\end{array}
$$

is commutative.

For simplicity's sake, we assume $\dim K^- < \infty$. In [16], it was shown that, for every K^- the universal graded Lie algebra $U(K^-)$ exists (and up to isomorphism is unique). Moreover, $U(K^-)$ can be realized as a subalgebra of the general Lie algebra of Cartan type $W(n)$, graded corresponding to the gradation of K^-. In particular, $W(n)$ is exactly the universal graded Lie algebra of depth one. And the contact Lie algebra $K(n)$ was shown to be the universal graded Lie algebra of Heisenberg type, i.e., with negative part isomorphic to the Heisenberg algebra (of dimension n).

Similarly, the universal Lie algebra $U(K^-, K_0)$ with a given nonpositive part $K^- \oplus K_0$ was considered. When K^- is of depth one, $U(K^-, K_0)$ is simply called the universal graded Lie algebra of type K_0 which, when realized as a subalgebra of $W(n)$, is just the extension of K_0 in $W(n)$ (cf. [35]). It was shown that the Cartan type Lie algebras $S(n)$ and $H(n)$ are precisely the universal graded Lie algebras of type $K_0 = \mathfrak{sl}(n)$ and $\mathfrak{sp}(n)$ respectively, and are the only nontrivial cases when K_0 is isomorphic to a classical Lie algebra.

When Char $F = p > 0$, the universal graded Lie algebra with a given p-flag $\mathcal{F} : 0 \subset \mathcal{F}_1 \subset \ldots \subset \mathcal{F}_k = K^-$ ($\mathcal{F}$ is the p-flag of a graded Lie algebra L of type K^- if $\mathcal{F}_i = \{x \in L^- \mid (\mathrm{ad}_L x)^{p^i} = 0\}$) and the universal restrictedly graded Lie algebras were defined and discussed, by which the graded Lie algebras of Cartan type and the restricted Lie algebras of Cartan type are respectively characterized.

In the discussion, not only the existence but also the explicit constructions of the universal graded Lie algebra are obtained. It should be noted that from the universal properties things comes out naturally. For instance, the notions of the differential operators and the general Lie algebra $W(n)$ emerge naturally from the discussion and it is not necessary to presuppose any knowledge of them. Concepts which are familiar to the differential geometers such as the exterior differential forms and the exterior differential operation can be derived and treated naturally from an abstract Lie algebra theoretical standpoint of view (for a detailed discussion, cf. [40]).

4. Simple Lie algebras of low characteristics

By the verification of GKŠC, the classification problem of simple Lie algebras over an algebraically closed field of characteristic $p > 7$ is essentially solved. It is estimated that there exists no difficuties in principle to obtain the classification for $p = 7$ and 5: GKŠC still holds for $p = 7$; for $p = 5$, alongside with the classical and generalized Cartan type algebras, the addition of the Melikyan algebras will complete the list of simple algebras (cf. [44]). On the contrary, the prospects for the cases $p = 2$ and $p = 3$ are quite obscure at present. It is desirable that, by the discover of more new simple Lie algebras, a pattern might be shaped and a conjecture for the classification formulated.

By the methods illustrated in [16], when a negatively graded Lie algebra K^- is given, the universal graded Lie algebra $U(K^-)$ can in principle be constructed by mechanical computations. As seen in Section 3, the Cartan type Lie algebras are realizations of universal graded Lie algebras of depthes at most 2. Now we consider $\Gamma^- = \Gamma_{-1} \oplus \Gamma_{-2} \oplus \Gamma_{-3}$ where $\Gamma_{-1} = \langle z_1, z_2 \rangle$, $\Gamma_{-2} = \langle z_3 \rangle$ and $\Gamma_{-3} = \langle z_4, z_5 \rangle$ with the multiplication table

$$\begin{cases} [z_1, z_2] = z_3, \ [z_1, z_3] = z_4, \ [z_2, z_3] = z_5, \\ [z_i, z_j] = 0, \text{ otherwise, } 1 \leq i, j \leq 5, \end{cases}$$

which is one of the simplest negatively graded Lie algebras of depth 3. Let α and β be respectively the short and long roots of the classical Lie algebra G_2, x_μ be the root vector belonging to the root μ. Let G_2 be graded by letting $\deg x_{i\alpha+j\beta} = i$. Then G_2^- is isomorphic to Γ^- if Char $F \neq 2$, 3. The universal graded Lie algebra $U(\Gamma^-)$ was explicitly constructed in [41]. When Char $F = 0$ or Char $F > 5$, $U(\Gamma^-)$ is isomorphic to G_2. When Char $F = 5$, $U(\Gamma^-)$ turns out to be the universal Melikyan algebra $g(\infty)$ whose finite-dimensional "sections" $g(m_1, m_2)$ are the only known simple Lie algebras of characteristic 5 which are neither classical nor of generalized Cartan type. When Char $F = 2$, $U(\Gamma^-)$ is a nonrestricted simple Lie algebra of dimension 14 which is new and in fact is the Lie algebra obtained from an integral but non-Chevalley basis of G_2 by modulo 2 process. When Char $F = 3$, $U(\Gamma^-)$ is not simple.

As a graded Lie algebra of depth 2 (by letting $\deg x_{i\alpha+j\beta} = j$), G^- is isomorphic to the Heisenberg algebra of dimension5 and G_2 can be gradedly imbedded in the contact algebra $K(5)$ with $(G_2)_0$ isomorphic to $\mathfrak{sl}(2)$. Let $V^- = K(5)^-$. It was shown in [41] that G_2 is precisely the universal graded Lie algebra $U(V^-, (G_2)_0)$. This situation can be generalized by choosing different subalgebras V_0 of $K(5)_0$ which are, or nearly are, isomorphic to $\mathfrak{sl}(2)$. In case Char $F = 2$, this leads to families of new simple Lie algebras (of dimension 14, 15 and 16) which , more or less, bear resemblance to G_2 and are called the variations of G_2. In some cases, $U(V^-, V_0)$ are not simple but admit simple subquotients (all of dimension 8) which are new and in a certain sense can be regarded as "variations" of $\mathfrak{sl}(3)$.

The Lie algebras of Cartan type can be defined in any prime characteristic p. However, systematic discussion of them are usually under the assumption $p > 2$ (or even $p > 3$), especially for the types H and K. In [51], it was shown that when Char $F = 2$, $H(n, \mathbf{m})$ is simple but $K(n, \mathbf{m})$, defined as usual by the differential form $\omega_K = dx_n + \sum_{i=1}^r (x_i dx_{i+r} - dx_{i+r} dx_i)$ $(n = 2r+1)$, is not simple. The nonsimplicity of $K(n, \mathbf{m})$ in characteristic 2 is in fact due to the incorrect choice of ω_K. L. Lin showed in [28] that if ω_K is modified to be $dx_n + \sum_{i=1}^r (\mu_i x_i dx_{i+r} - \nu_i x_{i+r} dx_i)$, $\mu_i + \nu_i = 1$, then the resultant Lie algebra $K(n, \mathbf{m}, \mu_i)$ is simple. (An explanation of why the modification should be made can be found in [16]). In [56], a detailed investigation of $K(n, \mathbf{m}, \mu_i)$ was given.

Detailed discussions of other graded Cartan type Lie algebras in low characteristics can be found in [54], [48], [55] and [50]. In particular, in all cases the derivation algebras were determined and the invariance of filtrations proved. In [53] and [57], *the Lie algebras of class Σ* defined in [33] were investigated in low characteristics and most of the fundamental results of [33] were generalized to low characteristic cases.

The Hamiltonian Lie algebra $H(n, \mathbf{m})$ is the extension of the classical Lie algebra $\mathfrak{sp}(2r)$ (cf. [35]) which is closely related to the nondegenerate alternating form of rank $2r$. When Char $F = 2$, the set P_0 of symmetric matrices is a subalgebra of $\mathfrak{gl}(n)$ which corresponds to the existence of the nondegenerate nonalternating antisymmetric form over F. L. Lin in [29] considered the extension $P'(n, \mathbf{m})$ of P_0 in $W(n, \mathbf{m})$. It turns out that $P(n, \mathbf{m}) := P'(n, \mathbf{m})^{(1)}$ (called the nonalternating Hamiltonian algebra) is simple unless $n < 4$ and $\mathbf{m} = \mathbf{1}$. The simple Lie algebras $G(n)$ constructed in [25] by I. Kaplansky were shown to be special cases of $P(n, \mathbf{m})$ when $\mathbf{m} = \mathbf{1}$ $(n \geq 4)$. The derivation algebra of $P(n, \mathbf{m})$ was determined. By proving the invariance of the natural filtration, $P(n, \mathbf{m})$ was shown to be new in general.

5. Representation

There has been (and is being) a vigorous study in the modules of modular classical Lie algebras, which in fact is generally regarded as comprised in the representation theory of algebraic groups. On the contrary, except the simplest case of the Witt algebra $W(1,1)$, of which a thorough description of the irreducible modules was given as early as in 1941 [3], systematic works on the representations of the Cartan type Lie algebras was comparatively few before mid 80's. The gradation structure plays a protruding role in the study of the graded Lie algebras $X(n, \mathbf{m})$, $X = W$, S, H and K, of Cartan type (as well as other important Lie algebras, e.g., the classical Lie algebras). A general discussion of the graded modules of a graded Lie algebra was given by G.-Y. Shen in [36]. Let $L = \oplus_{i \in \mathbf{Z}} L_i$ be a graded Lie algebra (of any charcteristic and possibly infinite-dimensional) and $V = \oplus_{i \geq 0} V_i$ be a positively (resp. negatively) graded module of L (i.e., satisfying $L_i V_j \subseteq V_{j+i}$ (resp. $L_i V_j \subseteq V_{j-i}$)) where V_0 is assumed to be nonzero if $V \neq 0$ and is called the base (resp. top) space of V. Let $L^+ = \oplus_{i > 0} L_i$, $L^- = \oplus_{i < 0} L_i$ and $B = L_0 + L^+$. It was shown that $V \mapsto V_0$ is a bijective map of the set of isomorphism classes of irreducible graded modules of L onto the set of isomorphism classes of the irreducible modules of L_0. A finite-dimensional irreducible module L is isomorphic to a graded module if and only if the elements of L^+ and L^- act nilpotently on V. It follows that if L is restricted and centerless, every irreducible restricted module of L is graded and the isomorphism classes of the restricted modules of L are in one-one correspondence with those of L_0. The duality relations between the positively and the negatively graded modules were clarified.

Let $\mathfrak{A}$ be a commutative associative algebra and $D_1, \ldots, D_n$ a set of mutually commutative derivations of $\mathfrak{A}$. Then $K = \{\sum a_i D_i \mid a_i \in \mathfrak{A}\}$ is a subalgebra of $\text{Der}\mathfrak{A}$. Let E_{ij} be the $n \times n$ matrix whose (k, l)-component is $\delta_{ki} \delta_{lj}$ and L_0 a subalgebra of $\mathfrak{gl}(n)$. Then $L = \{\sum a_i D_i \in K \mid \sum D_i(a_j) \otimes E_{ij} \in \mathfrak{A} \otimes L_0\}$ is a Lie subalgebra of K which is called the extension of L_0 in K. In particular, the finite-dimensional (resp. infinite-dimensional) Lie algebras of Cartan type $W(n, \mathbf{m})$, $S(n, \mathbf{m})$ and $H(n, \mathbf{m})$ (resp. $W(n)$, $S(n)$ and $H(n)$) are the extensions of $\mathfrak{gl}(n)$, $\mathfrak{sl}(n)$ and $\mathfrak{sp}(n)$ in $W(n, \mathbf{m})$ (resp. $W(n)$) respectively. In [35] it was shown that if L is an extension of L_0 then for an arbitrary module V_0 of L_0, an L-module structure can be defined on the space $\mathfrak{A} \otimes V_0$ which is called the mixed product of $\mathfrak{A}$ and V_0 and is denoted by $\mathfrak{A} \rtimes V_0$. The essential point is that the actions of the elements of L on $\mathfrak{A} \rtimes V_0$ are explicitly defined. In particular, for an irreducible module V_0 of $\mathfrak{gl}(n)$, $\mathfrak{sl}(n)$ or $\mathfrak{sp}(n)$, a positively graded module

$\widetilde{V}_0 := \mathfrak{A}(n, \mathbf{m}) \rtimes V_0$ is obtained for $L = W(n, \mathbf{m})$, $S(n, \mathbf{m})$ or $H(n, \mathbf{m})$ respectively. The irreducible graded module with V_0 as base space is isomorphic to the unique minimal submodule of $\widetilde{V}_0$. It turns out [38] that, except V_0 is a module $V_0(\lambda_i)$ with a highest weight which is a fundamental weight λ_i, $\widetilde{V}_0$ is itself irreducible. In [38], the decompositions of the $\widetilde{V}_0(\lambda_i)$ were completely described. Thus, all irreducible positively graded modules of $X(n, \mathbf{m})$, $X = W$, S and H, and in particular all irreducible restricted modules of $X(m, \mathbf{1})$ were determined via the irreducible modules of the classical Lie algebras $\mathfrak{gl}(n)$, $\mathfrak{sl}(n)$ and $\mathfrak{sp}(n)$. The results generalize to the infinite-dimensional $X(n)$, $X = W$, S and H, as well. By duality, the irreducible negatively graded modules were also determined. It was later found out that actually $\widetilde{V}_0 \cong \hom_{\mathcal{U}(\mathcal{B})}(\mathcal{U}(L), V_0)$ (in the category of graded modules) for $L = X(n)$ (cf. [36, Thm.3.1.]) and $\widetilde{V}_0 \cong \hom_\theta(\mathcal{U}(L), V_0)$ for $L = X(n, \mathbf{m})$ where θ is a certain subalgebra of $\mathcal{U}(L)$ containing $\mathcal{U}(\mathcal{B})$ (cf. [14]), here $\mathcal{U}(L)$ denotes the universal enveloping algebra of L. In the final part of [38], the irreducible positively and negatively filtered modules of L were described.

For the remianing case, the contact Lie algebra $L = K(2r + 1, \mathbf{m})$, the situation is more complicated. The mixed product applies no longer. However, N.-H. Hu in [19] obtained an explicit formula of the actions of the elements of L on $V = \hom_\theta(\mathcal{U}(L), V_0)$ where V_0 is an irreducible module of $L_0 \cong \mathfrak{sp}(2r)$. It was shown that V is irreducible but for a few exceptional V_0. Furthermore, in [20], Hu, under the condition $p > r + 1$, realized the irreducible graded modules with exceptional base spaces and described the decomposition of V. Hu's results generalized those obtained by R. Holmes in [17] and [18] where the restricted $K(2r + 1, \mathbf{1})$ was considered.

In [8], the indecomposable modules of the restricted Lie algebra $L = X(m, \mathbf{1})$, $X = W$, S, H or K, were considered. Exploiting certain techniques of the representation theory of reductive algebraic groups, in particular Jantzen's $u(L^e)$-T^e-modules, S. Chiu obtained properties of the principal indecomposable modules of $u(L)$ (the universal restricted enveloping algebra of L) which parallel closely to those of classical Lie algebras. For instances, a reciprocity theorem was proved and linkage conditions were obtained. And it was shown that the decopmosition of the Cartan matrix $C = BD$ is also valid for the restricted Cartan type Lie algebras.

6. Cohomology

It is a classical results that if L is a simple Lie algebra of characteristic zero, then the cohomology groups $H^i(L, M) = 0$, $i = 1$, 2, for any
irreducible module M of L. Counterexamples show that this is not the
case for simple Lie algebras of characteristic $p > 0$. In particular, little
was known of the cohomology of Lie algebras of Cartan type until recently. A. S. Dzhumadil'daev determined the structure of $H^1(W(1, \mathbf{n}), V)$
for every irreducible module V [10]. In [9], Chiu and Shen showed that if
$L = X(n, \mathbf{m})$, $X = W$, S or H, and V_0 is an irreducible module of L_0, then
$H^*(L, \tilde{V}_0) = 0$ (where $\tilde{V}_0 = \mathfrak{A} \rtimes V_0$ as in Section 5) unless V_0 is an integral
highest weight module. Since $M(V_0)$, the irreducible L-module with base
space V_0, is the minimal submodule of $\tilde{V}_0$, the assertion can be shown to
be valid also for $H^*(L, M(V_0))$. By reducing to the cohomology of Lie
algebras of classical groups, $H^1(L, \tilde{V}_0)$ and $H^1(L, M(V_0))$ were completely
determined for $L = W(n, \mathbf{m})$, $n = 2$, 3, $p \geq n + 2$, and $L = H(2, \mathbf{m})$. In
[5], $H^1(L, \tilde{V}_0)$ and $H^1(L, M(V_0))$ were also computed for $L = S(3, \mathbf{m})$.

In [10], [11] and [12], A. S. Dzhumadil'daev and R. Farnteiner deter-
mined the structures of $H^2(L, F)$ for $L = W(n, \mathbf{m})$, $p \geq 3$, $S(3, \mathbf{m})$, $p > 3$,
$H(n, \mathbf{m})$, $p > 3$ and $K(n, \mathbf{m})$, $p > 3$. In [7], Chiu gave a new unifying
approach based mainly on the computation of $H^1(L, L^*)$. A description of
the dual adjoint module L^* was given by means of the mixed product and
the coinduced module. The computation of $H^1(L, L^*)$ was then reduced
to the computation of the cohomology of L_0 which is the Lie algebra of a
certain reductive algebraic group. Exploiting techniques of the representation theory of reductive algebraic groups, Chiu was able to determine the
structures of $H^1(L, L^*)$ and $H^2(L, F)$ for $L = W(n, \mathbf{m})$, $p > 0$, $S(n, \mathbf{m})$,
$n \geq 3$ and $p > 2$, $H(n, \mathbf{m})$, $p > 2$, and $K(n, \mathbf{m})$, $p > 2$ and $n + 3 \not\equiv 0$
(mod p).

References

[1] R. Block and R. L. Wilson, *The restricted simple Lie algebras are of classical or
    Cartan type*, Proc. Natl. Acad. Sci. USA, (1984), 5271–5274.

[2] E. Cartan, *Les groupes de transformations continues, infinis, simples*, Ann. Sci.
    École Norm. Sup., **26** No. 3 (1909), 93–161.

[3] H.-J.Chang, *Über Wittsche Ringe*, Abh. Math. Sem. UnivHamburg, **14** (1941),
    151–184.

[4] S. Chiu, *Second cohomology of the Witt algebra*, Chin. Ann. of Math., **9A** No. 5 (1988), 524–529. (Chinese)

[5] ——, *Cohomology of graded Lie algebras of Cartan type $S(n, \mathbf{m})$*, Chin. Ann. of Math., **10B** No. 1 (1989), 105–114.

[6] ——, *Derivations of the graded Lie algebras of Cartan type*, Chin. Ann. of Math., **13B** No. 2 (1992), 196–204.

[7] ——, *Central extensions and $H^1(L, L^*)$ of the graded Lie algebras of Cartan type*, J. Algebra, **149** (1992), 46-67.

[8] ——, *Principal indecomposable representations for restricted Lie algebras of Cartan type*, J. Algebra, **155** No. 1 (1993), 142–160.

[9] S. Chiu and G.-Y. Shen, *Cohomology of graded Lie algebras of Cartan type of characteristic p*, Abh. Math. Sem. Univ. Humburg, **57** (1987), 139–156.

[10] A. S. Dzhumadil'daev, *On cohomology of modular Lie algebras Russian*, Mat. Sb., **119** (1989), 132–149.

[11] ——, *Central extensions of the Zassenhaus algebras and their irreducible representations*, Mat. Sb., **126** (1985), 473–489. (Russian)

[12] R. Farnteiner, *Central extensions and invariant forms of graded Lie algebras*, Algebras, Group and Geometries, **3** (1986), 431–455.

[13] ——, *Dual space derivations and $H^2(L, F)$ of modular Lie algebras*, Canad. J. Math., **39** (1987), 1078–1106.

[14] ——, *Extension functors of modular Lie algebras*, Math. Ann., **288** (1990), 713–730.

[15] Q.-Y. Fei, *On new simple Lie algebras of Shen Guangyu*, Chin. Ann. of Math., **10B** No. 4 (1989), 448–457.

[16] Q.-Y. Fei and G.-Y. Shen, *Universal graded Lie algebras*, J. Algebra, **152** (1992), 439–453.

[17] R. Holmes, *Simple restricted modules for the restricted contact Lie algebras*, Proc. Amer. Math. Soc., **116** (1992), 329–337.

[18] ——, *Dimensions of the simple restricted modules for the restricted contact Lie algebra* Preprint.

[19] N.-H. Hu, "*The graded modules for the graded contact Cartan algebras*," Doctoral Dissertation, East China Normal University, 1993.

[20] ——, *Irreducible constituents of graded modules for graded contact Cartan algebras* (to appear).

[21] N. Jacobson, "*Lie Algebras*, Interscience, New York, 1962.

[22] N. Jin, *ad-nilpotent elements, quasi-nilpotent elements and invariant filtrations of infinite dimensional Lie algebras of Cartan type*, Science in China, **Ser. A 35** (1992), 1192–1200.

[23] N. Jin and G.-Y. Shen, *Invariant filtrations of Lie algebras of Cartan type over a commutative ring* (to appear).

[24] V. G. Kac, *Description of filtered Lie algebras with which graded Lie algebras of Cartan type are associated*, Izv. Akad. Nauk SSSR Ser. Mat., **38** (1974), 800–834. (Russian)

[25] I. Kaplansky, *Some simple Lie algebras of characteristic 2*, LNM 933, Springer, 1982.

[26] A. I. Kostrikin and I. R. Šafarevič, *Cartan pseudogroups and Lie p-algebras,* Dokl. Akad. Nauk SSSR, **186** (1966), 740–742. (Russian)

[27] ——, *Graded Lie algebras of finite characteristic,* Izv. Akad. Nauk SSSR Ser. Mat., **33** (1969), 251–322. (Russian)

[28] L. Lin, *Lie algebras $K(\mathcal{F}, \mu_i)$ of Cartan type of characteristic $p = 2$ and their subalgebras,* J. East China Normal Univ. (Natural Science Ed.) No. 1 (1988) 16–23. (Chinese)

[29] ——, *Non-alternating Hamiltonian algebra $P(n, \mathbf{m})$ of characteristic two,* Comm. Alg., **21** (1993), 399–411.

[30] R. Ree, *On generalized Witt algebras,* Tran. Amer. Math. Soc., **83** (1956), 510–546.

[31] S. Serconek and R. L. Wilson, *Forms of restricted simple Lie algebras,* Comm. Alg., **19** (1991), 1603–1628.

[32] G.-Y. Shen, *An intrinsic property of the Lie algebra $K(m, \mathbf{n})$,* Chin. Ann. of Math., Eng. Issue, **2** (1981), 105–115.

[33] ——, *New simple Lie algebras of characteristic p,* Chin. Ann. of Math., **4B** No. 3 (1983), 329–346.

[34] ——, *On Lie algebras associated with nodal noncommutative Jordan algebras,* Acta Math. Sinica, New Ser, **2** No. 1 (1986), 14–24.

[35] ——, *Graded modules of graded Lie algebras of Cartan type,* I , Scientia Sinica, Ser. A, **29** No. 6 (1986), 570–581.

[36] ——, *Graded modules of graded Lie algebras of Cartan type,* II, Scientia Sinica, Ser A, **29** No. 10 (1986), 1009–1019.

[37] ——, *Notes on Lie algebras $\Sigma(n, \mathbf{m}, r, G)$,* Chin. Ann. of Math., **8B** No. 3 (1987), 329–331.

[38] ——, *Graded modules of graded Lie algebras of Cartan type,* III, Chin. Ann. of Math., **9B** No. 4, (1988), 404–417.

[39] ——, *Realization of irreducible modules of $\mathfrak{sl}(2)$ of characteristic p,* Northeast Math. J., **6** No. 2 (1990), 151–156.

[40] ——, *Alegbraic explanation of exterior differential,* Chin. Quaterly J. of Math., **6** No. 1 (1991), 41-49.

[41] ——, *Variations of the classical Lie algebra G_2 in low characteristic,* to appear in Nuovo J. of Algebra and Geometry.

[42] B. Shu, *"The forms of Lie algebras of Cartan type,* Doctoral Dissertation, East China Normal Univ., 1993.

[43] S. M. Skryabin, *Modular Lie algebras of Cartan type over algebraically nonclosed field,* I, Comm. Alg., **19**, (1991), 1629–1741.

[44] H. Strade, *Die Klassifikation der einfach Lie Algebren über Körpern mit positiver Characteristik: Methoden und Resultate,* Jber. der Dt. Math. Verein, **95** (1993), 28–46.

[45] H. Strade and R. Farnsteiner, *"Modular Lie Algebras and Their Representations,* Marcel Dekker, New York and Basel, 1988.

[46] H. Strade and R. L. Wilson, *Classification of simple Lie algebras over algebraically closed fields of prime characteristic,* Bull. Amer. Math. Soc., **24** (1991), 357–362.

[47] W. C. Waterhouse, *Automorphism and twisted forms of generalized Witt Lie algebras,* Trans. Amer. Math. Soc., **327** (1991), 185–200.

[48] Y. Wang and Y.-Z. Zhang, *The derivation algebra of Lie algebra $H(n, \mathbf{m})$ of Cartan type of characteristic $p = 2$* (to appear).

[49] R. L. Wilson, *A structural characterization of the simple Lie algebras of generalized Cartan type over fields of prime characteristic*, J. Algebra, **40** (1976), 418–465.

[50] L.-Y. Xue, *On the invariance of filtrations of Cartan type Lie p-algebra of characteristic 2*, Northeastern Math. J., **2** (1986), 150–163. (Chinese)

[51] X.-X. Xue" *The structure of Cartan type Lie algebras $H_n(\mathcal{F})$ and $K_n(\mathcal{F})$ when the characteristic equals to two*, Masterial Thesis, Chin. Univ. of Science and Technology.

[52] Y.-Z. Zhang, *Lie algebra $K(m, \mathbf{n})$ of characteristic 3*, J. of East China Normal Univ. (Natural Science Ed.) No. 4 (1990), 22–26. (Chinese)

[53] ——, *On the Lie algebras of class Σ of Shen Guangyu*, to appear in Acta Math. Sinica. (Chinese)

[54] ——, *Derivation algebra of Cartan type Lie algebra $S(n, \mathbf{m})$* (to appear). (Chinese)

[55] ——, *Filtrations of the finite-dimensional Lie algebras of Cartan type in low characteristics* (to appear).

[56] Y.-Z. Zhang and L. Lin, *Lie algebra $K(n, \mathbf{m}, \mu_i)$ of Cartan type of characteristic $p = 2$*, Chin. Ann. of Math., **13B** No. 3 (1992), 315–326.

[57] Y.-Z. Zhang and Y. Wang, *On the Lie algebra $\overline{\Sigma}$ of characteristic 2 of Shen Guangyu* (to appear)

Left Cells in Certain Coxeter Groups

Jian-yi Shi

Department of Mathematics, East China Normal University
Shanghai 200062, The People's Republic of China

A survey is given on the achievements of KL-cell theory of certain
Coxeter groups. Some techniques applied in that theory are introduced.
Also, we propose several open problems for further study.

In order to construct the representations of a Coxeter group W and
the associative Hecke algebra, D. Kazhdan and G. Lusztig defined certain
equivalence classes of W called left, right and two-sided cells. Thus the
description of cells of W and the structural study of these cells become
interesting and also important in the representation theory of groups and
algebras. In the present paper, we shall make a survey on the achievements
of studying left cells of W. According to the definition, the description
of left cells might involve complicated computation of Kazhdan-Lusztig
polynomials and is hard even by a computer when the order of W is
getting larger. Thus we shall introduce some methods to simplify our
work. They reduce the computation in significant rate so that sometimes
we can reach our goal only by hand even when W is in some infinite case.
We shall see that the study of cells of W involves some combinatorial
techniques and has to invoke some other mathematical theory. We also
propose some related open problems for further study.

The content of this paper is organized as below. In Section 1, we make
a historical review on the cells of a Coxeter group, introduce the definition
of cells by Kazhdan and Lusztig and some related concepts. Then we state
some results of Lusztig concerning cells of Coxeter groups with properties

* Supported by the National Natural Science Foundation of China and by the Science
Foundation of the University Doctoral Program of CNEC.

2.3, (a), (b), mainly of affine Weyl groups. A survey is given in Section 3 on the achievements for the description of left cells of Weyl groups, affine Weyl groups and some other finite Coxeter groups. We construct an algorithm in Section 4 for finding a representative set of left cells in any given two-sided cell of Coxeter groups with properties 2.3, (a), (b). Several conjectures concerning left cells of W are scattered in Sections 2 to 4. Finally in Section 5, we state some more conjectures on left cells of affine Weyl groups.

§1. Kazhdan-Lusztig cells

1.1 The concept of cells originally came from combinatorial theory. Robinson [26] defined a map from the symmetric group S_n to the set of pairs of standard Young tableaux of the same shape and of rank n: $\sigma \longrightarrow (\phi(\sigma), \phi(\sigma^{-1}))$. Then Schensted [27] proved that this map is bijective. This map is called the Robinson-Schensted map. Hence a left cell of S_n is, by definition, the set of elements of S_n corresponding to the set of such pairs (P, Q) with P fixed. A two-sided cell of S_n is defined to be the set of elements of S_n corresponding to the set of such pairs (P, Q) with P, Q of fixed shape. Thus there is 1-1 correspondence between the set of two-sided cells of S_n and the set of partitions of n. This is the prototype of cells which applied to any Coxeter group.

Then Vogan and Joseph defined the concept of left cells in the Weyl group W in terms of primitive ideals in the enveloping algebra of a complex semisimple Lie algebra. For $w \in W$, let J_w be the annihilator of the irreducible module of the enveloping algebra with highest weight $-w\rho - \rho$, where ρ is half the sum of positive roots. Then w, w' are said to be in the same left cell precisely when $J_w = J_{w'}$. This definition of left cells and the corresponding Weyl group representations involves some deep results about the multiplicities of the composition factors of the Verma modules with highest weight $-w\rho - \rho$, [44], [14].

In 1979, Kazhdan and Lusztig [15] gave the definition of cells for an arbitrary Coxeter group. Their definition is elementary but it gives rise not only to the representations of the Coxeter group, but also to that of the corresponding Hecke algebra. This makes possible applications of the results on cells to more general representation theory. On the other hand, the definition of Kazhdan-Lusztig cells coincides with that of Vogan and Joseph in the case of Weyl groups.

We adopt Kazhdan-Lusztig's definition of cells.

1.2 Let (W, S) be a Coxeter group with S its simple reflection set. Let $\leq$ be the Bruhat order of W and let $\ell(x)$ be the length of an element $x \in W$. Let $\mathcal{A} = Z[u, u^{-1}]$ be the ring of Laurent polynomials in an indeterminate u with integer coefficients. There exists an associative algebra $\mathcal{H} = \mathcal{H}(W)$ over $\mathcal{A}$ with $\{T_w \mid w \in W\}$ and $\{C_w \mid w \in W\}$ its two free $\mathcal{A}$-bases. Its multiplication rule in terms of T_w's is given by

$$\begin{cases} T_w T_{w'} = T_{ww'}, & \text{if } \ell(ww') = \ell(w) + \ell(w'); \\ (T_s - u^{-1})(T_s + u) = 0, & \text{for } s \in S. \end{cases}$$

The relation between these two bases is as below.

$$C_w = \sum_{y \leq w} u^{\ell(w)-\ell(y)} P_{y,w}(u^{-2}) T_y, \quad \text{for } w \in W.$$

where the $P_{y,w}$'s are Kazhdan-Lusztig polynomials in $Z[u]$, which satisfy the conditions: $P_{y,w} = 0$ if $y \not\leq w$; $P_{w,w} = 1$ and $\deg P_{y,w} \leq (1/2)(\ell(w) - \ell(y) - 1)$ if $y < w$.

1.3 For $y, w \in W$ with $\ell(y) \leq \ell(w)$, we denote by $\mu(y, w)$ or $\mu(w, y)$ the coefficient of $u^{(1/2)(\ell(w)-\ell(y)-1)}$ in $P_{y,w}$. We say that y and w are joint, written $y\!-\!w$, if $\mu(y, w) \neq 0$. To any $x \in W$, we associate two subsets of S:

$$\mathcal{L}(x) = \{s \in S \mid sx < x\} \quad \text{and} \quad \mathcal{R}(x) = \{s \in S \mid xs < x\}.$$

We have the following relations: for any $x \in W$ and $s \in S$,

$$C_s C_x = \begin{cases} (u^{-1} + u) C_x, & \text{if } s \in \mathcal{L}(x); \\ \displaystyle\sum_{\substack{y-x \\ sy<y}} \mu(x, y) C_y, & \text{if } s \notin \mathcal{L}(x); \end{cases} \tag{1.3.1}$$

1.4 For any $x, y \in W$, we denote $x \underset{L}{\leq} y$ (resp. $x \underset{R}{\leq} y$), if there exists a sequence of elements $x_0 = x, x_1, \cdots, x_r = y$ in W with some $r \geq 0$ such that for every i, $1 \leq i \leq r$, $x_{i-1}\!-\!x_i$ and $\mathcal{L}(x_{i-1}) \not\subseteq \mathcal{L}(x_i)$ (resp. $\mathcal{R}(x_{i-1}) \not\subseteq \mathcal{R}(x_i)$). We denote $x \underset{LR}{\leq} y$, if there exist elements $x_0 = x, x_1, \cdots, x_r = y$ in W such that either $x_{i-1} \underset{L}{\leq} x_i$ or $x_{i-1} \underset{R}{\leq} x_i$ holds for $1 \leq i \leq r$. We write $x \underset{L}{\sim} y$ (resp. $x \underset{R}{\sim} y$, resp. $x \underset{LR}{\sim} y$), if the relation $x \underset{L}{\leq} y \underset{L}{\leq} x$ (resp. $x \underset{R}{\leq} y \underset{R}{\leq} x$, resp. $x \underset{LR}{\leq} y \underset{LR}{\leq} x$) holds. These are equivalence relations on W, and the equivalence classes of W with respect to $\underset{L}{\sim}$ (resp. $\underset{R}{\sim}$, resp. $\underset{LR}{\sim}$) are called the left (resp. right, resp. two-sided) cells of W. The preorders $\underset{L}{\leq}$, $\underset{R}{\leq}$ and $\underset{LR}{\leq}$ on elements of W induce partial orders on the corresponding cells of W.

1.5 Each cell of W provides a representation of W and of the associated Hecke algebra. Suppose that Γ is a left cell of W. Let $I_{\leq\Gamma}$ (resp. $I_{<\Gamma}$) be the $\mathcal{A}$-submodule of $\mathcal{H}$ spanned by $\{C_w \mid w \underset{L}{\leq} x$, for some $x \in \Gamma\}$ (resp. $\{C_w \mid w \underset{L}{\leq} x$, for some $x \in \Gamma$ but $w \notin \Gamma\}$). Then $I_{\leq\Gamma}$ (resp. $I_{<\Gamma}$) is a left ideal of $\mathcal{H}$ by 1.2.1. Thus the quotient $I_\Gamma = I_{\leq\Gamma}/I_{<\Gamma}$ is a left $\mathcal{H}$-module. It becomes a W-module if we specialize $u = 1$ for $\mathcal{A}$. So studying cells of W is of significance in the representation theory of groups and algebras. In principle, when W is a finite Coxeter group, all the cells of W could be described explicitly by a finite step of computation. For example, let $W = D_m = \langle s, t \mid o(s) = o(t) = 2, o(st) = m \rangle$ be the dihedral group of order $2m$, $m > 1$. Then we have $P_{x,y} = 1$ iff $x \leq y$. This implies that there are three two-sided cells $\{e\}$, $\{w_0\}$ and $\Gamma_s \bigcup \Gamma_t$, where e, w_0 are the identity element and the longest element of D_m, respectively, and

$$\Gamma_s = \{s,\ ts,\ sts,\ \cdots,\ \overbrace{\cdots sts}^{m-1 \text{ factors}}\ \}$$

$$\Gamma_t = \{t,\ ts,\ tst,\ \cdots,\ \overbrace{\cdots tst}^{m-1 \text{ factors}}\ \}.$$

The first two two-sided cells themselves are left cells. The third one is a union of two left cells Γ_s and Γ_t.

Kazhdan and Lusztig gave explicit description of cells for some Coxeter groups of lower ranks [15], D. Alvis described all the left cells of the Coxeter group of type H_4 [1], and K. Takahashi did the same thing for the Weyl group of type F_4 [42]. The most of their results were obtained by making direct computation of the related KL-polynomials. In Alvis's case, more than one million KL-polynomials were calculated by a computer. So their methods are not effective for the Coxeter groups of higher orders or of infinite orders. We must search some new methods, directly or indirectly.

1.6 Given an element $x \in W$, We define the set $M(x)$ of all the elements y such that there exists a sequence of elements $x_0 = x, x_1, \ldots, x_r = y$ in W with some $r \geq 0$, where for every i, $1 \leq i \leq r$, the conditions $x_{i-1}^{-1}x_i \in S$ and $\mathcal{R}(x_{i-1}) \underset{\not\subseteq}{\not\supseteq} \mathcal{R}(x_i)$ are satisfied. To any $x \in W$, we associate a graph $\mathfrak{M}(x)$ as follows. Its vertex set is $M(x)$. Its edge set consists of all two-element subsets $\{y, z\} \subset M(x)$ with $y^{-1}z \in S$ and $\mathcal{R}(y) \not\subseteq \mathcal{R}(z)$. By a path in the graph $\mathfrak{M}(x)$, we mean a sequence of vertices $z_0, z_1, \ldots, z_t$ in $M(x)$ such that $\{z_{i-1}, z_i\}$ is an edge of $M(x)$ for any i, $1 \leq i \leq t$. Two elements $x, x' \in W$ are said to have the same *generalized τ-invariant* if for any path $z_0 = x, z_1, \ldots, z_t$ in the graph $M(x)$, there exists a path $z'_0 = x', z'_1, \ldots, z'_t$ in $M(x')$ with $\mathcal{R}(z'_i) = \mathcal{R}(z_i)$ for every i, $0 \leq i \leq t$, and if the same condition holds when interchanging the roles of x and x'.

Note that our definition of a generalized τ-invariant is slightly different from the one given by D. Vogan (see [44]). It is known that if two elements of W are in the same left cell of W then they have the same generalized τ-invariant and that the converse is not true in general, i.e. it might happen that two elements having the same generalized τ-invariant belong to different left cells of W. However, when $W = S_n$, we have

Proposition (see [29]) $x \underset{L}{\sim} y$ in S_n iff x, y have the same generalized τ-invariant.

1.7   Given a Coxeter group W, define a V-left cell of W to be the set of all the elements of W having the same generalized τ-invariant. We use the terminology " a V-left cell " because when W is a Weyl group, it is precisely a left cell defined by D. Vogan (see [44]). As pointed out in the above, a V-left cell of W is a union of some left cells of W defined by Kazhdan and Lusztig. R. Bédard described the V-left cells of all the crystallographic, compact, hyperbolic groups of rank 3 [5].

§2. Some results of Lusztig on cells in certain Coxeter groups

All the results in the present section are due to Lusztig unless otherwise specified. These results are important not only on the cell theory itself but also on its application to the other mathematical fields. Their proofs are based on the very deep theory of the intersection cohomology, the algebraic K-equivariance and the character sheaves.

2.1   For any $x, y, z \in W$, we define $h_{x,y,z} \in \mathcal{A}$ by

$$C_x C_y = \sum_z h_{x,y,z} C_z. \qquad (2.1.1)$$

Define a function $a \colon W \longrightarrow N$ by setting $a(z)$ to be the smallest integer k satisfying the condition

$$u^k h_{x,y,z} \in Z[u], \qquad \text{for all } x, y \in W.$$

For $z \in W$, where we stipulate $a(z) = \infty$ if no such an integer k exists.

2.2   Two families of Coxeter groups are more interesting to us: Weyl groups and affine Weyl groups. This is because the cell theory of these groups is closely related with the representations of connected reductive algebraic groups over the complex field and over a p-adic field. These Coxeter groups have nice properties: their a-functions are upper-bounded, and the coefficients of their KL-polynomials and their Laurent polynomials $h_{x,y,z}$ are non-negative. These two properties strongly effect the cell structure of these groups.

2.3   Let W be a Coxeter group satisfying the following two properties:

(a) its a-function is upper bounded;

(b) the coefficients of the KL-polynomials and the Laurent polynomials $h_{x,y,z}$ associating with its elements are non-negative.

Then the following results on W hold.

(1) The function a is constant on each two-sided cell of W. So we may define the a-value $a(\Gamma)$ on a (left, right or two-sided) cell Γ of W by $a(x)$ for any $x \in \Gamma$ [20].

(2) Let $\delta(z) = \deg P_{e,z}$ for $z \in W$, where e is the identity of the group W. Then the inequality $i(z) := \ell(z) - 2\delta(z) - a(z) \geq 0$ holds for any $z \in W$. The set $\mathcal{D}_0 = \{w \in W \mid i(w) = 0\}$ consists of involutions (called distinguished involutions by Lusztig). Each left (resp. right) cell of W contains a unique element of $\mathcal{D}_0$ [21]. Let W' be a standard parabolic subgroup of W. Then a consequence of the above result is that the intersection of any left cell of W with W' is either empty or a single left cell of W'.

2.4   Let W_a be an irreducible affine Weyl group and let G be the connected reductive algebraic group over C associated to W_a. Then the following result of Lusztig establishes some connection between the set of two-sided cells of W_a and the set of unipotent conjugacy classes of G.

Theorem (see [23])   *There exists a bijection* $\mathbf{u} \mapsto \mathbf{c}(\mathbf{u})$ *from the set of unipotent conjugacy classes in* G *to the set of two-sided cells in* W_a. *This bijection satisfies the equation* $a(\mathbf{c}(\mathbf{u})) = \dim \mathfrak{B}_u$, *where* u *is any element in* $\mathbf{u}$, *and* $\mathfrak{B}_u$ *is the variety of Borel subgroups of* G *containing* u.

This is a result concerning two-sided cells of an affine Weyl group W_a. We also have a result concerning left cells of W_a: $\mathcal{D}_0$ is a finite set of involutions in W_a [21]. This particularly implies that the number of left cells of an affine Weyl group is finite.

2.5 We know the number of two-sided cells of an affine Weyl group but not the number of left cells of such a group in general. This is because we know the number of unipotent conjugacy classes of G (see [6]) but we don't know the number of distinguished involutions of W_a in general. However, Lusztig proposed the following

Conjecture (see [2]) *The number of left cells of W_a contained in the two-sided cell corresponding to a unipotent element $u \in G$ is equal to $\sum_i (-1)^i \dim H^i(\mathfrak{B}_u)^{A(u)}$, where $A(u)$ is the group of connected components of the centralizer of u in G, and $H^i(X)$ is the ith étale cohomology space of the variety X with values in the constant sheaf.*

Now we assume the above conjecture. Suppose that **u** is a unipotent class of G containing a regular unipotent element of a Levi subgroup L. Then the number of left cells in the corresponding two-sided cell **c(u)** is equal to the number of left cosets of W with respect to W_L, provided that the centralizer $C_G(u)$ of u in G is connected, where u is any element in **u**, and W_L is the standard parabolic subgroup of W determined by L.

The above conjecture is supported in all the cases where the numbers of left cells of W_a have been calculated. For example, the author showed it for W_a of type $\tilde{A}_{n-1}$. In that case, the group $A(u)$ is trivial for any unipotent element $u \in G$, and any unipotent class of G contains a regular unipotent element of some Levi subgroup L (see [6]). On the other hand, the two-sided cells of $W_a(\tilde{A}_{n-1})$ are in one-to-one correspondence with the partitions of n, where a partition of n is by definition a sequence of integers $\lambda_1 \geq \lambda_2 \geq \cdots \geq \lambda_r > 0$ with $\sum_{i=1}^r \lambda_i = n$. The number of left cells of $W_a(\tilde{A}_{n-1})$ in the two-sided cell corresponding to the partition $\lambda_1 \geq \lambda_2 \geq \cdots \geq \lambda_r$ is equal to $\dfrac{n!}{\prod_{j=1}^m \mu_j!}$, where $\mu_j = \#\{i \mid \lambda_i \geq j\}$ (see [29]).

§3. A survey on the achievements for describing left cells

3.1 Although it has taken about fifteen years to study, the left cells for the most Coxeter groups are still far from being known. However, some significant progresses have been made on the explicit description of all left cells of certain Coxeter groups satisfying the properties 2.3,(a), (b) , which are listed as follows.

(i) Weyl groups of types A (Kazhdan & Lusztig, see [15]), B, C, D

(Barbasch-Vogan, see [3], Garfinkle, Devra, see [13]), E_6 (Tong, see [43]), F_4 (K. Takahashi, see [42]) and G_2 ($\cong D_6$);

(ii) Affine Weyl groups of types $\tilde{A}$ (Shi, see [29]), $\tilde{B}_i$ ($i = 3, 4$) (Du and Zhang, [10] [46]), $\tilde{C}_j$ ($j = 2, 3, 4$) (Lusztig, Bédard and Shi, see [20] [4] [40]), $\tilde{D}_4$ (Chen and Shi, see [8] [39]) and $\tilde{G}_2$ (Lusztig, see [20]).

(iii) The Coxeter groups of types H_3 and H_4 (D. Alvis, see [1]), and all the dihedral groups D_m, $m \geq 1$, (see 1.4).

3.2   The description of the left cells of the affine Weyl groups $W_a(\tilde{A}_{n-1})$ is the most successful work among all in the above list.

A Young tableau is called *quasi-standard* if the numbers in each of its columns increase downwards, and is *standard* if it is quasi-standard and the numbers in each of its rows increase from left to right. Let $\mathfrak{C}_n$ (resp. $\mathfrak{G}_n$) be the set of all quasi-standard (resp. standard) Young tableaux of rank n.

Recall the Robinson-Schensted map ϕ from the symmetric group S_n to the set $\mathfrak{G}_n$ (see 1.1). The author constructed a surjective map from the affine Weyl group $W_a(\tilde{A}_{n-1})$ to the set $\mathfrak{C}_n$. This map, when restricted on S_n (note that S_n could be regarded as a maximal standard parabolic subgroup of $W_a(\tilde{A}_{n-1})$), is exactly the Robinson-Schensted map ϕ. So we call this map a generalized Robinson-Schensted map. The significance of this map is that it induces a bijection from the set of all left cells of $W_a(\tilde{A}_{n-1})$ to the set $\mathfrak{C}_n$. Two left cells are in the same two-sided cell iff the shapes of the corresponding Young tableaux are the same (see [29] [36]).

3.3   Left cells of the group $W_a(\tilde{A}_{n-1})$ could also be characterized by the generalized τ-invariant (see 1.6): two elements of $W_a(\tilde{A}_{n-1})$ are in the same left cell iff they have the same generalized τ-invariant (see [29]). This result generalizes Proposition 1.6. Left cells of $W_a(\tilde{A}_{n-1})$ have even more nice properties (see 3.6).

3.4   Garfinkle defined another kind of generalized Robinson-Schensted map. She associated each signed permutation to a domino Young tableau, by which she got a surjective map from the set of all the elements in the Weyl group of type B_n or D_n to a set of certain standard domino Young tableaux. Then she concluded that the fibers of such a map should be exactly all the left cells of the corresponding Weyl group (see [13]).

3.5   Let W_a be an affine Weyl group. Let Φ be the associated root system with $\{\alpha_j \mid 1 \leq j \leq \ell\}$ a choice of simple root system and $-\alpha_0$

its highest short root. Denote $s_i = s_{\alpha_i}$, $1 \leq i \leq \ell$, the reflection with respect to α_i, and denote $s_0 = T_{-\alpha_0} s_{\alpha_0}$, where $T_{-\alpha_0}$ is the translation $\lambda \mapsto \lambda - \alpha_0$ in the euclidean space spanned by Φ. Let $w \mapsto \bar{w}$ be the natural map $W_a = N \rtimes W \longrightarrow W_a/N \cong W$, where N is the maximal normal abelian subgroup of W_a consisting of all translations and W is the Weyl group of Φ. To each element $w \in W_a$, we can associate a unique Φ-tuple $(k(w, \alpha))_{\alpha \in \Phi}$ over Z subject to the following conditions:

$$\begin{cases} k(e, \alpha) = 0, & \text{for all } \alpha \in \Phi, \\ k(ws_i, \alpha) = k(w, (\alpha)\bar{w}) + k(s_i, \alpha), & \text{for all } s_i \in S, \end{cases}$$

where e is the identity of W_a and $k(s_i, \alpha)$ is defined by

$$k(s_i, \alpha) = \begin{cases} 0, & \text{if } \alpha \neq \pm\alpha_i, \\ -1, & \text{if } \alpha = \alpha_i, \\ 1, & \text{if } \alpha = -\alpha_i. \end{cases}$$

$(k(w, \alpha))_{\alpha \in \Phi}$ is called the alcove form of an element $w \in W_a$ (see [30]).

3.6   A Φ-sign type is a Φ-tuple $(X_\alpha)_{\alpha \in \Phi}$ with $X_\alpha \in \{+, -, \bigcirc\}$. We can associate an element $w \in W_a$ to a Φ-sign type $(X_\alpha)_{\alpha \in \Phi}$ by

$$X_\alpha = \begin{cases} +, & \text{if } k(w, \alpha) > 0, \\ -, & \text{if } k(w, \alpha) < 0, \\ \bigcirc, & \text{if } k(w, \alpha) = 0. \end{cases}$$

By abuse of terminology, we call the set of all elements of W_a corresponding to any given Φ-sign type also by " a sign type of W_a " provided that this set is non-empty [31]. Our result asserts

Theorem (see [29])   *Each left cell of the group $W_a(\widetilde{A}_{n-1})$ is a union of some sign types.*

Since any Weyl group could be regarded as a standard parabolic subgroup of some affine Weyl group, one can also define the alcove form and the sign type of an element of a Weyl group. Note that the above theorem is trivially valid in the case when $W_a(\widetilde{A}_{n-1})$ is replaced by a Weyl group since each sign type of a Weyl group consists of exactly one element. But this result does not hold for an arbitrary affine Weyl group. A counterexample could be found in the case of $W_a(\widetilde{B}_2)$ (see [20]). However, one may expect the truth of the following

Conjecture (see [34])   *If W is an irreducible affine Weyl group with simply-laced Dynkin diagram (i.e. W has type $\widetilde{A}$, $\widetilde{D}$ or $\widetilde{E}$), then each left cell of W is a union of some sign types of W.*

The results on the left cells of the affine Weyl group $W_a(\tilde{D}_4)$ support this conjecture.

3.7 Sometimes we are unable to describe all the left·cells of certain Coxeter groups, but we can describe all the left cells of those groups in certain two-sided cells. This is the case in the following two-sided cells Ω of affine Weyl groups·W_a:

(1) Ω is the lowest two-sided cell W_ν of W_a with respect to the partial order $\underset{LR}{\leq}$ (see [32] [33]);

(2) $a(\Omega) \leq 3$ (see [17] [18] [7] [28]);

(3) W_a has type $\tilde{B}_n$, $\tilde{C}_n$, or $\tilde{D}_n$ ($n \geq 5$), and, $a(\Omega) = 4$ (see [9]).

3.8 Let us explain the results of 3.7,(1). The lowest two-sided cell W_ν of W_a could be described as below.

$$W_\nu = \{w \in W_a \mid k(w, \alpha) \neq 0, \text{ for all } \alpha \in \Phi\}$$
$$= \Big\{w \in W_a \;\Big|\; w = xw_Jy, \text{ for some } x, y \in W_a \text{ and } J \subseteq S$$
$$\text{with } \ell(w) = \ell(x) + \ell(y) + \frac{1}{2}|\Phi|\Big\}.$$

where w_J is the longest element in the subgroup of W_a generated by J. Let

$$M = \{w \in W_\nu \mid sw \notin W_\nu \text{ for all } s \in \mathcal{L}(w)\}.$$

Then each element $w \in M$ could be written uniquely in the form $w = w_Jx$ for some $x \in W_a$ and $J \subseteq S$ with $\ell(w) = \ell(x) + \frac{1}{2}|\Phi|$ (call it the standard expression of w). There exists a bijective map from the set M to the set of all the left cells of W_a in W_ν by sending $w \in M$ to the set

$$\Gamma_w = \{yw \mid y \in W_a, \ell(yw) = \ell(y) + \ell(w)\}.$$

The last set forms a left cell of W_a, which is also a sign type of W_a (see [32] [33]).

The distinguished involutions of W_a in W_ν could be described as below (see [33]).

$$\mathcal{D}_0(W_\nu) = \{x^{-1}w_Jx \mid w_Jx \in M \text{ standard expression }\}. \tag{3.8.1}$$

Now we want to give another expression of this set. Let x, $y \in W_a$. In the product $T_xT_y = \sum_z f_{x,y,z}T_z$ ($f_{x,y,z} \in \mathcal{A}$), there exists a unique element $w \in W_a$ with $f_{x,y,w} \neq 0$ such that if z satisfies $f_{x,y,z} \neq 0$ then $z \leq w$ (see

[33]). Denote this element w by $\lambda(x, y)$. Then the set (3.8.1) could be reformulated as below.

$$\mathcal{D}_0(W_\nu) = \{\lambda(x^{-1}, x) \mid x \in M\}.$$

From this, one may expect the following more general result.

Conjecture 3.9 (see [34]) *Let $x \in W_a$ be a shortest element in the left cell of W_a containing it. Then $\lambda(x^{-1}, x) \in \mathcal{D}_0$. Conversely, any element $d \in \mathcal{D}_0$ has the form $d = \lambda(x^{-1}, x)$ for any shortest element x in the left cell of W_a containing d.*

Note that there may exist more than one shortest elements in a left cell of W_a. But this fact does not conflict with the above conjecture. For example, in the affine Weyl group

$$W_a(\tilde{A}_3) = \langle s_i \mid 0 \le i \le 3 \rangle \quad \text{with} \quad o(s_i s_j) = 3 \quad \text{for} \quad j \equiv i \pm 1 (\mathrm{mod}\, 4),$$

let $x = s_1 s_2 s_1 s_3$ and $y = s_2 s_3 s_2 s_1$. Then $x \underset{L}{\sim} y$ and, both x and y are shortest elements in the left cell containing them. We have

$$\lambda(x^{-1}, x) = \lambda(y^{-1}, y) = s_1 s_2 s_3 s_2 s_1 \in \mathcal{D}_0.$$

The above conjecture has been verified in all the left cells Γ of W_a with $a(\Gamma) \le 3$ (see [7] [28]) and in all the left cells Γ of W_a of classical types with $a(\Gamma) = 4$ (see [8] [9]).

3.10 The canonical left cells of an affine Weyl group, which will be defined shortly, are closely related to the spherical representations of the corresponding Hecke algebra [23, §9.]. The description of these left cells were given by Lusztig and Xi [24].

It is well known that for any x, y in a Coxeter group W,

$$x \underset{L}{\sim} y \Longrightarrow \mathcal{R}(x) = \mathcal{R}(y) \qquad (\text{see [15]}). \qquad (3.10.1)$$

Thus we can use the notation $\mathcal{R}(\Gamma)$ for a left cell Γ of W, which is $\mathcal{R}(x)$ for any $x \in \Gamma$. But the converse is not true in general:

$$\mathcal{R}(x) = \mathcal{R}(y) \nRightarrow x \underset{L}{\sim} y.$$

The weaker one is still not true in general:

$$x \underset{LR}{\sim} y \text{ and } \mathcal{R}(x) = \mathcal{R}(y) \nRightarrow x \underset{L}{\sim} y.$$

Call a subset $I \subseteq S$ to be special if $|I| = 1$ and $|W_{S-I}| = \max\{|W_{S-\{t\}}| \mid t \in S\}$, where the notation $|X|$ stands for the cardinality of a set X. For $J \subseteq S$, let Y_J be the set of all the elements $x \in W_a$ satisfying $\mathcal{R}(x) = J$. Then the fact (3.10.1) tells us that for any subset $J \subseteq S$, the intersection of a two-sided cell of W_a with the set Y_J is either empty or a union of some left cells of W_a. Lusztig and Xi showed the following stronger result in certain circumstance.

Proposition (see [24])   *For any special $I \subseteq S$, the intersection of a two-sided cell $\Omega \neq \{e\}$ of W_a with the set Y_I is exactly a single left cell of W_a (i.e. it is neither empty nor a union of more than one left cells).*

This tells us that for x, y in an affine Weyl group,

$$x \underset{LR}{\sim} y \text{ with } \mathcal{R}(x) = \mathcal{R}(y) \text{ special} \implies x \underset{L}{\sim} y.$$

A left cell Γ of W_a is *canonical* if either $\Gamma = \{e\}$ or that $\mathcal{R}(\Gamma)$ is special. An easy consequence is that the number $\kappa(W_a)$ of the canonical left cells of W_a is equal to $gh + 1 - g$, where g is the number of special subsets in S, and h is the number of two-sided cells of W_a, both of which are relatively well known. For example, when W_a is of type $\tilde{A}_{n-1}$, we have $g = n$ and $h = p_n$, the number of partitions of n. So $\kappa(W_a) = n \cdot p_n + 1 - n$.

§4. Algorithm for finding a representative set of left cells

Although we have succeeded in describing so many left cells, the whole figure about left cells of Coxeter groups is still far from being exposed. Thus we must describe more left cells of Coxeter groups whenever it is possible. To do so, it is desirable to design some algorithms. Here we introduce an algorithm which is effective for any Coxeter group with properties 2.3, (a), (b). [38]

In the present section, we always assume W to be a Coxeter group with properties 2.3, (a), (b).

4.1   For $x \in W$, let $\Sigma(x)$ be the set of all left cells Γ satisfying: there exists some $y \in \Gamma$ with $y \text{---} x$, $\mathcal{R}(y) \not\subseteq \mathcal{R}(x)$ and $a(x) = a(y)$. Then the author showed.

Theorem (see [38])   *If $x \underset{L}{\sim} y$ in W, then $\mathcal{R}(x) = \mathcal{R}(y)$ and $\Sigma(x) = \Sigma(y)$.*

Remark 4.2   (1) The author conjectured that the converse of the above result should also be true (see [38]), i.e.

$$x \underset{L}{\sim} y \iff \mathcal{R}(x) = \mathcal{R}(y) \text{ and } \Sigma(x) = \Sigma(y). \qquad (4.2.1)$$

The truth of the conjecture would characterize the left cells of W. This conjecture has been verified in the cases when W is a Weyl group and when W is an irreducible affine Weyl group with few cases excepted in $W_a(\tilde{F}_4)$ (see [41]).

(2) It is easily seen that if $x, y \in W$ satisfy $\mathcal{R}(x) = \mathcal{R}(y)$ and $\Sigma(x) = \Sigma(y)$ then x, y have the same generalized τ-invariant (see 1.6 for the definition).

Remark 4.3 A subset $K \subset W$ is called *a representative set of left cells* (or *an l.c.r. set* for brevity) of W (resp. of W in a two-sided cell Ω), if $|K \cap \Gamma| = 1$ for any left cell Γ of W (resp. of W in Ω). The following is a criterion for an l.c.r. set of W in a given two-sided cell.

Theorem (see [38]) *Let Ω be a two-sided cell of W. Assume that a non-empty subset $M \subset \Omega$ satisfies the following conditions.*

(1) $x \underset{L}{\nsim} y$ for any $x \neq y$ in M;

(2) Given an element $y \in W$. If there exists some element $x \in M$ such that $y \text{—} x$, $\mathcal{R}(y) \nsubseteq \mathcal{R}(x)$ and $a(y) = a(x)$. Then there exists some $z \in M$ with $y \underset{L}{\sim} z$.

Then M is an l.c.r. set of W in Ω.

In principle, this theorem provides us a method to find an l.c.r. set of W in any given two-sided cell Ω from a non-empty subset of Ω, and hence of W itself provided that one could find at least one element from each two-sided cell of W.

Remark 4.4 A subset $P \subset W$ is said to be *distinguished* if $P \neq \emptyset$ and $x \underset{L}{\nsim} y$ for any $x \neq y$ in P.

Now assume that P is a non-empty subset of a two-sided cell Ω of W. We introduce the following two processes.

(A) *Take a subset Q of the largest possible cardinality from set $\bigcup_{x \in P} M(x)$ with Q distinguished.*

(B) *For each $x \in P$, find elements $y \in W$ such that $y \text{—} x$, $\mathcal{R}(y) \supsetneq \mathcal{R}(x)$ and $a(y) = a(x)$, add these elements y on the set P to form a set P' and then take a largest possible subset Q from P' with Q distinguished.*

Note that the resulting sets in both Processes **(A)** and **(B)** are automatically in the two-sided cell Ω.

Now we are ready to introduce an algorithm to find an l.c.r. set of W in a given two-sided cell Ω.

Algorithm 4.5   (see [38])

(1) Find a non-empty subset P of Ω (usually we take P to be distinguished for avoiding unnecessary complication if possible);

(2) Perform Processes (**A**) and (**B**) alternately on P until the resulting distinguished set can not be further enlarged by both processes.

Remark 4.6   The above algorithm has been applied by several persons to classify the left cells of affine Weyl groups W in the following cases.

(1) For W of type $\tilde{D}_4$, by the author [39] (The author understand that Chen Chengdong also did this but his method is different from the author [8]).

(2) For W of type $\tilde{C}_4$, by the author [40].

(3) For W of type $\tilde{B}_4$, by Zhang [46].

(4) For all the left cells with their a-values equal to 3 in any irreducible affine Weyl group W, by Rui [28].

(5) For all the left cells with their a-values ≤ 5 in W of type $\tilde{F}_4$, by the author [38].

Remark 4.7   As mentioned in 2.3,(2), the set $\mathcal{D}_0$ forms an l.c.r. set of W. Thus one can also classify the left cells of W by first finding the set $\mathcal{D}_0$. Chen made this approach in his papers [7], [8] and [9]. In general, it is more difficult to find the set $\mathcal{D}_0$ directly than to find an arbitrary l.c.r. set of W by algorithm 4.5. On the other hand, suppose that one has got an l.c.r. set of W by algorithm 4.5. Then by applying the result [35, Proposition 5.12], one can find the set $\mathcal{D}_0$ considerably easier.

§5. Some more open problems

In this section, we assume that W is either a Weyl group or an affine Weyl group. We want to state two more conjectures proposed by Lusztig. One is concerning the connectedness of left cells of W. The validity of this conjecture would ensure certain good behavior for the left cell representations of the corresponding Hecke algebra, such as cyclicity. The other is concerning the cells of the affine Weyl groups of classical types. That conjecture is based on the belief that there should exist certain very strong combinatorial background for the cells of these groups.

Remark 5.1 Connectedness of left cells of W.

A subset $K \subseteq W$ is *connected*, if, for all $x, y \in K$, there exists a sequence of elements $x_0 = x$, x_1, $\cdots$, $x_r = y$ in K with some $r \geq 0$ such that for all i, $1 \leq i \leq r$, we have $x_{i-1}x_i^{-1} \in S$. Lusztig proposed:

Conjecture (see [2]) *When W is either a Weyl group or an affine Weyl group, each left cell of W is connected.*

This conjecture is supported in all the cases where the left cells of W have been described explicitly (see 3.1). On the other hand, Xi and Du made some progress in the proof of this conjecture by showing the following result individually: It is finite for the number of connected components in each left cell of an affine Weyl group (see [45] [12]). Comparing with the result of Xi-Du, the above conjecture asserts that this number is one.

By 2.3, it is sufficient to verify the above conjecture in the cases of affine Weyl groups. Note that the truth of Conjecture 3.8 might be helpful in the verification of the present conjecture.

Remark 5.2 Left cells in the affine Weyl group of type $\tilde{B}_n$, $\tilde{C}_n$ or $\tilde{D}_n$.

Define a permutation group on the integer set Z as follows.

$$\mathcal{A}_n = \left\{ \sigma : Z \mapsto Z \;\middle|\; (i+n)\sigma = (i)\sigma + n, \quad \forall i \in Z; \; \sum_{i=1}^{n}(i)\sigma = \sum_{i=1}^{n} i \right\}.$$

Then it is known that $\mathcal{A}_n$ is isomorphic to the affine Weyl group of type $\tilde{A}_{n-1}$ with its simple reflection set $S = \{s_t \mid 0 \leq t \leq n - 1\}$, where

$$s_t(i) = \begin{cases} i, & \text{if } i \not\equiv t, t+1 \pmod n, \\ i+1, & \text{if } i \equiv t \pmod n, \\ i-1, & \text{if } i \equiv t+1 \pmod n. \end{cases}$$

The symmetric group S_n could be described as a subgroup of $\mathcal{A}_n$:

$$S_n = \{\sigma \in \mathcal{A}_n \mid 1 \leq (i)\sigma \leq n, \text{ for all } 1 \leq i \leq n\}.$$

The affine Weyl groups of types $\tilde{B}_\ell$, $\tilde{C}_\ell$ and $\tilde{D}_\ell$ could also be described as permutation groups on Z.

(a) Let $\sigma : \; s_i \mapsto s_{2\ell+1-i}$, for all $0 \leq i \leq 2\ell$, be the involutive automorphism in $\mathcal{A}_{2\ell+1}$ with the convention that $s_{2\ell+1+i} = s_i$ for $i \in Z$. Then we have $W_a(\tilde{B}_\ell) \cong \mathcal{A}_{2\ell+1}^\sigma$, the latter is the subgroup of $\mathcal{A}_{2\ell+1}$ consisting of all the σ-fixed elements (similar notations will be used in (b) and (c)).

(b) Let $\tau : \ s_i \mapsto s_{2\ell+1-i}$, for all $0 \leq i \leq 2\ell + 1$, be the involutive automorphism in $\mathcal{A}_{2\ell+2}$ with the convention that $s_{2\ell+2+i} = s_i$ for $i \in Z$. Then we have $W_a(\widetilde{C}_\ell) \cong \mathcal{A}_{2\ell+2}^\tau$.

(c) Let $\eta : \ s_i \mapsto s_{2\ell-i}$, for all $0 \leq i \leq 2\ell - 1$, be the involutive automorphism in $\mathcal{A}_{2\ell}$ with the convention that $s_{2\ell+i} = s_i$ for $i \in Z$. Then $W_a(\widetilde{D}_\ell) \cong \mathcal{A}_{2\ell}^\eta$.

We have even simpler description for $W_a(\widetilde{C}_\ell)$ as follows.

$$W_a(\widetilde{C}_\ell) = \{w \in \mathcal{A}_{2\ell+2} \mid (-i)w = -(i)w, \ \text{for all} \ i \in Z\}.$$

Note that this description is slightly different from the one by R. Bédard (see [4]). Our description of $W_a(\widetilde{C}_\ell)$ has the advantage of exposing more group-theoretic symmetry in element form.

We can also give some similar descriptions for $W_a(\widetilde{B}_\ell)$ and $W_a(\widetilde{D}_\ell)$, however they are slightly complicated.

Lusztig suggested the following

Conjecture *Each left cell of $W_a(\widetilde{X})$ ($X \in \{B_\ell, \ C_\ell, \ D_\ell\}$) has the form*

$$\Gamma = W_a(\widetilde{X}) \cap \left(\bigcup_{i \in I} \Gamma_i \right),$$

where the Γ_i ($i \in I$) are some left cells of $\mathcal{A}_m$ (m is determined by X as above).

Remark 5.3 The author would like to take this opportunity to thank Lusztig for telling me the above conjecture by private communication. This conjecture has been verified by the author in the case of $\ell \leq 3$ (unpublished).

Remark 5.4 Description of cells in terms of partitions.

To each element $w \in \mathcal{A}_n$, we associate a sequence of integers $d_1 \leq d_2 \leq \cdots \leq d_t = n$ as follows.

$$d_k = \max \left\{ |X| \ \middle| \ X = \bigcup_{i=1}^{k} X_i \subset Z; \ u \not\equiv v (\mathrm{mod}\, n), \qquad \forall \, u \neq v \ \text{in} \ X: \right.$$
$$\left. \text{and} \ u < v \ \text{in some} \ X_i \ \text{implies} \ (u)w > (v)w \right\}$$

Then we have $d_1 \geq d_2 - d_1 \geq d_3 - d_2 \geq \cdots \geq d_t - d_{t-1}$, which is a partition of n. We denote it by $\psi(w)$. This defines a map $\psi : \mathcal{A}_n \longrightarrow \Lambda_n$, where Λ_n is the set of all partitions of n. This map is compatible with the map mentioned in 3.2, and so it induces a bijection from the set of two-sided cells of $\mathcal{A}_n$ to the set Λ_n. It would be interesting to define an analogous map for the group $W_a(\widetilde{X})$ ($X \in \{B_\ell, C_\ell, D_\ell\}$) such that one can describe all the two-sided cells (and hence all the left cells) in the element level. That is, to associate any element of $W_a(\widetilde{X})$ to a pair of partitions of some fixed integer with certain properties such that the fibers of the corresponding map coincide with two-sided cells of $W_a(\widetilde{X})$.

Note that some progress in this direction has been made on the group $W_a(\widetilde{C}_n)$ by the author (see [37]). We omit the detail.

References

[1] D. Alvis, *The left cells of the Coxeter group of type H_4*, 1987 J. Algebra **107** (1987), 160-168.

[2] T. Asai et al., *Open problems in algebraic groups*, Proc. Twelveth International Symposium, Tôhoku Univ., Japan (1983), 14.

[3] D. Barbasch and D. Vogan, *Primitive ideals and Orbital integrals in complex classical groups*, Math. Ann., **259** (1982), 153-199.

[4] R. Bédard, *Cells for two Coxeter groups*, Comm. Algebra14 (1986), 1253–1286.

[5] R. Bédard, *Left V-cells for hyperbolic Coxeter groups*, Comm. Algebra **17** (1989), 2971-2997.

[6] R. W. Carter "*Finite Groups of Lie Type: Conjugacy Classes and Complex Characters*" Wiley Series in Pure and Applied Mathematics, John Wiley London (1985).

[7] Chen Chengdong, *Two-sided cells in affine Weyl groups*, Northeastern Math. J., **6** (1990) 425–441.

[8] Chen Chengdong, *Cells of the affine Weyl group of type $\widetilde{D}_4$*, J. Algebra (to appear).

[9] Chen Chengdong, *Two-sided cells ($a = 4$) in the affine Weyl group of type $\widetilde{B}_n$*, J. London Math. Soc. (to appear).

[10] Du Jie, *The decomposition into cells of the affine Weyl group of type $\widetilde{B}_3$*, Comm. Algebra **16** (1988), 1383–1409.

[11] Du Jie, 384–404, *Cells in the affine Weyl group of type $\widetilde{D}_4$*, J. Algebra (1990)**128** (1990) .

[12] Du Jie, *Sign types and Kazhdan-Lusztig cells*, Chin. Ann. Math., **12B(1)** (1991), 33-39.

[13] Garfinkle, Devra, *On the classification of primitive ideals for complex classical Lie algebra, I*, Compositio Math. **75**, (1990), 135-169 .

[14] A. Joseph, *Goldie rank in the enveloping algebra of a semisimple Lie algebra, I*, J. Algebra, **65** (1980), 269-283.

[15] D. Kazhdan & G. Lusztig, 165–184, *Representations of Coxeter groups and Hecke algebras*, Invent. Math. **53** (1979).

[16] D. Kazhdan & G. Lusztig, *Schubert varieties and Poincaré duality*, Proc. Sympos. Pure Math., Amer. Math. Soc., Providence, R. I.**36** (1980), 185-203.

[17] G. M. Lawton, *On cells in affine Weyl groups*, Ph.D. Thesis, MIT, 1986.

[18] G.Lusztig, 623–653, *Some examples in square integrable representations of semisimple p-adic groups*, Trans. of the AMS., **277** (1983).

[19] G. Lusztig, *The two-sided cells of the affine Weyl group of type $\widetilde{A}_n$*, in "Infinite Dimensional Groups with Applications", (V. Kac, ed.), Springer-Verlag (1985), 275–283.

[20] G. Lusztig, *Cells in affine Weyl groups*, in "Algebraic Groups and Related Topics" (R. Hotta, ed.), Advanced Studies in Pure Math., Kinokuniya and North Holland (1985), 255–287.

[21] G. Lusztig, 536–548, *Cells in affine Weyl groups, II*, J. Algebra, **109** (1987).

[22] G. Lusztig, 223–243, *Cells in affine Weyl groups, III*, J. Fac. Sci. Univ. Tokyo Sect. IA. Math., **34** (1987).

[23] G. Lusztig, 297–328, *Cells in affine Weyl groups, IV*, J. Fac. Sci. Univ. Tokyo Sect. IA. Math., **36** (1989).

[24] G. Lusztig & Xi Nan-hua, *Canonical left cells in affine Weyl groups*, Adv. in Math., **72** (1988), 284-288 .

[25] Hiroshi, Naruse, *On an Isomorphism between Specht module and left cell of S_n*, Tokyo J. Math., **12** (1989), 247-267

[26] G. de B. Robinson, *On the representations of the symmetric groups*, Amer. J. Math., **60** (1938), 745–760; **69** (1947), 286–298; **70** (1948), 277–294.

[27] C. Schensted, *Longest increasing and decreasing subsequences*, Canad. J. Math., **13** (1961), 179-191.

[28] Rui Hebing, *Some cells in affine Weyl groups of types other than $\widetilde{A}_n$ and $\widetilde{G}_2$*, preprint.

[29] Jian-yi Shi" *The Kazhdan-Lusztig cells in certain affine Weyl groups*, Lect. Notes in Math., Springer-Verlag, **1179** (1986).

[30] Jian-yi Shi, 42–55, *Alcoves corresponding to an affine Weyl group*, J. London Math. Soc. **35** (1987).

[31] Jian-yi Shi, *sign types corresponding to an affine Weyl group*, (1987), J. London Math. Soc., **35** (1987), 56-74.

[32] Jian-yi Shi, 407–420, *A two-sided cell in an affine Weyl group*, J. London Math. Soc., **36** (1987).

[33] Jian-yi Shi, 253–264, *A two-sided cell in an affine Weyl group, II* J. London Math. Soc., **37** (1988).

[34] Jian-yi Shi, *A survey on the cell theory of affine Weyl groups*, Advances in Science of China, Mathematics **3** (1988), 79–98.

[35] Jian-yi Shi, *The joint relations and the set $\mathcal{D}_1$ in certain crystallographic groups*, Adv. in Math., **81(1)** (1990), 66-89

[36] Jian-yi Shi, *The generalized Robinson-Schensted algorithm on the affine Weyl group of type $\widetilde{A}_{n-1}$*, J. Algebra, **139** (1991), 364-394.

[37] Jian-yi Shi, *Some results relating two presentations of certain affine Weyl groups*, J. Algebra (to appear).

[38] Jian-yi Shi, *Left cells in affine Weyl groups*, Tôhoku Math. J. (to appear).

[39] Jian-yi Shi, *Left cells of the affine Weyl group $W_a(\widetilde{D}_4)$*, Osaka J. Math. (to appear).

[40] Jian-yi Shi, *The classification of left cells of the affine Weyl group of type $\widetilde{C}_4$*, in preparation.

[41] Jian-yi Shi, *The verification of a conjecture on the left cells of certain Coxeter groups*, Hiroshima J. Math. (to appear).

[42] K. Takahashi, *The left cells and their W-graphs of Weyl group of type F_4* **13** (1990), Tokyo J. Math., 327-340.

[43] Tong Chang-qing, preprint.

[44] D. Vogan, 209–224, *A generalized τ-invariant for the primitive spectrum of a semisimple Lie algebra*, Math. Ann., **242** (1979),

[45] Xi Nan-hua, *An approach to the connectedness of the left cells in affine Weyl groups*, Bull. London Math. Soc., **21** (1989), 557-561.

[46] Zhang Xin-fa, *Cells decomposition in the affine Weyl group $W_a(\widetilde{B}_4)$*, Comm. Algebra (to appear).

Some Works on Modular Representations of Finite Groups [*]

Shi Shengming

Department of Mathematics, The Capital Normal University, Beijing

Zhang Jiping

Department of Mathematics, Peking University, Beijing

Professor Hsio-Fu Tuan is the first Chinese mathematician who made important contributions to modular representation theory of finite groups in the 1940s. Ten years ago, some Chinese mathematicians began to study the recent development in this area and have made some progress. Here we present some of their results.

§1. Problems on defect groups

Defect groups play a key role in the block theory of finite groups and are the most important objects which connect group theoretical properties and representation theoretical ones. As is well-known, the problem "When a given p-subgroup D is a defect group for some p-block of a finite group G? If D is such a subgroup, in terms of group theoretical properties, count the number of p-blocks with D as a defact group" is of great importance in the modular representation theory of finite groups, which was posed as Problem 19 in [8] by R. Brauer and as Problem 5 in [16] by W. Feit. Brauer and Fowler (1955), Tsushima (1977), Iizuka and Watanabe (1972) and W.Willems (1978) all made contributions. In 1983 G.R. Robinson obtained in [30] a precise formula for the number of p-blocks of G with defect group D under the assumption that D is normal in G. In his formula, the number is equal to the rank of a certain matrix which is related with the set of (P, P) double-cosets in G, where P is a Sylow p-subgroup of G. S.M. Shi gave in [33] a generalized formula and many

---

[*] Supported partly by the National Natural Science Foundation of China

consequence without the assumption $D \triangleleft G$. The key point of his work is the socalled p-power homomorphism. Of course the number is by no means easy to calculate. So for a concrete p-subgroup D, counting the number of p-blocks with D as a defect group or finding the necessary and sufficient conditions that such p-blocks exist is still an important task.

For $D = <1>$, such a p-block is called a defect zero p-block. In 1986, the following theorems were proved.

Theorem 1.1   ([44]) *A finite group G with a cyclic Sylow p-subgroup P has a p-block of defect zero if and only if $O_p(G) = 1$. (A sketched proof of this theorem was given in [43]).*

Theorem 1.2   ([52]) *A finite group G with a T.I. Sylow p-subgroup P has a p-block of defect zero if and only if P is not nomal in G.*

Also J.P. Zhang proved independently in [47] the following general result. (cf. [27],[41])

Theorem 1.3   *Let G be a finite simple group of Lie type, then G has a p-block of defect zero for any prime p.*

The following theorem solved a conjecture of J. Alperin [1]. (The case $p = 2$ was due to Robinson [31].)

Theorem 1.4   ([46]) *If a finite group G has a strongly p-embedded subgroup then G has a p-block of defect zero.*

The following corollary to Theorem 1.4 generalizes a result of Alperin [1]. We refer the reader to [1] for the notation and terminology.

Corollary 1.5   *If B is the principal p-block of a finite group G with $l(B) = l(b)$, then B is a controlled p-block.*

The defect zero problem for sporadic simple groups is easy to solve by checking character tables, and for alternating groups, considerable progress has also been made by Atkin and Olsson (cf. [27]). But in contrast, the problem for finite solvable group G is not well understood. In 1950s Ito obtained some results and the following is the best to cite ([16]).

Theorem   *Let G be a finite solvable group of odd order such that $O_{p'}(G)$ and $G/O_{p'}(G)$ are nilpotent for some prime p. Then G has a p-block of defect zero if and only if $O_p(G) = 1$.*

Note that examples show that the theorem is not true for groups of even order and the nilpotence of $O_{p'}(G)$ can not be discarded. So improvements on this theorem are difficult.

In 1991, J.P. Zhang made new progress on defect-zero problem for finite solvable groups by using some new and deep results on socalled regular-orbit problem. In order to formulate his results we need first fix some notations.

Let V be an elementary abelian q-group of order q^{pn} such that p divides $q^n - 1$. Consider V as the additive group of the Galois field $F_{q^{pn}}$ of q^{pn} elements. Let α be an element of the Galois group $\mathrm{Gal}(F_{q^{pn}}/F_q)$ of order p, and N, a subgroup of the multiplicative group $F_{q^{pn}}^*$ of order $(q^{pn}-1)/(q^n-1)$. Now $N :< \alpha >$ acts naturally on V. Set $E(p,q,n) = V : (N :< \alpha >)$. Then $E(p,q,n)$ is uniquely determined up to isomorphism by the three parameters p, q and n. It is easy to see now that a Sylow p-subgroup of $E(p,q,n)$ is non-cyclic of order p^2 and $O_p(E(p,q,n)) = 1$. But $E(p,q,n)$ has no p-blocks of defect zero.

Theorem 1.6 ([45]) *Let G be a finite group of odd order and p a prime divisor of the order of G such that $O_p(G) = 1$. If G is $E(p,q,n)$-free for any prime q and any positive integer n then there exists an element x in $O_{p'}(G)$ such that $C_G(x)$ is a p'-subgroup of G. Therefore G has a p-block of defect zero.*

An immediate corollary to Theorem 1.6 is the following, which generalizes Ito's theorem.

Corollary 1.7 ([45]) *If G is a finite group of odd order such that both $O_{p'}(G)$ and $G/O_{p'}(G)$ are supersolvable for some prime p then $O_p(G)$ is a defect group for some p-block. Thus G has a defect-zero p-block if and only if $O_p(G) = 1$.*

During the course of the proof of Theorem 1.6 J.P. Zhang posed the following open problem on regular orbits.

Open Problem 1.8 *Let G be a finite group acting on a vector space V over a field such that for any prime p there exists an element v_p in V with $C_G(v_p)$ a p'-subgroup. Then there exists an vector $v \in V$ such that $C_G(v) = 1$.*

About the number of p-blocks with defect zero, H. Blau and G. Michler obtained the following result.

Theorem ([5]) *Let G be a finite group with a T.I. Sylow p-subgroup P. Then the number of p-blocks with defect zero is equal to the difference of the number of the conjugacy classes of G and the one of $N_G(P)$.*

Brauer gave a perfect result for the case that D is a Sylow p-subgroup of G.

Theorem *Let D be a Sylow p-subgroup of G. Then the number of* *p-blocks with D as a defect group is equal to the number of p'-conjugacy classes of G with D as a defect group.*

Using the p-power homomorphism S.M. Shi proved in [33] the following theorem, which is a generalization of the above theorem of Brauer.

Theorem 1.9   *Let G be a finite group and D be a Sylow p-subgroup of $DC_G(D)$. For any p-subgroup D_1 of G, if D is conjugate to a subgroup of D_1 (we denote this by $D \overset{<}{G} D_1$), then the number of p-blocks with D_1 as a defect group is equal to the number of p'-classes of G with D_1 as a defect group.*

In [35], such a p-subgroup D is called a constrained p-subgroup. In fact D is a constrained p-subgroup if and only if any p-element of G which commutes with all the elements in D is contained in D. It is obvious that if D is a constrained p-subgroup and $D \overset{<}{G} P_o$, so is P_0. Besides the Sylow p-subgroups of G, the typical examples of constrained p-subgroups are the Sylow p-subgroups of $O_{p'p}(G)$, where G is a p-constrained group. [35] proved the following theorems.

Theorem 1.10   *Let G be a p-constrained group and $Q \in S_p(O_{p'p}(G))$. Then Q is a constrained p-subgroup. For any p-subgroup P_0 which satisfies $Q \overset{<}{G} P_0$, the number of p-blocks of G with P_0 as a defact group is equal to the number of p'-classes of G with P_o as a defect group.*

Theorem 1.11   *Let G be a p-constrained group. Q and P_0 are as in Theorem 1.10. Then the number of p-blocks of G with P_0 as a defect group is equal to the number of p'-classes of G with P_0 as a defect group which are contained in $O_{p'}(G)$.*

Corollary (P. Fong)   *Let G be a p-constrained group and $O_{p'}(G) = 1$. Then G has only the principal block.*

Generally speaking, it is difficult to give the number of p-blocks with a given defect group. We turn to study the existence of such p-blocks.

A maximal Sylow p-intersection D of a finite group G is a maximal element in the set $\{P \cap P^x \mid x \in G \setminus N_G(P)\}$, where $P \in S_p(G)$. A p-subgroup D of a finite group G is called cocyclic if for all $P \in S_p(N_G(D))$, P/D is cyclic. [34] proved the following theorems.

Theorem 1.12   *Let D be a p-subgroup of G such that D is not a Sylow p-subgroup of $DC_G(D)$ and is cocyclic. Then G has a p-block with D as a defect group if and only if $D = O_p(DC_G(D))$.*

Theorem 1.13   *Let D be a p-subgroup of G such that D is not a Sylow p-subgroup of $DC_G(D)$ and is a maximal Sylow p-intersection of G.*

Then the following are equivalent,

(i) G *has a p-block with D as a defect group,*

(ii) $D = O_p(DC_G(D))$,

(iii) $DC_G(D)$ *is not p-closed.*

Theorem 1.14 *Let P_1 be a maximal subgroup of a Sylow p-subgroup P of G and P_1 is not a Sylow p-subgroup of $P_1C_G(P_1)$. Then the following are equivalent,*

(i) G *has a p-block with P_1 as a defect group,*

(ii) *There exists an element x of G and a Sylow p-subgroup P such that $P^x \cap P = P_1$,*

(iii) P *is not normal in $N_G(P_1)$.*

J.P. Zhang introduces the following concept. A p-subgroup D of G is called strong radical if $N_G(D)/D$ has a strongly p-embedded subgroup ([46]). He solved the problem when a strongly radical p-subgroup is a defect group for some p-block, which generalizes the above results.

Theorem 1.15 *Let G be a finite group. Then a strong radical p-subgroup D of G is a defect group for some p-block of G if and only if there exists a p-regular element x in G such that D is a Sylow p-subgroup of $C_G(x)$.*

Theorem 1.16 *If D is a strong radical p-subgroup of a finite group G such that D is not a Sylow p-subgroup of $DC_G(D)$ then D is a defect group for some p-block of G.*

The following corollary to the two theorems above is really surprising.

Corollary 1.17 *Let G be a finite group with an abelian Sylow p-subgroup. Then every strong radical p-subgroup of G is a defect group for some p-block of G. In particular every maximal Sylow p-intersection of G is a defect group for some p-block of G.*

The following result was conjectured by G. Robinson.

Theorem 1.18 *Suppose G is a finite p-solvable group such that $O_p(G)$ is a strong radical p-subgroup of G. Then for any p-subgroup D of G, D is a defect group for some p-block of G if and only if there exists a p-regular element x in $O_{p'}(G)$ such that $D \in S_p(C_G(x))$.*

As is well-known, a finite group G has at least one p-block, the principal block. G may have only one p-block and a group theoretical characterization of such groups G remained an open problem (posed by Brauer) for some years. Until 1984, M. Harris solved this problem in [21]. The principal block is of course of the highest defect.

It may be the case that a finite group G has only p-blocks of highest defect. A finite group G is said to be full p-defective if all p-blocks of G have Sylow p-subgroups of G as defect groups. The group-theorectical characterization of full p-defective groups is a much studied problem. Harada [20] (1967) proved that full 2-defective groups with an abelian Sylow 2-subgroup P are 2-closed, i.e. P is normal in G. Herzog (1976) [22] improved Harada's result by weaken the assumption on Sylow 2-subgroups. Further results can be found in Tsushima [39] (1977).

J.P. Zhang in [47] succeeded in proving the following

Theorem 1.19 *Let G be an arbitrary finite group, P a Sylow p-subgroup of G, and $M = O_{p'}(C_G(P))$. Then G is full p-defective if and only if the following two statements hold:*

(i) $O_{p'}(G) = \cup_{x \in G} M^x$,

(ii) $F^*(G/O_{p'}(G)) = O_p(G)O_{p'}(G)/O_{p'}(G)$, *and for $p = 2$ we may have besides*

$$F^*(G/O_{2'}(G)) = (N \times O_{2'}(G))/O_{2'}(G),$$

where N is a normal subgroup of G such that all components of N are of type either M_{22} or M_{24}.

Corollary 1.20 *If G is a finite group with an Abelian Sylow p-subgroup P. Then G is full p-defective if and only if P is normal in G.*

The corollary generalizes the main result of Harada [20].

In [47], J.P. Zhang proved also Theorem 1.2 independently.

For further details or generalizations on full p-defective groups we refer the reader to [47] and [29].

We generalize the concept of full defect to the following one. Let P_0 be a p-subgroup of G. If the defect groups of all the p-blocks of G contain a conjugacy of P_0, we call G a P_0-defective group. If an element y of G has the property $P_0 \overset{<}{G} C_G(y)$, we call y P_0-defective.

[35] proved the following result.

Theorem 1.21 *Let G be p-constrained and Q, P_0 be as in Theorem 1.10. Then G is P_0-defective if and only if $O_{p'}(G)$ consists of P_0-defective elements. In this case, $O_{p'p}(G) = O_p(G) \times O_{p'}(G)$. In particular, $O_p(G) \neq 1$.*

It generalizes a corollary of [39].

§2. Finite linear groups

The study of linear groups of small degree over a field dates at least from Jordan (for the field C of complex numbers) and from Dickson (for finite fields). The investigation of linear groups of degree $d < p$ over C may be traced through Blichfeld [4], Brauer [11], Tuan [40], Feit-Thompson [17] and Feit [13], [15], among others. Brauer noted the great significance of linear groups G of small degree over a given field ([8], Problem 39 and 40) and Feit [14] provided some key methods for their analysis when G has a Sylow p-subgroup of order p.

The classification of the finite simple groups has been used over the past decade to solve previously intractable problems in linear groups. J.P. Zhang [49] characterized the finite complex linear groups of degree $d \leq p-1$, which not only extended the most of previous results but also, more importantly, offered an affirmative answer to Problem 39 of Brauer in a more general way. J.P. Zhang [50] determined the finite complex linear groups G of degree $d \leq |P| - 1$ with the assumption that G has a cyclic or T.I. Sylow p-subgroup P, which extended an important result of H. Blau [2]. In [51], J.P. Zhang was able to classify the finite linear groups G in characteristic p with degree $d < p - 2$ under the assumption that G has an abelian Sylow p-subgroup P, and in a joint work with H. Blau, he extended the result to the groups of degree $d < p$ and without any restriction on the structure of P, which solved essentially Problem 40 of Brauer.

Now we go into details.

Theorem 2.1 ([49]) *Let G be a finite complex linear group of degree d at least 7 and with a nontrivial Sylow p-subgroup P of order p^a. If $d \leq p-1$ then there exists a normal subgroup H in G with $C_G(H) \leq H$ such that one of the following conditions occurs.*

(1) $H = P(C_G(P)$, *that is, P is normal in G.*

(2) $H = Q_0 Z(G)$, $p = 1 + 2^m$, *where Q_0 is an extra-special 2-group of order 2^{2m+1}, $Q_0 O_2(Z(G)) \in Syl_2(O_{p'}(G))$, G is irreducible of degree $d = p - 1$ and $|P/O_p^{\cdot}(Z(G))| = p$.*

(3) $H = H_0 M$, *where $M = O_p(G) \times O_{p'}(G)$, H_0 is a normal subgroup of G with $H_0/Z(H_0) \cong PSL(2,p)$, $PSL(2,p) \times PSL(2,p)$ or $p = 7$ with $H_0/Z(H_0) \cong A_7$.*

(4) $H/Z(H) \cong SL(2, 2^m)$, $p = 1 + 2^m$.

(5) G is irreducible of degree $d = p-1$, and $H/Z(H) \cong PSL(2,q)$, $p = (q+1)/2$; $PSL(n,q), p = (q^n-1)/(q-1)$; $PSU(n,q), p = (q^n+1)/(q+1)$; $PSp(2n,q), p = (q^n+1)/2$; $^2G_2(q), p = q^2-q+1$ or A_p.

(6) G is irreducible of degree $d = p-1$, and $H/Z(H) \cong PSL(3,4)$, $PSU(4,3)$ or J_2 for $p = 7$; M_{11}, M_{12} or M_{22} for $p = 11$; $G_2(4)$ or Suz for $p = 13$; J_3 for $p = 19$; M_{23} for $p = 23$; Ru for $p = 29$ or A_p.

Note that together with the list of finite complex linear groups of degree at most 6 (cf. Tuan [40] and Blau [2]), Theorem 2.1 determines all finite complex linear groups of degree at most $p-1$ provided $p||G|$. The following corollary generalizes a result of Feit [13].

Corollary 2.2 ([49]) *Suppose that G is a finite complex linear group with a non-normal Sylow p-subgroup P, where p is a prime at least 7. If G has degree $p-2$ then $G \cong SL(2,2^m) \times Z(G)$ with $p = 1 + 2^m$.*

The following theorem offers more than what was asked by Brauer in his Problem 39 [8].

Theorem 2.3 ([49]) *Let G be a finite complex linear group of degree $m > 5$. If G has a nontrivial Sylow p-subgroup P with $p > m+1$, then P is normal in G unless $G \cong SL(2,p-1) \times Z(G)$ or $G = O_{pp'}(G)M$, where M is a normal subgroup of G such that $M/Z(M) \cong PSL(2,p)$.*

By assuming some conditions on Sylow p-subgroups, the above results can be generalized.

Theorem 2.4 ([50]) *Suppose that G is a finite group having a T.I. Sylow p-subgroup P of degree at least 8. Set $H = O^{p'}(G)$. If G has a faithful complex representation ρ of degree at most $|P|-1$ then one of the following is true:*

(1) P is normal in G.

(2) P is generalized quaternion, G is solvable, and $\rho(1) \geq |P|/2 - 1$.

(3) P is cyclic; ρ is irreducible of degree $|P|-1$; $O_{p'}(G) = QZ(G)$, where Q is an extra-special q-subgroup of order q^{2r+1} and $Z(Q) = Z(G) \cap Q$; and $|P| = 1 + q^r$ (hence $r = 1$ if $p = 2$ and $|P| = p$ or 9 if $q = 2$). Furthermore, if G is not solvable then $q = 2$, $|P| = 1+2^r$ and $H/O_{p'}(H) \cong PSp(2n,2^m)$, $PSO^-(2n,2^m)'$ or $PSL(2,p)$, where $mn = r$.

(4) $|P| = p = 1 + 2^n$, $n \geq 3$; $G \cong SL(2,2^n) \times Z(G)$ and $\rho(1) \geq p-2$.

(5) ρ is irreducible of degree $|P|-1$; $H/Z(H) \cong PSL(3,4)$ if $|P| = 9$ or $Aut(S_Z(32))$ if $|P| = 125$.

(6) ρ is irreducible of degree $|P|-1$; P is cyclic; $H/Z(H) \cong PSL(n,q)$ with $n \geq 3$ and $|P| = (q^n-1)/(q-1)$, $PSU(n,q)$ with n odd and $|P| =$

$(q^n + 1)/(q + 1)$ or $PSp(2n, q)$ with $|P| = (q^n + 1)/2$.

(7) $H/Z(H) \cong PSL(2, q)$ with $|P| = (q + 1)/(2, q - 1)$, q or $q + 1$ and $\rho(1) \geq (q - 1)/(2, q - 1)$; $PSU(3, q)$ with $|P| = q^3$ and $q \geq 3$ and $\rho(1) \geq q(q - 1)$; $^2G_2(q)$ with $|P| = q^3$ and $\rho(1) \geq q^2 - q + 1$; $^2B_2(q)$ with $|P| = q^2$ and $\rho(1) \geq (q - 1)\sqrt{2q}$ or M_C with $|P| = 125$ and $\rho(1) \geq 22$.

(8) $|P| = p$, ρ is irreducible of degree $p - 1$; $H/Z(H) \cong M_{11}$, M_{12} or M_{22} for $p = 11$; $G_2(4)$ or Suz for $p = 13$; J_3 for $p = 19$; M_{23} for $p = 23$; Ru for $p = 29$ or A_p.

Theorems 2.1 and 2.4 depend on the classification of the finite simple groups, but one of the main difficulties in the proofs occurs in p-solvable case. For the case a deep group-theoretical result due to R. Braner [9] is needed, and it seems that the result does not receive much attention.

The following result extended the main result of Michler and Landrock [26] on Leonard conjecture.

Theorem 2.5   ([48]) *Let G be a finite group with a T.I. Sylow p-subgroup P. If G has a faithful complex character χ such that $\chi(1) \geq \sqrt[3]{|P|}(\sqrt[3]{|P|} - 1) - 1$, then P is normal in G.*

Finite linear groups of small degree in characteristic p are more difficult to determine. In [3] Blau classified the linear groups in characteristic p with degree $d \leq |P| - 2$ under the assumption of P cyclic. In [51], J.P. Zhang proved the following very concise result.

Theorem 2.6   ([51]) *Suppose that G is a finite linear group in characteristic $p(> 5)$ with a nontrivial abelian Sylow p-subgroup P. Set $H = O^{p'}(G)$. If G has degree $d < p - 2$ then one of the following holds:*

(a) *P is normal in G,*

(b) *$H/Z(H) \cong \oplus_{i \leq t} PSL(2, p^{n_i})$, where n_i and t are positive integers with $2t < p - 2$,*

(c) *$p = 7$ or 11 and $H \cong 2 \cdot A_7$ or J_1, respectively, and $d \geq p - 4$.*

In [6], J.P. Zhang and H. Blau generalize the above theorem by raising the bound on the degree from $p - 3$ to $p - 1$ and discarding the assumption on P, which solves essentially the Problem 40 of R. Brauer.

Theorem 2.7   ([6]) *Let G be a finite group and F, a field of characteristic p. Suppose that there is a faithful FG-module V with $\dim V = d < p$. Set $\overline{G} = G/O_p(G)$. Assume that $G = O^{p'}(G)$ and $P \in Syl_p(G)$. Then one of the following must be true:*

(a) *$G = P$.*

(b) *$|P| = p = 1 + 2^n$ for some n; $O_{p'}(G) = RZ(G)$ where R is a*

Sylow 2-subgroup of $O_{p'}(G)$; $R/Z(R)$ is an elementary abelian 2-group of order 2^{2n}; $Z(R) = R \cap Z(G)$; P acts regularly and irreducibly on $R/Z(R)$; $C_G(P) = P \times Z(G)$; $R = [P,R]Z(R)$ with $[P,R]$ extra-special of order 2^{1+2n} : $\dim V = p-1$ and $V|_{[P,R]}$ is absolutely irreducible; $G/O_{p'}(G) \cong P$, $PSp(2a, 2^b)$ or $PSO^-(2a, 2^b)'$ where $ab = n$, $PSL(2, 17)$ where $p = 17$, A_5 or A_6 where $p = 5$.

    (c) *$|P| = p$, $C_G(P) = P \times Z(G)$, $\dim V = p-1$, V is absolutely irreducible, and $G/Z(G) \cong PSL(2, q)(q \text{ odd})$, $p = (q+1)/2$; $PSU(n, q)$ (n odd prime, $n \nmid (q+1)$), $p = (q^n + 1)/(q+1)$; $PSp(2n, q)$ ($n = 2^a$ for some $a > 0$, q odd), $p = (q^n + 1)/2$; A_5, $p = 3$; A_7, $p = 5$; $PSL(3, 4)$, $PSU((4, 9)$, J_2 or A_7, $p = 7$; M_{11}, M_{12} or M_{22}, $p = 11$; Suz or $G_2(4)$, $p = 13$; J_3, $p = 19$; Ru, $p = 29$.*

    (d) *$|P| = p$, $C_G(P) = P \times Z(G)$, $\dim V = p-1$, V is absolutely irreducible. and $G/Z(G) \cong A_n$ (n prime, $n \geq 7$),$p = n$; $PSL(n, q)$ (n prime, $n \nmid (q-1)$), $p = (q^n - 1)/(q-1)$; A_6 or A_7, $p = 5$, M_{11}, $p = 11$; or M_{23}, $p = 23$.*

    (e) *V has two irreducible constituents: the trivial module V_0 of dimension 1 and, say, W of dimension $p-2$; $\ker W = O_p(G)$ and $\overline{G}$ satisfies conclusion (d).*

    (f) *$p = 7$ (rep. $p = 11$), V has an absolutely irreducible constituent W of dimension 4 (rep. dimension 7), $\ker W = O_p(G)$, all other constituents (if any) are isomorphic to V_0 and $\overline{G}/Z(\overline{G}) \cong A_7$ (resp. J_1).*

    (g) *$p = 7$ (resp. $p = 11$), V has absolutely irreducible constituents of dimension 4 (resp. 7 and 2) and has no other irreducible constituents (unless $p = 11$ and V_0 appears), and for some a, $\overline{G}/(Z(\overline{G}) \cong A_7 \times PSL(2, 7^a)$ (resp. $J_1 \times PSL(2, 11^a)$.*

    (h) *$p = 11$, V has absolutely irreducible constituents of dimensions 7 and 3, and for some a, $\overline{G}/Z(\overline{G}) \cong J_1 \times PSL(2, 11^a)$, $J_1 \times PSL(3, 11^a)$ or $J_1 \times PSU(3, 11^a)$.*

    (i) *$\overline{G}/Z(\overline{G}) \cong \oplus_{i \leq t} L_i$ (a direct product), where L_i is a simple group of Lie type defined over a field of characteristic p, with $\sum_{i \leq t} R_p(L_i) \leq d < p$, where $R_p(L_i)$ is the minimal integer s such that L_i is embedded in $PGL(s, F')$ for some field F' of characteristic p.*

§3. The number of irreducible characters in a p-block

In 1956, R. Brauer posed a conjecture that the number $k(B)$ of ordinary irreducible characters of a finite group G in a p-block B is bounded by the order $|D|$ of a defect group D of B. ([10]). It is called the Brauer's $k(B)$-conjecture. In [8] Brauer put it as problem 20 and as problem 10 by Feit in [16].

This problem is very difficult. Under specific conditions on G, one can verify the conjecture. When G is a finite group with a T.I. Sylow p-subgroup P and P is not normal in G, H. Blau and G. Michler proved that for a p-block B with P as a defect group, the $k(B)$-conjecture is true ([5]). For G being the general linear groups, the unitary groups and the symmetric groups, G. Michler and J.B. Olsson verified the same assertion.

When the defect group D is not complicated, one can also verify the conjecture. For example, for the cases, $|D| = p^d$ and $d \leq 2$, or $|D| \leq 8$, or D is cyclic, the $k(B)$ conjecture is true.

In general case, Brauer and Feit proved that $k(B) \leq \frac{1}{4}|D|^2 + 1$.

In the case of p-solvable groups, Nagao used the result of Fong ([18]) to reduce the problem to the following question:

Let V be an elementary abelian finite p-group on which a p'-group G acts faithfully and irreducibly. Is it then true that the number $k(GV)$ of conjugacy classes of the semidirect product $G \ltimes V$ is bounded by the order $|V|$ of V? (i.e. $k(GV) \leq |V|$?) We called it the $k(GV)$-problem.

R. Knörr ([24]) proved that $k(GV) \leq |V|$ when G is supersolvable, a much better result than any obtained previously. More importantly, he introduced some highly effective and original method. Later D. Gluck ([19]) used his method to prove that $k(GV) \leq |V|$, when G is of odd order.

In [23], Y.M. Huang used the same methods and gave the following theorem which generalizes the result of Gluck.

Theorem 3.1 *Let G be a finite p'-group and $4 \nmid |G|$. If G acts faithfully and irreducibly on an elementary abelian finite p-group V, then $k(GV) \leq |V|$.*

But still using the methods of Knörr and Gluck, a further improvement to this theorem is very difficult. We turn to a new course. In [36], we obtained a formula to compute $k(GV)$.

Theorem 3.2 *Let V be an elementary abelian finite p-group and G be a p'-group of linear transformations on V. We denote by $\Pi_V(a_1, a_2)$ the number of the common fixed points of a_1 and a_2 on V for all $a_1, a_2 \in G$. then*

$$k(GV) = \frac{1}{|G|} \sum_{\substack{a_1, a_2 \in G \\ a_1 a_2 = a_2 a_1}} \Pi_V(a_1, a_2).$$

By using this formula we can easily prove $k(GV) \leq |V|$ if G is Abelian. We also get an interesting property of the maximal order of the Abelian p'-groups of linear transformations on V.

Corollary 3.3 *The maximal order of the Abelian p'-groups of linear transformations on V is $|V| - 1$.*

Under S.M. Shi's supervision, L.D. Xu proved the following result using the table of ordinary irreducible characters of $SL(2, F_0)$.

Theorem 3.4 ([42]) *Let G be $SL(2, F_0)$, where F_0 is a finite field of q elements and $q = p_0^n$, for a prime p_0. Let F be a finite splitting field of G and its subgroups with characteristic p such that $(p, |G|) = 1$. If V is a finite dimensional F-vector space which offers a faithful and irreducible representation of G, then $k(GV) \leq |V|$.*

Furthermore, S.M. Shi proved the following theorem without the restriction that F is a splitting field.

Theorem 3.5 ([38]) *Let G be $SL(2, F_0)$, where F_0 is a finite field of q elements and $q = p_0^n$, for a prime p_0. Let V be a finite dimensional F-vector space such that $(char\, F, |G|) = 1$. If V offers a faithful and irreducible representation of G, then $k(GV) \leq |V|$.*

"What is the number $l(B)$ of irreducible modular characters in a p-block B of G with a given p-subgroup D as a defect group?" The problem is also studied. In [32], G.R. Robinson gave a computing formula for the number $l(B)$ and in [37] S.M. Shi gave independently a different formula.

In [7], Brauer proved that if a defect group D of a block ideal B of the group algebra kG with a splitting field k of characteristic p is Abelian and the inertial index is 1, then $k(B) = |D|$ and $l(B) = 1$. In [25], Külshammer proved that if B satisfies the above condition then $B \cong Mat_m(kD)$, for a positive integer m. G. X. Zhang generalizes Kïshammer's result. In [52], he proved the following

Theorem 3.6 *Let G be a finite group and k a splitting field of G and all subgroups of G with characteristic p. Let B be a block ideal of kG with an abelian defect group D. Suppose that there is only one irreducible modular character in B and its dimension is m. Then $B \cong Mat_m(kD)$.*

References

[1] J.L. Alperin, *Weights for finite groups*, Proc. Symposia in Pure Math., 47 (1987).

[2] H.I. Blau, *On linear groups with a cyclic or T.I. Sylow subgroup*, J. Algebra, 114 (1988).

[3] H.I. Blau, *Minimal modular character degrees with a cyclic Sylow subgroup*, J. Algebra 131 (1990).

[4] H.F. Blichfeldt, "Finite Collineation Groups", University of Chicago Press, Chicago, 1917.

[5] H. Blau and G. Michler, *Modular representation theory of finite groups with T.I. Sylow p-subgroups*, Trans. Amer. Math. Soc., 319 (1990).

[6] H. Blau and J.P. Zhang, *Linear groups of small degree over fields of finite characteristic*, to appear in J. Algebra.

[7] R. Brauer, *Some applications of the theory of blocks of characters of finite groups*, IV, J. Algebra, 17 (1971).

[8] R. Brauer, *Representations of finite groups*, in "Lectures on Modern Mathematics", Vol.1, John Wiley & Sons, New York, 1963.

[9] R. Brauer, *On finite projective groups*, in "Contributions to Algebra: A Collection of Papers Dedicated to Ellis Kolchin", Academic Press, New York, 1977.

[10] R. Brauer, *Number theoretical investigations on groups of finite order*, Proceedings of the International Symposium on Algebraic Number Theory, Tokyo, 1956.

[11] R. Brauer, *On groups whose order contains a prime number to the first power*, I, II, Amer. J. Math., 64 (1942).

[12] R. Brauer and W. Feit, *On the number of irreducible characters of finite groups in a given block*. Proc. Nat. Acad. Sci. USA, Vol., 45 (1959).

[13] W. Feit, *Groups which have a faithful representation of degree less than $p-1$*, Trans. Amer. Math. Soc., 112 (1964).

[14] W. Feit, *Groups with a cyclic Sylow subgroup*, Nagoya Math. J., 27 (1966).

[15] W. Feit, *On finite linear groups*, I, J. Algebra 5 (1967), II, J., Algebra 30 (1974).

[16] W. Feit, "The Representation Theory of Finite Groups", North-Holland, 1974.

[17] W. Feit and J.G. Thompson, *Groups which have a faithful representation of degree less than $(p-1)/2$*, Pacific J. Math., 11 (1961).

[18] P. Fong, *On the characters of p-solvable groups*, Trans. Amer. Math. Soc., 98 (1961).

[19] D. Gluck, *On the $k(GV)$-problem*, J. Algebra, 89 (1984).

[20] K. Harada, *On groups all of whose 2-blocks have the highest defects*, Nagoya Math. J., 33 (1967).

[21] M. Harris, *On the p-deficiency class of a finite group*, J. Algebra 94 (1985).

[22] M. Herzog, *On groups with extremal blocks*, Bull. Austral. Math. Soc., 14 (1976).

[23] Y.M. Huang, *On Brauer's $k(B)$-conjecture (in Chinese)*, Acta Math. Sinica, 36 (4) (1993).

[24] R. Knörr, *On the number of characters in a p-block of a p-solvable group*, Illinois J. Math., 28 (2) (1984).

[25] B. Külshammer, *On the structure of block ideals in group algebras of finite groups*, Comm. in Algebra, 8:19 (1980).

[26] P. Landrock and G. Michler, *Minimal degrees of faithful characters of finite groups with a T.I. Sylow p-subgroup*, Proc. Amer. Math. Soc., 99 (1987).

[27] G. Michler, *Modular representation theory and the classification of finite simple groups*, Proc. Symposia in Pure Math., 47 (1987).

[28] H. Nagao, *On a conjecture of Brauer for p-solvable groups*, J. Math. Osaka City Univ., 13 (1962).

[29] G. Pazderski and G. Tiedt, *On groups with extremal p-blocks*, Arch. Math., 57 (1991).

[30] G.R. Robinson, *The number of blocks with a given defect group*, J. Algebra, 84 (1983).

[31] G.R. Robinson, *The Frobinius-Schur indicator and projective modules*, J. Algebra, 126 (1989).

[32] G.R. Robinson, *Counting modular irreducible characters*, J. Algebra, 90 (1984).

[33] S.M. Shi, *On the number of p-blocks with given defect groups and some applications of the p-power homomorphisms*, Lecture Notes in Math., 1185 (1986).

[34] S.M. Shi, *Some results on defect groups*, J. Algebra, 142 (1991).

[35] S.M. Shi, *The constrained p-subgroups and some results on defect groups of p-constrained groups*, Comm. in Algebra, 20 (3) (1992).

[36] S.M. Shi, Y. Fan and H.L. Ling, *On the number of conjugacy classes in semidirect products of certain finite groups (abstract)*, Advances in Mathematics (China), 21 (3) (1992).

[37] S.M. Shi, *The number of irreducible modular characters lying in a p-block (in Chinese)*, Dongbei Shuxue, 4 (1987).

[38] S.M. Shi and L.D. Xu, *A result on k(GV)-problem*, in preparation.

[39] Y. Tsushima, *On the weakly regular p-blocks with respect to $O_{p'}(G)$*, Osaka J. Math., 14 (1977).

[40] H.F. Tuan, *On groups whose orders contain a prime number to the first power*, Ann. of Math., 45 (1944).

[41] W. Willems, *Blocks of defect zero in finite simple groups of Lie type*, J. algebra, 113 (1988).

[42] L.D. Xu, *A result on k(GV)-problem*, Master Thesis, University of Science and Technology, Beijing, 1993.

[43] J.P. Zhang, *A condition for the existence of p-blocks of defect zero*, Proc. Symposia in Pure Math., 47 (1987).

[44] J.P. Zhang, *On the existence of p-blocks of defect zero (in Chinese)*, Acta Math. Sinica, 30 (1987).

[45] J.P. Zhang, *p-Regular orbits and p-blocks of defect zero*, Comm. in Algebra, 20 (1993).

[46] J.P. Zhang, *Studies on defect groups*, to appear in J. Algebra.

[47] J.P. Zhang, *On finite groups all of whose p-blocks are of the highest defect*, J. Algebra, 118 (1989).

[48] J.P. Zhang, *Minimal degrees of faithful representations of finite groups with a T.I. Sylow p-subgroup (in Chinese)*, Acta Math. Sinica, 31 (1989).

[49] J.P. Zhang, *Complex linear groups of degree at most p − 1*, Contemp. Math., 82 (1989).

[50] J.P. Zhang, *On linear groups of degree at most $|P| - 1$*, to appear.

[51] J.P. Zhang, *On linear groups over finite fields*, Proc. Amer. Math. Soc., 110 (1990).

[52] L.W. Zhang, *Some studies on block theory (in Chinese)*, Doctoral Thesis, Peking University 1987.

[53] G.X. Zhang, *On the block algebras having only one irreducible module*. Chinese Ann. of Math., 14 B (1993).

The Quantitative Structure of Groups and Related Topics

W. J. Shi*

Department of Mathematics, Southwest Teachers University
Beibei, Chongqing, 630715, P. R. China

The concepts of the order of a group and its element orders are the most fundamental in group theory. But they play an important role in the quantitative structure of groups. For example, Sylow's theorems and Lagrange's theorem are very well-known in the groups of finite order.

In 1902 W. Burnside put forward the celebrated problem [15]: Let G be a finitely generated group. If for every $a \in \pi_e(G)$, where $\pi_e(G)$ denotes the set of element orders in G, a is finite, whether G is necessarily finite? Although the problem of Burnside had been given the negative solution [25], it stressed the role of the orders of elements in the structure of groups.

In 1937 B.H.Neumann discussed such groups G with $\pi_e(G) = \{1, 2, 3\}$ [35]. After 20 years G. Higman researched the finite groups G with $\pi_e(G) = \{p^n, q^m, \cdots\}$ [30], where $p, q, \cdots$ are primes. For convenience, we call such groups finite EPPO-groups. The main result in [30] is that if G is a finite solvable EPPO-groups then $|\pi(G)| \leq 2$, where $\pi(G)$ denotes the set of all prime divisors of the order $|G|$ of G

In 1962 M. Suzuki classified all finite simple EPPO-groups [68]. They are the following groups: $L_2(q), q = 5, 7, 8, 9$ or 17, $L_3(4)$, $Sz(8)$ and $Sz(32)$.

1. On element orders

In [7] and [64], the authors go further into the finite EPPO-groups.

*The author gratefully acknowledges the support of K. C. Wong Education Foundation, Hong Kong.

Considering the special case of EPPO-groups, the authors of [63] and [24] discussed finite EPO-groups, i.e. the groups in which every nontrivial element has prime order. We prove the following conclusion [63].

Theorem 1.1 *Let G be a finite EPO-group. The one of the following cases occurs:*

(1) *G is a p-group of exponent p;*

(2) *$G = Q \ltimes P$, where the Sylow p-subgroup P is normal in G and the Sylow q-subgroup Q is of order q ($q \neq p$). Moreover $G(|G| = p^n q)$ has the chief factors $q, p^b, \cdots, p^b$, $n = kb$, where b is the exponent of $p(\mathrm{mod}\, q)$, and the class of P is not more than k; or*

(3) *$G \cong A_5$.*

Question 1.2 *What are the infinite EPO-groups and the infinite EPPO-groups? (For the locally finite EPPO-groups we may refer to [66])*

A finite EPO p-group is a regular p-group, and so p-abelian. But it is more difficult to determine all isomorphic classes of EPO p-groups of order p^m for $m \geq 8$ (see [77]).

We derive easily a pure quantitative characterization of A_5 from Theorem 1.1 (3). (also see [39])

Corollary 1.3 *Let G be a finite group. Then $G \cong A_5$ if and only if $\pi_e(G) = \{1, 2, 3, 5\}$.*

In Sec.5 we will discuss the similar characterization for many other simple groups. Denote $\pi_e(G) = \{1\} \cup \pi'_e(G) \cup \pi''_e(G)$, where $\pi'_e(G)$ is the subset of prime number and $\pi''_e(G)$ is the subset of composite number in $\pi_e(G)$.

If G is finite and $\pi''_e(G) = \phi$, then G is an finite EPO-group. For $|\pi''_e(G)| = 1$, we have the following theorems [62].

Theorem 1.4 *Let G be a finite simple group with $|\pi''_e(G)| \leq 1$. Then G is one of the following groups:*

(1) *Z_p, p prime,*

(2) *$L_2(q)$, $q = 5, 7, 8, 9, 11, 13$ or 16,*

(3) *$L_3(4), Sz(8)$,*

(4) *$L_2(3^n)$, where $(3^n - 1)/2$ and $(3^n + 1)/4$ are primes, or*

(5) *$L_2(2^n)$, where $2^n - 1$ and $(2^n + 1)/3$ are primes.*

Theorem 1.5 *Let G be a finite nonsolvable group with $|\pi''_e(G)| \leq 1$. Then one of the following holds:*

(1) *G is one of the groups listed in Theorem 1.4, or*

(2) *$G/N \cong A_5$, where N is an elementary Abelian 2-group, and $\pi_e(G) =$*

$\{1, 2, 3, 4, 5\}$.

Corollary 1.6 *Let G be a finite group and $|\pi_e''(G)| = k$. If $k = 0$ then $|\pi(G)| \leq 3$, and if $|\pi(G)| = 3$ then G is simple. If $k = 1$ then $|\pi(G)| \leq 4$, and if $|\pi(G)| = 4$ then G is simple.*

Question 1.7 *Let G be a finite group. Whether there exists a general relation between $|\pi(G)|, |\pi_e''(G)|$ and the simplicity of G?*

A C_{22} group (or CIT-group) is of even order and the centralizer of any involution is a 2-group [69]. Let $|G|$ be even and $2 \nmid a$ for all $a \in \pi_e''(G)$. For such special C_{22} groups we proved the following results [48].

Theorem 1.8 *Let G be a finite group. If $2 \nmid a$ for all $a \in \pi_e''(G)$, then G is one of the following groups:*

(1) *G is of odd order,*

(2) *$G = P_2O(G)$, where $|P_2| = 2$ and $O(G)$ is abelian,*

(3) *$G = H \ltimes P_2$, where $P_2 \in Syl_2G$, P_2 is elementary abelian, and G is a Frobenius group with Frobenius kernel P_2 and complement H, or*

(4) *$G \cong L_2(2^m)$, $m \geq 2$.*

Corollary 1.9 *Let G be a finite group satisfying the condition of Theorem 1.8. If $4 \mid |G|$ and G is not 2-closed, the $G \cong L_2(2^m)$, $m \geq 2$.*

Corollary 1.10 *Let G be a finite group. Then $G \cong L_2(2^m)$, $m \geq 2$, if and only if $\pi_e(G) = \pi_e(L_2(2^m))$, $m \geq 2$.*

Similarly we showed the following theorem by using the result of C_{33} groups [3].

Theorem 1.11 *[55] Let G be a finite group. If $3 \nmid a$ for all $a \in \pi_e''(G)$, then G is one of the following groups:*

(1) *G is a 3'-group,*

(2) *$G = Z_3 \ltimes N$ or $G = S3 \ltimes N$, where N is a nilpotent normal 3'-subgroup of G,*

(3) *$G = H \ltimes P_3$, where $P_3 \in Syl_3G$, $exp(P_3) = 3$ and G is a Frobenius group with kernel P_3 and complement H,*

(4) *$G/H \cong L_2(5)$, where N is a normal elementary abelian 2-subgroup of G,*

(5) *G is simple and $G \cong L_2(5), L_2(7), L_3(4)$ or $L_2(3^n)$, $n > 1$, or*

(6) *G has a normal subgroup N, $N \cong L_2(3^n)$, $n > 1$ and $|G : N| = 2$.*

Corollary 1.12 *[55] Let G be a finite group. Then $G \cong L_2(3^m)$, $m > 2$, if and only if $\pi_e(G) = \pi_e(L_2(3^m))$, $m > 2$.*

Searching for the converse of the well-known Dickson's theorem, it was proved that the above conclusion is also true for $L_2(q)$, q odd and $q \neq 9$

(see Sec.4).

Question 1.13 *Suppose G is a finite group and $p \nmid a$ (p is a given prime, $p \neq 2,3$) for all $a \in \pi_e''(G)$, what is the structure of G?*

In Sec.6 we will discuss such groups whose $\pi_e(G)$ takes some special numbers.

2. On orders of simple groups

As early as 1940s R. Brauer and H.F. Tuan characterized some projective special linear groups using the conditions of the group order $|G|$ and the simplicity of G [13]. Especially they proved the following theorem:

Theorem 2.1 *Let G be a simple group of order prq^b , where p, q and r are three different primes, and b is a positive integer. Then $G \cong L_2(5)$ or $L_2(7)$.*

Then many mathematicians characterized many simple groups from various point of view. For example some mathematicians characterized the simple groups G using the conditions of $|G|$ and the simplicity of G (see [65], [36] and [14]). Some characterization using the decomposition of prime factors of $|G|$ and the simplicity we refer to [11], [28], [1], [2], [70] and [71].

Using the number $|\pi(G)|$ we have the following result [29].

Theorem 2.2 *Let G be a finite simple group. If $|\pi(G)| = 3$, then G is isomorphic to one of the following groups: $A_5, L_2(7), L_2(8), A_6, L_2(17), L_3(3), U_3(3) or U_4(2)$.*

D. Gorenstein called above eight simple groups as simple K_3-groups and pointed out "that classification is obtained only as a corollary of the complete classification of all finite simple groups"[26]. We determined all simple K_4-groups using the classification theorem [40].

Theorem 2.3 *Let G be a simple K_4-group. Then G is isomorphic to one of the following groups: A_n, $n = 7, 8, 9, 10$; $M_{11}, M_{12}, J_2; L_2(q)$, $q = 16, 25, 49, 81$; $L_3(q)$, $q = 4, 5, 7, 8, 17$; $L_4(3)$; $O_5(q)$, $q = 4, 5, 7, 9$; $O_7(2)$, $O_8^+(2), G_2(3); U_3(q), q = 4, 5, 7, 8, 9; U_4(3); U_5(2); {}^3D_4(2); {}^2F_4(2)'$; $Sz(8), Sz(32)$; and $L_2(r)$, r being prime and satisfying the following equation:*

$$r^2 - 1 = 2^a 3^b u^c, \tag{1}$$

where $a \geq 1, b \geq 1, c \geq 1$, u prime, $u > 3$; $L_2(2^m)$ and satisfying the *following equations:*

$$\begin{cases} 2^m - 1 = u \\ 2^m + 1 = 3t^b \end{cases} \qquad (2)$$

where $m \geq 1$, u, t primes, $t > 3, b \geq 1$; $L_2(3^m)$ and satisfying the following equations:

$$\begin{cases} 3^m + 1 = 4t \\ 3^m - 1 = 2u^c \end{cases} \qquad (3)$$

or

$$\begin{cases} 3^m + 1 = 4t^b \\ 3^m - 1 = 2u \end{cases} \qquad (4)$$

where $m \geq 1$, u, t odd primes, $c \geq 1, b \geq 1$.

Remark 2.4   There is an error in [23]. Perhaps the equation $3^m - 1 = 2u^c$, where $c > 1$ and u prime, has no other solution except $m = 5, u = 11, c = 2$ and $m = u = c = 2$. But I have not proved yet.

Question 2.5   *The number of simple K_4-groups is determined by the number of solution of equations (1)–(4). But it is unknown whether the number of solution is finite or infinite. In other words, is the number of simple K_4-groups finite or infinite?*

If G is a simple K_4-group and we know some $p \in \pi(G)$, where $p \neq 2, 3$, then we can determine all such groups by the equations(1)–(4).For example we have the following corollary [41].

Corollary 2.6   *Let G be a simple group of order $2^a 3^b 5^c 7^d$, $abcd \neq 0$. Then G is isomorphic to one of the following groups: A_n, $n = 7, 8, 9, 10$; J_2; $L_2(49)$; $L_3(4)$; $O_5(7)$; $O_7(2)$; $O_8^+(2)$; $U_3(5)$ and $U_4(3)$.*

Although we do not know the number of simple K_4-groups, we have the following conclusion from Theorem 2.3.

Corollary 2.7   *Let G be a simple K_4-group. Then the order of G contains a prime to the first power except $G \cong S_4(7)$.*

Problem 2.8   (H.F. Tuan) Determine all finite simple groups whose order does not contain a prime to the first power.

For the orders of finite simple groups Z.M. Chen proved the following theorem [19].

Theorem 2.9   *The order of any finite simple group G is not the kth power, $k \geq 3$. The order of G is square if and only if $G \cong S_4(p)$, where p is a prime satisfying the following condition:*

$$p = 1 + 2C_{2n+1}^2 + 2^2 C_{2n+1}^4 + \cdots + 2^n C_{2n+1}^{2n}, \qquad n \geq 1 \qquad (5)$$

If $p < 10^{15}$, *then the above equation has only solutions* $p = 7, 41, 239, 9369319,$ 63018038201. [67]

Question 2.10 *How many primes satisfy the equation (5)?*

In 1955 E. Artin studied the orders of the known types of finite simple groups. He pointed out that it is of interest to study simple groups of order $|G|$ which contain a Sylow subgroup of order greater than $|G|^{1/3}$ [4].The simplest question of this type is the following one which E. Artin asked R. Brauer in a letter: "What are the simple groups G of order $|G|$ which is divisible by a prime $p > |G|^{1/3}$?" R. Brauerand W.F. Reynolds solved the simplest case and proved the following theorem [12].

Theorem 2.11 *Let G be a finite simple non-cyclic group whose order $|G|$ is divisible by a prime $p > |G|^{1/3}$. Then G is isomorphic either to $L_2(p)$ where $p > 3$ is a prime or to $L_2(p-1)$ where $p > 3$ is a Fermat prime.*

Using the classification theorem of all finite simple groups we solved E. Artin's problem and proved the following theorems [42]:

Theorem 2.12 *Let G be a finite simple non-cyclic group. If $p^k|||G|$ (i.e. $p^k|||G|$ and $p^{k+1} \nmid |G|$) where p is an odd prime, and $|G| < p^{3k}$, then G is isomorphic to one of the following groups:*

(1) *A simple group of Lie type of characteristic p;*

(2) A_5 *($k = 1, p = 5$),* A_6 *($k = 2, p = 3$),* A_9 *($k = 4, p = 3$); or*

(3) $L_2(p-1)$ *($k = 1$, p is a Fermat prime),* $L_2(8)$ *($k = 2, p = 3$),* $U_5(2)$ *($k = 5, p = 3$).*

Theorem 2.13 *Let G be a finite simple non-cyclic group. If $2^k|||G|$ and $|G| < 2^{3k}$, then G is isomorphic to one of the following groups:*

(1) *A simple group of Lie type of characteristic 2 or $^2F_4(2)'$;*

(2) $L_2(r)$ *where r is a Fermat prime or a Mersenne prime; or*

(3) A_6 *($k = 3$),* $U_3(3)$ *($k = 5$),* $A_8, A_9, M_{12}, A_{10}, M_{22}, J_2$ *($k = 7$),* HS *($k = 9$),* M_{24} *($k = 10$),* Suz *($k = 13$),* Ru *($k = 14$),* Fi_{22} *($k = 17$),* Co_2 *($k = 18$),* Co_1 *($k = 21$),* B *($k = 41$).*

Making use of the above theorems we proved that almost all finite simple groups can be characterized by the conditions of the simplicity of G and $|G|$ [43] and [32].

Theorem 2.14 *Let G be a simple group and $|G| = |M|$, where M is a known finite simple group. Then one of the following conclusions holds.*

(1) *If $|M| = |A_8| = |L_3(4)|$, then $G \cong A_8$ or $L_3(4)$;*

(2) *If $|M| = |B_n(q)| = |C_n(q)|$ where $n \geq 3$ and q is odd ,then $G \cong$*

$B_n(q)$ or $C_n(q)$;or

(3) *If $|M|$ is not the above case (1) or (2), then $G \cong M$.*

By inspecting the orders of all the known finite simple groups we proved the following statement [43].

Theorem 2.15 *Let G be a finite simple non-cyclic group. Then G contains a proper subgroup H such that $|G| < |H|^2$.*

3. On quantitative characterization

Now we give the pure quantitative characterization for all finite simple groups. We replace the condition of the simplicity of G with a quantitative set $\pi_e(G)$. That is, we characterize the finite simple groups using only the sets $|G|$ and $\pi_e(G)$. We have proved the following theorem [44, 45, 46, 58, 59, 60].

Theorem 3.1 *Let G be a group and M one of the following simple groups:*

(1) *A cyclic simple group Z_p;*

(2) *An alternating group A_n, $n \geq 5$;*

(3) *A sporadic simple group;*

(4) *A Lie type group except $B_n, D_n, \ D_n, \ ^2A_n$ and ^2D_n; or*

(5) *A simple group with order $< 10^8$.*

Then $G \cong M$ if and only if (a) $\pi_e(G) = \pi_e(M)$, and (b) $|G| = |M|$.

Since we have not found any counter example, we have the following conjecture.

Conjecture 3.2 *Let G be a group and M a finite simple group. Then $G \cong M$ if and only if (a) $\pi_e(G) = \pi_e(M)$, and (b) $|G| = |M|$.*

We believe the above conjecture is correct and it gives an integral characterization to all finite simple groups. Since this characterization establishes the connection between the pure quantities and the finite simple groups, maybe it will bring such significance to some research and application.

For proving Theorem 3.1 we may first consider the set of orders of elements. Using the classification theorem, the results on the prime graph components [78] and the orders of centralizers of p-elements in group G we discuss the normal series of G which satisfy the conditions. We may also consider the order of group G first, and introduce some special number-theoretic function, again discuss the group-theoretic properties. For example, for characterizing A_n, we introduce the following function [58]

$$t_n(k) = \prod_{i=1}^{k} \left(\prod_{n/(i+1)<p_i \le n/i} P_i \right)^i$$

where p_j is a prime and defined as 1 if there is no prime between $n/(j+1)$ and n/j.

Let G denote the groups which satisfy the conditions $\pi_e(G_n) = \pi_e(A_n)$ and $|G_n| = |A_n|$. We have the following lemma [58].

Lemma 3.3 G_n is unsolvable for $n \ge 5$. If $n \ge 11$, then G_n has a normal series $G_n \ge H > N \ge 1$ where H/N is non abelian simple and $t_n(1) \mid |H/N|$. If $n \ge 46$,then $t_n(3) \mid |H/N|$. Again if $n \ge 83$, then $t_n(6) \mid |H/N|$.

This lemma is a key to the proof of A_n in Theorem 3.1.

Besides the finite simple groups we may characterize some unsolvable groups G using only the conditions of $\pi_e(G)$ and $|G|$. For example in [5], the author proved the following theorem.

Theorem 3.4 Let G be a group. Then $G \cong S_n$, $n \ge 3$ (S_n denotes the symmetric group of degree n) if and only if (a) $\pi_e(G) = \pi_e(S_n)$, and (b) $|G| = |S_n|$.

Generally we can not characterize the unsolvable groups, even p-groups, using only the conditions of $\pi_e(G)$ and $|G|$. From this view we may say the simple groups are simple.

There exist other pure quantitative characterizations for the finite simple groups. For instance there are the following results.

Theorem 3.5 [6] Let G be a group. Then $G \cong A_n$, $n \ge 5$, if and only if $\pi_n(G) = \pi_n(A_n)$, where $\pi_n(G) = \{|N_C(P)|;\ P \in Syl_pG,\ \text{for all}\ p \in \pi(G)\}$.

Theorem 3.6 [74, 75, 76, 79, 82] Let G be a group and M be one of the following groups:

(1) $L_2(q)$ or $SL_2(q)$;

(2) $Sz(2^{2m+1})$, $m \ge 1$;

(3) A sporadic simple group except $J_4, Fi_{23}, Fi'_{24}, TH, B$ and M; or

(4) A simple group with order less than 10^8.

Then $G \cong M$ if and only if $\pi_m(G) = \pi_m(M)$ where $\pi_m(G)$ is the set of orders of maximal subgroup in G.

From theorem 3.6 we put forward the following conjecture:

Conjecture 3.7 *Let G be a group and M a finite simple group. Then $G \cong M$ if and only if $\pi_m(G) = \pi_m(M)$.*

4. Characterization only using $\pi_e(G)$

In Theorem 3.1 if we use only the condition (b), then only Z_n can be characterized. Generally we obtain finite different groups with the same order. Now we are interested in comparing groups by the set $\pi_e(G)$ rather than by the orders of the groups them selves. Thus we are interested in the following function h defined on subsets of positive integers.

Definition 4.1 *For a set m of positive integers let $h(m)$ be the number of isomorphism classes of finite group G such that $\pi_e(G) = m$.*

From the classification of finite EPPO-groups we proved the following theorem:

Theorem 4.2 *Let G be a finite simple EPPO-group. Then $(\pi_e(G)) = 1$ except $G \cong A_6$. Moreover $h(\pi_e(A_6)) = \infty$.*

In other words, we proved the following results:

Let G be a finite group. Then

$G \cong A_5$ if and only if $\pi_e(G) = \{1, 2, 3, 5\}$ [39];

$G \cong L_2(7)$ if and only if $\pi_e(G) = \{1, 2, 3, 4, 7\}$ [47];

$G \cong L_2(8)$ if and only if $\pi_e(G) = \{1, 2, 3, 7, 9\}$ [48];

$G \cong L_2(17)$ if and only if $\pi_e(G) = \{1, 2, 3, 4, 8, 9, 17\}$ [7];

$G \cong L_3(4)$ if and only if $\pi_e(G) = \{1, 2, 3, 4, 5, 7\}$ [49];

$G \cong Sz(8)$ if and only if $\pi_e(G) = \{1, 2, 4, 5, 7, 13\}$ [50];

$G \cong Sz(32)$ if and only if $\pi_e(G) = \{1, 2, 4, 5, 25, 31, 41\}$ [50].

Moreover if $\pi_e(G) = \pi_e(A_6) = \{1, 2, 3, 4, 5\}$, then $G \cong A_6$ or $G \cong N \rtimes Q$, where $Q \cong A_5$ and N is an elementary abelian 2-group and a direct sum of natural $SL(2, 4)$-modules [8].

Furthermore we have proved the following conclusion [34, 39, 8, 51, 37, 52, 53, 54, 56, 61, 9, 10, 48, 49, 50, 85].

Theorem 4.3 *Let G be one of the following groups:*

(1) A_n, $n = 5, 7, 8, 9, 11, 13$, or S_7;

(2) A sporadic simple groups except J_2, and Co_1; or

(3) $L_2(q)$, $q > 3$ and $q \neq 9$, $L_3(4)$, $U_3(4)$, $U_4(3)$, $U_5(2)$, $U_6(2)$, $Sz(2^{2m+1})$, $m \geq 1$, $R(3^{2m+1})$, $m \geq 1$, M_{10} or $L_3(4) \langle \beta \rangle$, where β is a unitary automorphism of $L_3(4)$.

Then $h(\pi_e(G)) = 1$. Moreover $h(\pi_e(J_2)) = h(\pi_e(S_8)) = \infty$.

However we do not know of any set m for which $h(m) \notin \{0, 1, \infty\}$. We make the following conjecture:

Conjecture 4.4 *For all subsets m of positive integers, $h(m) \in \{0, 1, \infty\}$.*

Question 4.5 *For $q = p^k \geq 5$, where p is a prime and $k \geq 1$, is $G \cong A_q$ characterized by the set $\pi_e(G)$, that is, is $h(\pi_e(A_q)) = 1$?*

Question 4.6 *We have proved $h(\pi_e(Co_1)) = 1$ or ∞ [56]. is $h(\pi_e(Co_1)) = 1$ or $h(\pi_e(Co_1)) = \infty$?*

Question 4.7 *We have proved $h(\pi_e(Sz(2^{2m+1}))) = 1$ for $m \geq 1$ and $h(\pi_e(R(3^{2m+1}))) = 1$ for $m \geq 1$. For $m \geq 1$, is $h(\pi_e({}^2F_4(2^{2m+1}))) = 1$?*

Since we proved $h(\pi_e(S_n)) = \infty$ for $n = 2, 3, 4, 5, 6$ and 8, and $h(\pi_e(S_7)) = 1$, we ask:

Question 4.8 *Is there an integer $n > 7$ such that S_n is characterized by $\pi_e(S_n)$? If so, what is the smallest such n?*

Z. M. Chen has proved the following lemma on conjecture 4.4 [56].

Lemma 4.9 Let N be a minimal normal subgroup of G and $\exp(N) \in \pi_e(N)$, where $\exp(N)$ is the exponent of N. Then $h(\pi_e(G)) = \infty$.

Corollary 4.10 *Let G be a finite solvable group. Then $h(\pi_e(G)) = \infty$.*

On the characterization of simple groups by only $\pi_e(G)$, our argument heavily use the following unpublished result of K. W. Gruenberg and O.H. Kegel [27]:

Theorem 4.11 *If G is a finite group whose prime graph has more than one component, then G has one of the following structure: (a) Frobenius or 2-Frobenius; (b) simple; (c) an extension of a π_1-group by a simple group; (d) simple by π_1-solvable; or (e) π_1 by simple by π_1, where π_1 denotes the connected component of the prime graph containing 2.*

The characteristic property of some simple groups by their element orders depends on the results of simple C_{22} groups and simple C_{33} groups. The following theorem gives a classification of simple C_{pp} groups, p is prime and $p = 2^\alpha 3^\beta + 1, \alpha \geq 0, \beta \geq 0$ [20].

Theorem 4.12 *Let p be a prime and $p = 2\alpha 3\beta + 1$, $\alpha \geq 0, \beta \geq 0$. Then any finite simple C_{pp} group is given by the following table.*

Question 4.13 *Let p be any prime. What is the classification of finite simple C_{pp} groups?*

For some infinite groups we may also characterize them by their element orders.

Theorem 4.14 *[31] The locally finite simple group G is linear if and* *only if there is a natural number n that does not appear as order of an element of G.*

Using the characterization of A_5 (i.e. a finite group $G \cong A_5$ if and only if $\pi_e(G) = \{1, 2, 3, 5\}$) we may obtain easily a symmetric graph H with 30 vertices such that $\text{Aut}(H) \cong A_5$ (see [83]).

p	finite simple C_{pp} groups
2	$A_5 A_6; L_2(q)$, where q is a Fermat prime, a Mersenne prime or $q = 2^n$, $n \geq 3$, $L_3(2^2), Sz(2^{2n+1}), n \geq 1$.
3	$A_5 A_6; L_2(q)$, where $q = 2^3, 3^{n+1}$ or $2 \cdot 3^n \pm 1$ is a prime, $n \geq 1$, $L_3(2^2)$.
5	$A_5 A_6, A_7; M_{11} M_{22}; L_2(q), q = 7^2, 5^n$, or $2 \cdot 5^n \pm 1$ is a prime $n \geq 1, L_3(2^2), S_4(q), q = 3, 7; U_4(3); Sz(q), q = 2^3, 2^5$.
7	$A_7 A_8; M_{22} J_1, J_2, HS; L_2(q), q = 2^3, 7^n$, or $2 \cdot 7^n - 1$ is a prime $n \geq 1, L_3(2^2), S_6(2), O_8^+(2), G_2(q), q = 3, 19; U_3(q); q = 3, 5, 19; U_4(3), U_6(2), Sz(2^3)$.
13	$A_{13} A_{14}, A_{15}; S_{UZ}, Fi_{22}; L_2(q), q = 3^3, 5^2, 13^n$, or $2 \cdot 13^n - 1$ is a prime, $n \geq 1, L_3(3), L_4(3), O_7(3), S_4(5), S_8(3), O_8^+(3), G_2(q), q = 2^2, 3; F_4(2), U_3(q), q = 2^2, 23, Sz(2^3), {}^3 D_4(2), {}^2 E_6(2), {}^2 F_4(2)'$.
17	$A_{17}, A_{18}, A_{19}; J_3, He, Fi_{23}, Fi_{24}'; L_2(q), q = 2^4, 17^n, 2 \cdot 17^n \pm 1$ is a prime, $n \geq 1, S_4(4), S_8(2), F_4(2), O_8^-(2), O_{10}^-(2), {}^2 E_6(2)$.
19	$A_{19}, A_{20}, A_{21}; J_1, J_3, O'N, Th, HN; L_2(q), q = 19^n, 2 \cdot 19^n - 1$ is a prime, $n \geq 1, L_3(7), U_3(2^3), R(3^3), {}^2 E_6(2)$.
37	$A_{37}, A_{38}, A_{39}; J_4, Ly; L_2(q), q = 37^n, 2 \cdot 37^n - 1$ is a prime, $n \geq 1, U_3(11), R(3^3), {}^2 F_4(2^3)$.
73	$A_{73}, A_{74}, A_{75}; L_2(q), q = 73^n, 2 \cdot 73^n - 1$ is a prime, $n \geq 1, L_3(2^3), S_6(2^3), G_2(q), q = 2^3, 3^2, F_4(3), E_6(2), E_7(2), U_3(3^2), {}^3 D_4(3)$.
109	$A_{109}, A_{110}, A_{111}; L_2(q), q = 109^n, 2 \cdot 109^n - 1$ is a prime, $n \geq 1, {}^2 F_4(2^3)$.
$2^{2^m} + 1$	$A_p, A_{p+1}, A_{p+2}; L_2(q), q = 2^{2^m}, p^n, 2p^n \pm 1$ is a prime, $n \geq 1, S_{2a+1}(2^{2^b}), a \geq 1, a + b = m, F_4(2^b), b \geq 1, 4b = 2^m, O_{2(2^m+1)}^-(2), m \geq 2, 0_{2a+1}^-(2^{2^b}), a \geq 2, a + b = m$.
other	$A_p, A_{p+1}, A_{p+2}; L_2(q), q = p^n, 2 \cdot p^n - 1$ is a prime, $n \geq 1$.

5. Thompson's two problems

A problem related to Conjecture 3.2 is the following open problem put forward by J.G. Thompson [72].

For each finite group G and each integer $d \geq 1, let G(d) = \{x \in G; x^d = 1\}$.

Definition 5.1 G_1 and G_2 are of the same order type if and only if $|G_1(d)| = |G_2(d)|, d = 1, 2, \cdots$.

Problem 5.2 Suppose G_1 and G_2 are groups of the same order type. Suppose also that G_1 is solvable. Is it true that G_2 is also necessarily solvable?

In Thompson's letter he pointed out that "The problem arose initially in the study of algebraic number fields, and is of considerable interest".

In fact the groups of the same order type are such groups whose "order equations" are the same. That is, according to the orders of elements as the equivalent relation, the group G is partitioned into the same order classes. The "order equation" is

$$|G| = n_1 + n_2 + \cdots + n_k,$$

where n_i denotes the number of elements of order i in G. Obviously the same order class is an union of several conjugacy classes and the partition of "order equation" is rougher than the class equation.

If G_1 and G_2 are finite groups of the same order type, then $\pi_e(G_1) = \pi_e(G_2)$ and $|G_1| = |G_2|$. Thus if we prove Conjecture 3.2, then we have the following corollary: Let G_1 and G_2 be finite groups of the same order type. If G_1 is solvable, then G_1 is not simple.

In [21], [22] and [38], the authors call the groups of the same order type conformal groups. In [38], the author dealt with the following problem.

Question 5.3 For which natural number n are any two conformal groups of order n isomorphic?

Another conjecture of Thompson which also aims at characterizing all finite simple groups by a number theoretic quantity is the following [73].

If G is a finite group, set $N(G) = \{n \in N; G$ has a conjugacy class C with $|C| = n\}$.

Conjecture 5.4 If G and M are finite groups and $N(G) = N(M)$, and if in addition, M is a nonabelian simple group while the center of G is 1, then G and M are isomorphic.

On Conjecture 5.4, G.Y. Chen has dealt with the situation that M is a finite simple group having a nonconnected prime graph. He has set up two important lemmas.

Lemma 5.5 Let G be a finite group with $Z(G) = 1$ and M a nonconnected prime graph nonabelian simple group such that $N(G) = N(M)$. Then $|G| = |M|$.

Lemma 5.6 *Let G and M be finite groups having same orders. If $N(G) = N(M)$, then the number of components of prime graph of G is equal to that of M. And the prime graph components of G are equal to those of M, where the prime graph components are viewed as the sets of vertices.*

By using above lemmas Chen has proved that Thompson's conjecture holds for all nonabelian simple groups except $A_n(q)$, $B_n(q)$, $C_n(q)$, $D_n(q)$, $^2A_n(q)$, $^2D_n(q)$, $E_7(q)$ and the alternating groups A_n (see [16],[17] and [18]).

6. Some related topics

(a) Finite groups whose element orders are consecutive integers.

Definition 6.1 *Let n be a positive integer. Then a group G is an OC_n group if every element of G has order $\leq n$ and for each $m \leq n$ there exists an element of G having order m.*

Using the classification of finite simple groups and the conclusion of any finite group having at most six prime graph components we proved the following theorem [8].

Theorem 6.2 *Let G be a finite OC_n group. Then $n \leq 8$.*

Moreover we have obtained a complete classification of finite OC_n groups. For example, let G be a finite OC_7 group then $G \cong A_7$.

Question 6.3 *For the infinite OC_n groups, what is the maximal number n? The infinite OC_n groups are periodic, are they locally finite?*

After [8] was published, D. Macttal posed the following problem: To classify all finite groups in which proper subgroup orders are consecutive integers. The following finite $OS_{n; \ m_1, \ m_2}$ groups have been investigated in generally. [33]

Definition 6.4 *A group G is called a $OS_{n; \ m_1, \ m_2}$ group is $\pi_s(G) = \{1, 2, \cdots, n, m_1, m_2\}$, where $m_2 > m_1 > n + 1$, and $\pi_s(G)$ is the set of all proper subgroup orders of G.*

Theorem 6.5 *Let G be a finite $OS_{n; \ m_1, \ m_2}$ group.   Then $n = 1, 2, 3, 4, 6$ and G is isomorphic to one of the following groups:*

(1) $n = 1, m_1 = t, m_2 = t^2$, G is isomorphic to one of the groups of order t^3, t is a prime;

(2) $n = 1, m_1 = t, m_2 = s$, G is isomorphic to one of the groups of order ts, $t < s$, both t and s are primes;

(3) $n = 2, m_1 = 4, m_2 = 8$, G is isomorphic to one of the groups of order 2^4;

(4) $n = 2, m_1 = 3, m_2 = 4$, G is isomorphic to A_4;

(5) $n = 3, m_1 = 6, m_2 = 9$, G is isomorphic to one of the groups of order 2.3^2.

(6) $n = 3, m_1 = 4, m_2 = 6$, G is isomorphic to $Z_4 \times Z_3, Z_6 \times Z_2$, $\left\langle a, b; a^6 = b^2 = 1, \ a^b = a^{-1} \right\rangle$ or $\left\langle a, b; a^6 = 1, b^2 = a^3, a^b = a^{-1} \right\rangle$;

(7) $n = 4, m_1 = 6, m_2 = 8$, G is isomorphic to $\langle a, b, c; a^4 = c^3 = 1,$ $b^2 = a^2, a^c = b, \ b^c = ab, a^b = a^{-1} \rangle$; or

(8) $n = 6, m_1 = 10, m_2 = 12$, G is isomorphic to A_5.

Theorem 6.6   Let G be a finite $OS_{n,m}$ groups. Then $n \leq 4$ and G is isomorphic to one of the following groups:

(1) $n = 1, m = t$, G is isomorphic to one of the groups of order t^2, t is a prime;

(2) $n = 2, m = 4$, G is isomorphic to one of the groups of order 2^3;

(3) $n = 2, m = s$, G is isomorphic to one of the groups of order $2s$, s is a odd prime;

(4) $n = 3, m = 4$, G is isomorphic to A_4; or

(5) $n = 4, m = 6$, G is isomorphic to $Z_4 \times Z_3, Z_6 \times Z_2, \langle a, b; a^6 = b^2 = 1,$ $a^b = a^{-1} \rangle$ or $\left\langle a, b; a^6 = 1, b^2 = a^3, a^b = a^{-1} \right\rangle$.

Theorem 6.7   Let G be a finite OS_n group. Then $n \leq 4$ and G is isomorphic to one of the following groups:

(1) $n = 1$, G is isomorphic to Z_p, p is a prime;

(2) $n = 2$, G is isomorphic to one of the groups of order 2^2;

(3) $n = 3$, G is isomorphic to one of the groups of order 6; or

(4) $n = 4$, G is isomorphic to A_4.

Corollary 6.8   Let G be a finite group. Then G is isomorphic to A_n if and only if $\pi_s(G) = \pi_s(A_n)$, $n = 4, 5$.

Also the following finite $OC_{n,p}$ -groups have been investigated. [80]

Definition 6.9   A group G is called $OC_{n,p}$ -group if its element orders set $\pi_e(G) = \{1, 2, \cdots, n; p_1, \cdots, p_s\}$, where n is a natural number, p_i is a prime and $p_i > n + 1$, $i = 1, 2, \cdots, s$.

Theorem 6.10   Let G be an $OC_{n,\ p}$ -group, $s \geq 1$. Then $1 \leq s \leq 2$ and $1 \leq n \leq 5$ or $n = 8$. Moreover

(1) If $1 \leq n \leq 2$, then G is a solvable group with all elements of prime order except '1;

(2) If $n = 3$, then $G \cong A_5$;

(3) *If $n = 4$, then $G \cong L_2(7)$;*

(4) *If $n = 5$, then $G \cong L_3(4)$;*

(5) *If $n = 8$, then $G \cong M_{22}$.*

(b) Finite groups in which the number of the maximal order elements is given

This topic is relative to the same order type groups. In [81], the author proved the following result:

Theorem 6.11 *Let G be a finite group and $M(G)$ the set of maximal order elements. If one of the following conditions holds, then G is solvable.*

(1) $2 \nmid |M(G)|$,

(2) $|M(G)| \leq 4$,

(3) $|M(G)| = 2p$, *p is prime, or*

(4) $|M(G)| = \phi(k)$, *where $\phi(k)$ is Eulerian function of the maximal order k. Moreover, if $|M(G)|$ is 2, 4 or odd, then G is supersolvable.*

(c)A quantitative theorem on "orders"

Let n be a positive integer and $n = p_1^a p_2^b \cdots p_k^c$ ($a \geq 1, b \geq 1, \cdots, c \geq 1, p_i \neq p_j$, if $i \neq j$), the prime-factor-decomposition of n. We put $|\pi(n)| = k$, $|\pi(G)| = |\pi(|G|)|$ and $\rho(G) = max\{|\pi(n)|; n \in \pi_e(G)\}$. If G is finite nilpotent, then $|\pi(G)| = \rho(G)$. While G is finite supersolvable we have proved [57]

$$|\pi(G)| \leq (1/2)\rho(G)(\rho(G) + 3).$$

In [84], the author proved the following:

Theorem 6.12 *Let G be an arbitrary finite group. Then*

$$|\pi(G)| \leq 8\rho(G)^3 (\rho(G) + 3) \exp(\rho(G)).$$

From the results of finite EPPO-groups perhaps the bound in Theorem 6.12 is not best.

Question 6.13 *What is the best bound in Theorem 6.12?*

(d)A general topic on $\pi_e(G)$

If $G = G_1 \times G_2$, then $\pi_e(G) = \{n_1 n_2 / (n_1, n_2); n_1 \in \pi_e(G_1), n_2 \in \pi_e(G_2)\}$. C.E. Praeger asked:

Question 6.14 *What properties does the set $X = \{m;$ there exist (finite) groups G with $\pi_e(G) = m\}$ have?*

There are some trivial closure properties, for example, for $m \in X$, if $k \in m$ then all divisors of k are in m. Consider lots of standard constructions for groups, for instance, $G_1 \times G_2, G_1 \rtimes G_2, G_1 wr G_2$, and $G_1 * G_2$, to find more properties on X.

In general, we believe that there are many problems in there searches of groups using the "orders". Some are interesting and some are difficult. We hope these works and results can be generalized and applied to other branch of algebra, combinatorics, computer and algorithm.

Acknowledgements

The author wants to express his profound gratitude to Professor Z.M. Chen and Professor J.G. Thompson for their kind and helpful guidance and encouragement. The author also would like to thank Professor H.F. Tuan, Professor O.H. Kegel, professor C.E. Praeger and Dr.R. Brandl for their many interesting suggestions and kind help.

References

[1] L.J. Alex, *Simple groups of order $2^a 3^b 5^c 7_p^d$*, Trans. AMS., 173 (1972), 389–398.

[2] L.J. Alex, *On simple groups of order $2^a 3^b 5^c 7_p^d$*, J. Algebra, 25 (1973), 113–124.

[3] Z. Arad and D. Chillag, *On centralizers of elements of odd order in finite groups*, J. Algebra, 61 (1979), 269–280.

[4] E. Artin, *The orders of the classical simple groups*, Comm. Pure Appl. Math, 8 (1955), 355–365.

[5] J.X. Bi, *A characterization of symmetric groups*, (in Chinese) Acta Math. Sinica, 33 (1990), 70–77.

[6] J.X. Bi, *On the characteristic property of the alternating groups*, (in Chinese) Chinese Sci. Bull., 34 (1989), 1117.

[7] R. Brandl, *Finite groups all of whose elements are of prime power order*, Boll. U. M. I. (5)18-A (1981), 491–493.

[8] R. Brandl and W.J. Shi, *Finite groups whose element orders are consecutive integers*, J. Algebra, 143 (1991), 388–400.

[9] R. Brandl and W. J. Shi, *The characterization of $PSL(2, q)$ by its element orders*, to be published in J. Algebra.

[10] R. Brandl and W. J. Shi, *A characterization of finite simple groups with abelian Sylow 2-subgroups*, Ricerche di Maematica, 42 (1993), 193–198.

[11] R. Brauer, *On simple groups of order $5 \cdot 3^a \cdot 2^b$*, Bull.AMS.,74 (1968), 900–903.

[12] R. Brauer and W.F. Reynolds, *On a problem of E. Artin*, Ann. Math., 68 (1958), 713–720.

[13] R. Brauer and H.F. Tuan, *On simple groups of finite order*, Bull.AMS., 51 (1945), 756–766.

[14] N. Bryce, *On the Mathieu group M_{23}*, J. Austral. Math.Soc., 12 (1971), 385–392.

[15] W. Burnside, *On an unsettled question in the theory of discontinuous groups*, Quart. J. Pure Appl. Math., 33 (1902), 230–238.

[16] G.Y. Chen, *On Thompson's conjecture*, (in Chinese) Proc. China Assoc. Sci. and Tech. First Academic Annual Meeting of Youths. Chinese Sci.and Tech. Press, Beijing, 1992, 1–6.

[17] G.Y. Chen, *On Thompson's conjecture-for simple groups having at least three prime graph components*, to appear.

[18] G.Y. Chen, *On Thompson's conjecture-for simple groups having exactly two prime graph components*, to appear.

[19] Z.M. Chen, *On the orders of the finite simple groups*, (in Chinese) Acta Math. Sinica, 30 (1987), 605–613.

[20] Z.M. Chen and W.J. Shi, *On simple C_{pp} groups*, (in Chinese) J. of Southwest-China Teachers University, 18:3 (1993), 249–256.

[21] C.D.H. Cooper, *Conformality and p-isomorphism infinite nilpotent groups*, J. Austral. Math. Soc., 7 (1975), 165–171.

[22] C.D.H. Cooper, *Words which give rise to another group operation for a given group*, Pro. Conf. Theory of Groups, Canberra 1973, 221–225.

[23] P. Crescenzo, *A diophantine equation which arise in the theory offinite groups*, Adv. Math., 17 (1975), 25–29.

[24] M. Deaconescu, *Classification of finite groups with all elements of prime order*, Proc. AMS., 106 (1989), 625–629; (Corrigendum and addendum), Proc. AMS., 117 (1993), 1205–1207.

[25] E.S. Golod, *On nil-algebras and residually finite p-groups*, Izv.Akad. Nauk. SSSR Ser. Mat., 28 (1964), 273–276.

[26] D. Gorenstein, *Finite Simple Groups*, Plenum Press, New York and London, 1982.

[27] K.W. Gruenberg and O. H. Kegel, *Unpublished material*, 1975.

[28] M. Herzog, *On groups of order $2^\alpha 3^\beta p^\gamma$ with a cyclic Sylow 3-subgroup*, Proc. AMS., 24 (1970), 116–118.

[29] M. Herzog, *On finite simple groups of order divisible by three prime only*, J. Algebra, 10 (1968), 383–388.

[30] G. Higman, *Finite groups in which every element has prime power order*, J. London Math. Soc., 32 (1957), 335–342.

[31] O.H. Kegel and W.J. Shi, *On the orders of elements of a locally finite simple group*, C. R. Math. Rep. Acad. Sci. Canada, 13 (1991), 253–254.

[32] W. Kimmerle, R. Lyons, R. Sandling and D.N. Teague, *Composition factors from the group ring and Artin's theorem on orders of simple groups*, Proc. London Math. Soc., (3)60 (1990), 89–122.

[33] J.H. Lee, *Finite groups whose proper subgroup orders are consecutive integers except two*, (in Chinese) J. of Southwest-China Teachers Univ., 18:4 (1993), 393–401.

[34] H. L. Li and W. J. Shi, *A characteristic property of some sporadic simple groups*, (in Chinese) Chin. Ann. Math., 14A:2 (1993), 144–151.

[35] B.H. Neumann, *Groups whose elements have bounded orders*, J. London Math. Soc., 12 (1937), 195–198.

[36] D. Parrott, *On the Mathieu groups M_{22} and M_{11}*, J. Austral. Math. Soc., 11 (1970), 69–81.

[37] C.E. Praeger and W.J. Shi, *A characterization of some alternating and symmetric groups*, to be published in Comm. in Algebra.

[38] R. Scapellato, *Finite groups with the same number of elements of each order*, Rendiconti di Matematica, Serie VII, 8 (1988).

[39] W.J. Shi, *A characteristic property of A_5*, (in Chinese) J. of Southwest-China Teachers University, 3 (1986), 11–14.

[40] W.J. Shi, *On simple K_4-groups*, (in Chinese) Chinese Science Bulletin, 36 (1991), 1281–1283.

[41] W. J. Shi, *The simple groups of order $2^a3^b5^c7^d$ and Janko's simple groups*, (in Chinese) J. of Southwest-China Teachers University, 4 (1987), 1–8.

[42] W.J. Shi, *On a problem of E. Artin*, (in Chinese) Acta. Math. Sinica, 35:2 (1992), 262–265.

[43] W.J. Shi, *On the orders of the finite simple groups*, (in Chinese) Chin. Sci. Bull., 38:4 (1993), 296–298.

[44] W.J. Shi, *A new characterization of the sporadic simple groups*, Group Theory, Proc. of the 1987 Singapore Conf. Walterde Gruyter. Berlin. New York, 1989, 531–540.

[45] W. J. Shi, *The pure quantitative characterization of finite simple groups (I)*, Progress in Natural Science, 4(1994), 316–326.

[46] W. J. Shi, *A new characterization of some simple groups of Lie type*, Contemporary Mathematics, 82 (1989), 171–180.

[47] W. J. Shi, *A characteristic property of $PSL_2(7)$*, J. Austral. Math. Soc. (Ser. A), 36 (1984), 354–356.

[48] W. J. Shi, *A characteristic property of J_1 and $PSL_2(2^n)$*, (in Chinese) Adv. in Math., 16 (1987), 397–401.

[49] W. J. Shi, *A characterization of some projective special linear groups*, J. of Math. (PRC), 5 (1985), 191–200.

[50] W. J. Shi, *A characterization of Suzuki's simple groups*, Proc. AMS., 114:3 (1992), 589–591.

[51] W.J. Shi, *A characteristic property of A_8*, Acta Math. Sin., New Ser.; 3 (1987), 92–96.

[52] W.J. Shi, *A characteristic property of Mathieu groups*, Chinese J. of Contemporary Mathematics, 9 (1988), 317–326.

[53] W. J. Shi, *A characterization of the Higman-Simsgroup*, Houston J. of Math., 16 (1990), 597–602.

[54] W.J. Shi, *A characterization of the Conway simple group Co_2*, (in Chinese) J. of Math. (PRC), 9 (1989), 171–172.

[55] W. J. Shi, *A class of special finite groups*, Northeast Math. J., 9:1 (1993), 1–4.

[56] W.J. Shi, *The characterization of the sporadic simple groups by their elements orders*, Algebra Colloquium, 1: 2 (1994), 316–326.

[57] W.J. Shi, *A quantitative theorem of the finite supersolvable groups*, to appear.

[58] W.J. Shi and J.X. Bi, *A new characterization of the alternating groups*, SEA Bull. Math., 16:1 (1992), 81–90.

[59] W.J. Shi and J.X. Bi, *A characteristic property for each finite projective special linear group*, Lecture Notes in Math.,1456 (1990), 171–180.

[60] W.J. Shi and J.X. Bi, *A characterization of Suzuki-Ree groups*, Science in China (Ser.A), 34 (1991), 14–19.

[61] W.J. Shi and H.L. Li, *A characteristic property of M_{12} and $PSU(6,2)$*, (in Chinese) Acta Math. Sinica, 32 (1989), 758–764.

[62] W.J. Shi and C. Yang, *A class of special finite groups*, Chinese Science Bulletin, 37 (1992), 252–253.

[63] W.J. Shi and W.Z. Yang, *A new characterization of A_5 and the finite groups in which every non-identity element has prime order*, (in Chinese) J. of Southwest-China Teachers College, 1 (1984), 36–40.

[64] W.J. Shi and W.Z. Yang, *The finite groups all of whose elements are of prime power order*, (in Chinese) J. Yunnan Educational College,1 (1986), 2–10.

[65] R.G. Stanton, *The Mathieu groups*, Canad. J. Math., 3 (1951), 164–174.

[66] A.I. Starostin, *On groups whose centralizers have partitions*, Izv, Akad. Nauk SSSR Ser. Mat. 29 (1965), 605–614.

[67] Q. Sun, *A unsolved diophantine equation which arises in the theory of finite groups*, (in Chinese) J. of Math. Res. and Exposition, 6 (1986), 20.

[68] M. Suzuki, *On a class of doubly transitive groups*, Ann. Math., 75 (1962), 105–145.

[69] M. Suzuki, *Finite groups with nilpotent centralizer*, Trans. AMS., 99 (1961), 425–470.

[70] K.B. Tchaberian, *A note on simple groups of order $2^a \cdot 3^b \cdot 5 \cdot p^c$*, C.R. Acad. Bulgare Sci., 33:8 (1980), 1037–1038.

[71] K. B. Tchaberian, *Groups containing a three-prime simple group*, C. R. Acad. Bulgare Sci., 33: 9 (1980), 1165 – 1167.

[72] J.G. Thompson, Private communication, (1987).

[73] J.G. Thompson, Private communication, (1988).

[74] D.J. Wang, *A characterizations of some finite simple groups using the set of the orders of their maximal subgroups*, (in Chinese), J.Southwest-China Teachers Univ. (Ser.B), 18:1 (1993), 18–21.

[75] D.J. Wang, *A characterizations of $L_2(q)$*, Chin. Ann. Math., 16A: 1(1995), 70 –76.

[76] D.J. Wang, *A characterizations of $SL(2,q)$*, Acta Math. Sinica, 37(1994), 601 – 606.

[77] D. Wilkinson, The groups of exponent p and order p7 (p any prime), J. Algebra, 118 (1988), 109–119.

[78] J.S. Williams, *Prime graph components of finite groups*, J. Algebra, 69 (1981), 487–513.

[79] J.Y. Xie and P. Yu, *A characterization of some sporadic simple groups*, (in Chinese) J. Southwest-China Teachers Univ. (ser.B), 18:3 (1993), 269–274.

[80] M.C. Xu, *Finite groups whose element orders are consecutive integers except some primes*, J. of Southwest-China Teachers Univ., 19 (1994), 116 –122.

[81] C. Yang, *Finite groups with a given number of the maximal orders elements*, (in Chinese) Chin. Ann. Math., 14A:5 (1993), 561–567.

[82] P. Yu, *A characterization of Suzuki's simple groups*, (in Chinese) J. Southwest-China Teachers Univ. (ser. B), 18:1 (1993), 15–17.

[83] J.S. Zhang and W. J. Shi, *An application of the quantitative characterization of A_5 in graph theory*, (in Chinese) J. of Southwest-China Teachers University, 16 (1991), 399–402.

[84] J. P. Zhang, *Arithmetical conditions on element orders and group structure*, to appear Proc. AMS,. 123:1 (1995), 39 –44.

[85] F. Zhou, *A characteristic property of some unitary groups*, (in Chinese) J. of Shanghai Jiaotong Univ., 28 (1994), 106 – 109.

Survey on Geometry of Classical Groups over Finite Fields and Its Applications

Zhe-xian Wan

The Chinese Academy of Sciences

1. The Germ of Our Study

Let $\mathbb{F}_q$ be a finite field with q elements, where q is a prime power, $\mathbb{F}_q^{(n)}$ be the n-dimensional row vector space over $\mathbb{F}_q$, and $GL_n(\mathbb{F}_q)$ be the *general linear group* of degree n over $\mathbb{F}_q$. $GL_n(\mathbb{F}_q)$ acts on $\mathbb{F}_q^{(n)}$ in the following way:

$$
\begin{aligned}
\mathbb{F}_q^{(n)} \times GL_n(\mathbb{F}_q) &\to \mathbb{F}_q^{(n)}, \\
((x_1, x_2, \cdots, x_n), T) &\mapsto (x_1, x_2, \cdots, x_n)T.
\end{aligned}
\tag{1}
$$

Let P be an m-dimensional subspace of $\mathbb{F}_q^{(n)}$ and $v_1, v_2, \cdots, v_m$ be a basis of P. Then

$$
\begin{pmatrix} v_1 \\ v_2 \\ \vdots \\ v_m \end{pmatrix}
\tag{2}
$$

is an $m \times n$ matrix of rank m over $\mathbb{F}_q$. We call the matrix (2) a *matrix representation* of the subspace P and use also the same letter P to denote the matrix (2) if no ambiguity arises. The action (1) of $GL_n(\mathbb{F}_q)$ on $\mathbb{F}_q^{(n)}$ induces an action on the set of subspaces of $\mathbb{F}_q^{(n)}$ such that $T \in GL_n(\mathbb{F}_q)$ carries the subspace P to PT. We may propose the following problems:

(i) What are the orbits of subspaces of $\mathbb{F}_q^{(n)}$ under the action of $GL_n(\mathbb{F}_q)$?

(ii) How many orbits are there ?

---

* Dedicated to Professor Hsio-Fu Tuan on his Eighty Second Birthday

(iii) What are the lengths of the orbits ?

(iv) What is the number of subspaces in an orbit contained in a given sub-space?

The answers to these four problems are well-known, they are:

i) Two subspaces belong to the same orbit if and only if their dimensions are equal.

ii) There are altogether $n + 1$ orbits.

iii) Denote the length of the orbit of m-dimensional subspaces ($0 \leq m \leq n$) by $N(m, n)$, then

$$N(m, n) = \frac{\prod\limits_{i=n-m+1}^{n} (q^i - 1)}{\prod\limits_{i=1}^{m} (q^i - 1)}. \tag{3}$$

iv) The number of k-dimensional subspaces contained in a given m-dimensional subspace ($0 \leq k \leq m \leq n$) is $N(k, m)$.

2. The Problems We Are Interested in

It is natural to propose the following problem.
Use any one of the other classical groups, such as the symplectic group $Sp_n(\mathbb{F}_q)$ (where $n = 2\nu$), the unitary group $U_n(\mathbb{F}_q)$ (where q is a square), or the orthogonal group $O_n(\mathbb{F}_q)$ (where $n = 2\nu + \delta$ and $\delta = 0, 1$, or 2) to replace $GL_n(\mathbb{F}_q)$, then we study Problems (i)–(iv).

Now let us introduce the definition of the other classical groups.

Let $n = 2\nu$. It is well-known that the cogredience normal form of $2\nu \times 2\nu$ nonsingular alternate matrices is

$$K = \begin{pmatrix} 0 & I^{(\nu)} \\ -I^{(\nu)} & 0 \end{pmatrix}. \tag{4}$$

Let

$$Sp_{2\nu}(\mathbb{F}_q) = \{T \in GL_{2\nu}(\mathbb{F}_q) | TK\,{}^tT = K\}. \tag{5}$$

Then $Sp_{2\nu}(\mathbb{F}_q)$ is a group with respect to the matrix multiplication, called the *symplectic group* of degree 2ν over $\mathbb{F}_q$.

Let $q = q_0^2$, where q_0 is a prime power. $\mathbb{F}_q = \mathbb{F}_{q_0^2}$ has an involutive automorphism

$$- : a \to \bar{a}, \tag{6}$$

whose fixed field is $\mathbb{F}_{q_0}$. Let

$$U_n(\mathbb{F}_q) = \{T \in GL_n(\mathbb{F}_q)|T\,^t\overline{T} = I^{(n)}\}. \tag{7}$$

Then $U_n(\mathbb{F}_q)$ is a group with respect to the matrix multiplication, called the *unitary group* of degree n over $\mathbb{F}_q$.

Let q be a power of an odd prime and z be a non-square element of $\mathbb{F}_q$. The cogredience normal forms of $n \times n$ nonsingular symmetric matrices over $\mathbb{F}_q$ are

$$S_0 = \begin{pmatrix} 0 & I^{(\nu)} \\ I^{(\nu)} & 0 \end{pmatrix}, \tag{8}$$

$$S_{1,d} = \begin{pmatrix} 0 & I^{(\nu)} \\ I^{(\nu)} & 0 \\ & & d \end{pmatrix}, \tag{9}$$

where $d = 1$ or z, and

$$S_2 = \begin{pmatrix} 0 & I^{(\nu)} \\ I^{(\nu)} & 0 \\ & & 1 \\ & & & -z \end{pmatrix}, \tag{10}$$

Corresponding to these four cases, n is equal to 2ν, $2\nu + 1$, $2\nu + 1$, and $2\nu + 2$, respectively. We use $n = 2\nu + \delta$ and $S_{\delta,d}$ to cover these four cases, where $\delta = 0, 1$, or 2, $d = 1$ or z when $\delta = 1$, and d disappears when $\delta = 0$ or 2. Let

$$O_{2\nu+\delta,d}(\mathbb{F}_q) = \{T \in GL_{2\nu+\delta}(\mathbb{F}_q)|TS_{\delta,d}\,^tT = S_{\delta,d}\}. \tag{11}$$

Then $O_{2\nu+\delta,d}(\mathbb{F}_q)$ is a group with respect to the matrix multiplication, called the *orthogonal group* of degree $2\nu+\delta$ over $\mathbb{F}_q$. It is easy to prove that $O_{2\nu+1,1}(\mathbb{F}_q)$ and $O_{2\nu+1,z}(\mathbb{F}_q)$ are isomorphic. Thus it is enough to consider the three orthogonal groups $O_{2\nu}(\mathbb{F}_q)$, $O_{2\nu+1,1}(\mathbb{F}_q)$, and $O_{2\nu+2}(\mathbb{F}_q)$. We write $O_{2\nu+1}(\mathbb{F}_q)$ simply for $O_{2\nu+1,1}(\mathbb{F}_q)$.

When $\mathbb{F}_q$ is of characteristic 2, there are also three types of orthogonal groups $O_{2\nu+\delta}(\mathbb{F}_q)$, where $\delta = 0, 1$, or 2, but their definitions are omitted.

3. The History of the Problems

In 1937 E. Witt [1] studied problem (i) for the orthogonal group over any field F of characteristic $\neq 2$. Let S be an $n \times n$ nonsingular symmetric matrix over F. The *orthogonal group* of degree n over F relative to S, denoted by $O_n(F, S)$, is defined to be

$$O_n(F, S) = \{T \in GL_n(F) | TS\,{}^tT = S\} \tag{12}$$

The famous Witt's Theorem asserts that two subspaces P_1 and P_2 of $\mathbb{F}_q^{(n)}$ belong to the same orbit of $O_n(F, S)$ if and only if $\dim P_1 = \dim P_2$ and $P_1S\,{}^tP_1$ and $P_2S\,{}^tP_2$ are cogredient. Later Witt's theorem was generalized to other classical groups by C. Arf [2], J, Dieudonné [3, 4], L. K. Hua [5], V. Pless [6], X. Feng and Z..Dai [7], and Z. Dai [8]. It is worth to mention that Hua [5] gave also a simple matrix proof of the generalized Witt theorem.

In 1948 B. Segre [9] studied the problem (iii) for the orthogonal group over $\mathbb{F}_q$, but he restricted himself to consider only the orbits of totally isotropic or totally singular subspaces corresponding to cases when q is odd or even, respectively. For simplicity we follow the notation of the previous paragraph, thus assume that $\mathbb{F}_q$ is of characteristic $\neq 2$. A subspace P of $F^{(n)}$ is called *totally isotropic*, if $PS\,{}^tP = 0$. By Witt's theorem totally isotropic subspaces of the same dimension of $F^{(n)}$ form an orbit under $O_n(F, S)$. Segre determined the lengths of the orbits of totally isotropic subspaces of the same dimension of $\mathbb{F}_q^{(2\nu+\delta)}$ under $O_{2\nu+\delta}(\mathbb{F}_q)$. He used geometric language to state his results as follows. The number of m-dimensional flats lying on a nondegenerate quadric in the $(n-1)$-dimensional projective space over $\mathbb{F}_q$, $PG(n-1, \mathbb{F}_q)$, is equal to

$$\frac{\prod\limits_{i=\nu-m}^{\nu} (q^i - 1)(q^{i+\delta-1} - 1)}{\prod\limits_{i=1}^{m+1} (q^i - 1)}, \tag{13}$$

where $-1 \leq m \leq \nu - 1$, $\nu = \frac{n-1}{2}$ when n is odd, and $\nu = \frac{n}{2}$ or $\frac{n}{2} - 1$ when n is even and the quadric is of the hyperbolic type or the elliptic type, respectively. He used geometric method to deduce this formula which holds also for the case of characteristic 2. In 1962 D.K.Raychaudhuri [10] obtained also Segre's formula (13).

In 1964 three students of mine at that time and myself [11–13, 7] studied the problem (iii) for the groups $Sp_{2\nu}(\mathbb{F}_q)$, $U_n(\mathbb{F}_q)$ (where q is a square), and $O_{2\nu+\delta}(\mathbb{F}_q)$ (where $\delta = 0, 1$, or 2). We determined not only the lengths of those orbits of totally isotropic or totally singular subspaces but also the lengths of all the orbits. We call these results the Anzahl theorems in the geometries of these classical groups. Our methods are algebraic and our results are compiled in our monograph [14].

In 1965 V. Pless [15] computed the lengths of the orbits of totally isotropic subspaces of $\mathbb{F}_q^{(2\nu)}$ under the group $Sp_{2\nu}(\mathbb{F}_q)$ and the number of totally isotropic subspaces of the same dimension of $\mathbb{F}_q^{(2\nu+\delta)}$ (where $\delta = 1$ or 2) with respect to a $(2\nu + \delta) \times (2\nu + \delta)$ nonsingular non-alternate symmetric matrix over $\mathbb{F}_q$ when q is even.

In 1965 B.Segre [16] and in 1966 R. C. Bose and I. M. Chakravarti [17] determined the lengths of the orbits of totally isotropic subspaces of $\mathbb{F}_q^{(n)}$ under the group $U_n(\mathbb{F}_q)$ (where q is a square of a prime power).

In 1966 the author studied problem (iv) for the group $Sp_{2\nu}(\mathbb{F}_q)$, $U_n(\mathbb{F}_q)$ (q is a square), and $O_{2\nu+\delta}(\mathbb{F}_q)$ (where $\delta = 0, 1$, or 2) and obtained closed formulas for the number of subspaces in an orbit under each of these groups contained in a given subspace. These results are also called Anzahl theorems and were compiled in [14] too.

4. Recent Results

In the early nineties I returned to the study of the geometry of classical groups over finite fields and obtained the following results.

1) Problems (i) and (ii) for the symplectic, unitary, and orthogonal groups over finite fields are studied [18–21]. Of course, Witt's theorem and its generalizations give a solution to problem (i), but we would like to use a set of numerical invariants to characterize an orbit and to derive the conditions satisfied by them that such an orbit exists, then the number of orbits can be computed.

Take the symplectic case as an example. Let (4)

$$K = \begin{pmatrix} 0 & I^{(\nu)} \\ -I^{(\nu)} & 0 \end{pmatrix}.$$

Then the symplectic group of degree 2ν is defined as (5)

$$Sp_{2\nu}(\mathbb{F}_q) = \{T \in GL_{2\nu}(\mathbb{F}_q) | TK\,{}^tT = K\}.$$

Let P be an m-dimensional subspace of $\mathbb{F}_q^{(2\nu)}$. Clearly $PK\,{}^tP$ is alternate, hence rank $PK\,{}^tP$ is even. Assume that rank $PK\,{}^tP = 2s$, then P

is said to be of *type* (m, s). From Dieudonné's generalization [3] of Witt's theorem it follows that two subspaces belong to the same orbit under $Sp_{2\nu}(\mathbb{F}_q)$ if and only if they are of the same type. It can be proved [18] that the type (m, s) of a subspace satisfies the inequality

$$2s \leq m \leq \nu + s \tag{14}$$

and for any pair of non-negative integers (m, s) satisfying (14) there exist subspaces of type (m, s). Thus the number of orbits of subspaces under $Sp_{2\nu}(\mathbb{F}_q)$ is equal to the number of pairs of non-negative integers (m, s) satisfying (14). We computed that the latter is equal to

$$\frac{1}{2}(\nu + 1)(\nu + 2). \tag{15}$$

By the way we mention that the length $N(m, s; 2\nu)$ of the orbit of subspaces of type (m, s) of $\mathbb{F}_q^{(2\nu)}$ given in [11] is

$$N(m, s; 2\nu) = q^{2s(\nu+s-m)} \frac{\prod\limits_{i=\nu+s-m+1}^{\nu} (q^{2i} - 1)}{\prod\limits_{i=1}^{s} (q^{2i} - 1)} \prod_{i=1}^{m-2s} (q^i - 1). \tag{16}$$

This is the solution to problem (iii) for the symplectic group.

2) The singular symplectic, unitary, and orthogonal groups are introduced and the problems (i)–(iv) are studied [22, 23].

Take the singular symplectic case as an example. Let

$$K_l = \begin{pmatrix} K & \\ & 0^{(l)} \end{pmatrix} \tag{17}$$

where K is the nonsingular alternate matrix (4). Define

$$Sp_{2\nu+l,\nu}(\mathbb{F}_q) = \{T \in GL_{2\nu+l}(\mathbb{F}_q) | TK_l{}^tT = K_l\}, \tag{18}$$

which is called the *singular symplectic group* over $\mathbb{F}_q$. Clearly, $Sp_{2\nu+l,\nu}(\mathbb{F}_q)$ acts on $\mathbb{F}_q^{(2\nu+l)}$ in an obvious way. Then problems (i)–(iv) can be studied for $Sp_{2\nu+l,\nu}(\mathbb{F}_q)$, and complete results are obtained.

Similarly, singular unitary and orthogonal groups over $\mathbb{F}_q$ can be defined, and complete results for problems (i)–(iv) are obtained.

A natural question arises. Why do we study the geometry of singular symplectic, unitary, and orthogonal groups over finite fields ?

The answer to problem (iv) for the general linear group $GL_n(\mathbb{F}_q)$ is easy: the number of k-dimensional subspaces contained in a given m-dimensional subspace $(0 \leq k \leq m \leq n)$ of $\mathbb{F}_q^{(n)}$ is $N(k,m)$. However, problem (iv) for the other classical groups is not so easy.

Take again the symplectic case as an example. Now assume that $Sp_{2\nu}(\mathbb{F}_q)$ acts on $\mathbb{F}_q^{(2\nu)}$. Given a subspace P of type (m,s), where (m,s) satisfies (14), we would like to compute the number of subspaces of type (m_1, s_1), where $2s_1 \leq m_1 \leq \nu + s_1$, contained in P. Denote this number by $N(m_1, s_1; m, s; 2\nu)$. We may choose a matrix representation of P, denoted by P again, such that

$$PK\,^tP = \begin{pmatrix} 0 & I^{(s)} & \\ -I^{(s)} & 0 & \\ & & 0^{(m-2s)} \end{pmatrix}. \tag{19}$$

Let P_1 be a subspace of type (m_1, s_1) contained in P. As an m_1-dimensional subspace of the m-dimensional space P, P_1 has a matrix representation, denoted by P_1 again, which is a $m_1 \times m$ matrix of rank m_1. Then as a subspace of $\mathbb{F}_q^{(2\nu)}$, the subspace P_1 has P_1P as a matrix representation. Similarly, we can choose the matrix P_1 such that

$$(P_1P)K\,^t(P_1P) = \begin{pmatrix} 0 & I^{(s_1)} & \\ -I^{(s_1)} & 0 & \\ & & 0^{(m_1-2s_1)} \end{pmatrix}. \tag{20}$$

Then

$$P_1 \begin{pmatrix} 0 & I^{(s)} & \\ -I^{(s)} & 0 & \\ & & 0^{(m-2s)} \end{pmatrix}\,^tP_1 = \begin{pmatrix} 0 & I^{(s_1)} & \\ -I^{(s_1)} & 0 & \\ & & 0^{(m_1-2s_1)} \end{pmatrix}. \tag{21}$$

Thus for any $T \in Sp_{2s+(m-2s),s}(\mathbb{F}_q), P_1TP$ is also a matrix representation of a subspace of type (m_1, s_1) and contained in P and as a subspace of P it is represented by the matrix P_1T. Therefore it is natural to introduce the singular symplectic group $Sp_{2s+(m-2s),s}(\mathbb{F}_q)$ and study how the subspaces of $\mathbb{F}_q^{(m)}$ are subdivided into orbits under $Sp_{2s+(m-2s),s}(\mathbb{F}_q)$, the length of each orbit, and what orbits are contained in P.

3) For pseudo-symplectic groups over finite fields of characteristic 2 problems (i)–(iv) are also studied [24, 25].

Now let $\mathbb{F}_q$ be a finite field of characteristic 2. Then any $n \times n$ non-singular non-alternate symmetric matrix over $\mathbb{F}_q$ is cogredient to either

$$S_1 = \begin{pmatrix} 0 & I^{(\nu)} & \\ I^{(\nu)} & 0 & \\ & & 1 \end{pmatrix}, \quad \text{when } n = 2\nu + 1 \text{ is odd} \tag{22}$$

or

$$S_2 = \begin{pmatrix} 0 & I^{(\nu)} & & \\ I^{(\nu)} & 0 & & \\ & & 0 & 1 \\ & & 1 & 1 \end{pmatrix}, \quad \text{when } n = 2\nu + 2 \text{ is even.} \tag{23}$$

We use S_δ ($\delta = 1$ or 2) to cover these two cases. Define the *pseudo-symplectic group* of degree $2\nu + \delta$ over $\mathbb{F}_q$ to be

$$Ps_{2\nu+\delta}(\mathbb{F}_q) = \{T \in GL_{2\nu+\delta}(\mathbb{F}_q) | TS_\delta\,{}^tT = S_\delta\}. \tag{24}$$

It was proved by Dieudonné [3] that $Ps_{2\nu+1}(\mathbb{F}_q) \simeq Sp_{2\nu}(\mathbb{F}_q)$ and $Ps_{2\nu+2}$ has a normal series with $Sp_{2\nu}(\mathbb{F}_q)$ as one of its factors and $\mathbb{F}_q$ as all the other factors. Thus from a group theory point of view the pseudo-symplectic group $Ps_{2\nu+\delta}(\mathbb{F}_q)$ is less interesting. However, its geometry is very peculiar. Let P be an m-dimensional subspace of $\mathbb{F}_q^{(2\nu+\delta)}$. Then $PS_\delta\,{}^tP$ is a symmetric matrix and is cogredient to one of the following normal forms

$$\begin{pmatrix} 0 & I^{(s)} & \\ I^{(s)} & 0 & \\ & & 0^{(m-2s)} \end{pmatrix}, \tag{25}$$

$$\begin{pmatrix} 0 & I^{(s)} & & \\ I^{(s)} & 0 & & \\ & & 1 & \\ & & & 0^{(m-2s-1)} \end{pmatrix}, \tag{26}$$

and

$$\begin{pmatrix} 0 & I^{(s)} & & \\ I^{(s)} & 0 & & \\ & & 0 & 1 & \\ & & 1 & 1 & \\ & & & & 0^{(m-2s-2)} \end{pmatrix}. \tag{27}$$

P is called a subspace of *type* $(m, 2s + \tau, s, \varepsilon)$ where $\tau = 0, 1$, or 2 corresponding to the above three normal forms (25), (26), or (27), respectively, $\varepsilon = 0$ or 1 corresponding to the cases $e_{2\nu+1} \notin P$ or $e_{2\nu+1} \in P$, respectively, and $e_{2\nu+1}$ is the $(2\nu + \delta)$-dimensional row vector whose $(2\nu + 1)$-th component is 1 and other components are all 0. It is proved that two subspaces of $\mathbb{F}_q^{(2\nu+\delta)}$ belong to the same orbit under $Ps_{2\nu+\delta}(\mathbb{F}_q)$ if and only if they are of the same type. It is also proved that subspaces of type $(m, 2s + \tau, s, \varepsilon)$ exist if and only if

$$(\tau, \varepsilon) = \begin{cases} (0,0), (1,0), (1,1), \text{ or } (2,0), & \text{when } \delta = 1, \\ (0,0), (0,1), (1,0), (2,0), \text{ or } (2,1), & \text{when } \delta = 2 \end{cases} \tag{28}$$

and

$$2s + \max\{\tau, \varepsilon\} \le m \le \nu + s + [(\tau + \delta - 1)/2] + \varepsilon. \tag{29}$$

Using conditions (28) and (29) we can compute the number of orbits of subspaces under $Ps_{2\nu+\delta}(\mathbb{F}_q)$, which is equal to

$$\frac{1}{2}(\nu + 1)((\nu + 4)\delta + 3\nu). \tag{30}$$

Denote the length of the orbit of subspaces of type $(m, 2s + \tau, s, \varepsilon)$ by $N(m, 2s + \tau, s, \varepsilon; 2\nu + \delta)$. Then

$$N(m, 2s + \tau, s, \varepsilon; 2\nu + \delta)$$
$$= q^{n_0 + 2(s + (2 - \delta)[\tau/2])(\nu + s - m + \delta[(\tau+1)/2] + (\delta-1)(\tau-1)(\tau-2)\varepsilon/2)}$$

$$\times \frac{\displaystyle\prod_{i=\nu+s-m+[(\tau+\delta-1)/2]+\varepsilon+1}^{\nu} (q^{2i} - 1)}{\displaystyle\prod_{i=1}^{s}(q^{2i} - 1) \prod_{i=1}^{m-2s-\max(\tau,\varepsilon)}(q^i - 1)}, \tag{31}$$

where $n_0 = 1$ when $\delta = 1$, and $n_0 = m, 0, 2(\nu + 1) - m, 2(\nu + 1) - m$, or $2(\nu + 1) - m$ corresponding to the cases $(\tau, \varepsilon) = (0, 0), (0, 1), (1, 0), (2, 0)$, or $(2, 1)$, respectively, when $\delta = 2$.

In order to study the problem (iv) for the pseudo-symplectic group $Ps_{2\nu+\delta}(\mathbb{F}_q)$ the singular pseudo-symplectic group is introduced and for which problems (i)–(iii) are studied [25].

4) The affine classification of quadrics over finite fields is obtained [26, 27].

The foregoing results together with our results obtained in the mid sixties are compiled in the monograph [28].

5. Application to Association Schemes and Designs

Why do we study problems (i)–(iv) for the classical groups over finite fields ? Of course, they are well-posed mathematical problems and were studied by several famous mathematicians before us. However, we have been interested in these problems mainly because they have interesting applications. In the sixties the geometry of classical groups over finite fields was used to construct association schemes and PBIB designs. I came back to this field in the early nineties because I found that it could be used to construct authentication codes.

A finite non-empty set X of cardinality v, whose elements are called *points*, together with a partition of the set of 2-element subsets of X into d non-empty classes $\Gamma_1, \Gamma_2, \cdots, \Gamma_d$ is called an *association scheme* with d classes [29] if the following two conditions are satisfied.

(a) Given $x \in X$, the number of points $y \in X$ such that $\{x, y\} \in \Gamma_i$ depends only on i, not on x. (So we write this number as n_i.)

(b) Given $x, y \in X$ with $\{x, y\} \in \Gamma_i$, the number of points $z \in X$ such that $\{x, z\} \in \Gamma_j$ and $\{z, y\} \in \Gamma_k$ depends only on i, j, k, not on x and y. (So we write this number as p_{jk}^i.)

We use the same letter X to denote the association scheme and the numbers $v, n_i (i = 1, 2, \cdots, d)$, and $p_{jk}^i (i, j, k = 1, 2, \cdots, d)$ are called its *parameters*. Points x and y are called the *i-th associates* if $\{x, y\} \in \Gamma_i$.

Let X be an association scheme with d classes and $\mathcal{B}$ a set of k-element subsets of X, called *blocks*. Then $(X, \mathcal{B})$ is called a PBIB(d) *design* [30] if the following two conditions are satisfied.

(a) Every point $x \in X$ belongs to the same number of blocks. (So we write this number as r.)

(b) Given $x, y \in X$ with $\{x, y\} \in \Gamma_i$, the number of blocks containing both x and y depends only on i, not on x and y. (So we write this number as λ_i.)

Let $|\mathcal{B}| = b$. Then the numbers $v, b, k, r, n_i, p_{jk}^i$, and λ_i $(i, j, k = 1, 2, \cdots, d)$ are called the *parameters* of the PBIB(d) design.

A PBIB(d) design with $d = 1$ is called a BIB *design* [31], and v, b, k, r, and $\lambda (= \lambda_1)$ are called its *parameters*.

The classical construction of BIB design by taking the set of 1-dimensional subspaces of $\mathbb{F}_q^{(n)}$ as the set of points and the set of m-dimensional subspaces of $\mathbb{F}_q^{(n)}$ for a fixed integer m satisfying $1 < m \leq n$, i.e., the set of subsets of 1-dimensional subspaces, each of which is a set of all 1-dimensional subspaces contained in an m-dimensional subspace, as the set of blocks is well-known. This BIB design has parameters

$$v = N(1, n), b = N(m, n), k = N(1, m),$$
$$r = N'(1, m, n), \lambda = N'(2, m, n), \tag{32}$$

where $N(m, n)$ is given by (3) and $N'(l, m, n)$ $(1 \leq l \leq m \leq n)$ denotes the number of m-dimensional subspaces of $\mathbb{F}_q^{(n)}$ containing a given

l-dimensional subspace and is given by

$$N'(l,m,n) = \frac{N(m,n)N(l,m)}{N(l,n)} = \frac{\prod\limits_{i=n-m+1}^{n-l}(q^i-1)}{\prod\limits_{i=1}^{m-l}(q^i-1)}. \qquad (33)$$

In 1954 W.H.Clatworthy [32] showed that a geometric configuration in $PG(3,\mathbb{F}_q)$ may be interpreted as a PBIB design. In our terminology, he took the set of 1-dimensional subspaces of the 4-dimensional symplectic space over $\mathbb{F}_q$ as the set of points and the set of 2-dimensional totally isotropic subspaces as the set of blocks. Two points are said to be the first associates (or second associates) if they span a 2-dimensional totally isotropic subspace (or nonisotropic subspace), respectively. A point is defined to be lie in a block if the 1-dimensional subspace as the point is contained in the 2-dimensional totally isotropic subspace as the block. Then a PBIB(2) design is obtained. Clatworthy also computed the parameters of the design.

In 1962 D.K.Ray-Chaudhuri [33] used the geometry of orthogonal groups, which was called the geometry of quadrics by him, to construct PBIB designs. He constructed several PBIB(2) designs with 1-dimensional or 2-dimensional totally isotropic (or singular) subspaces of the geometry of orthogonal groups over finite fields as points or blocks and computed their parameters. At that time only the length of the orbit of totally isotropic (or singular) subspaces of a given dimension under the orthogonal group over finite fields was known, so he naturally restricted himself to take only totally isotropic (or singular) subspaces as points and blocks in order to compute the parameters of the designs he constructed.

In the mid sixties after we had found the closed formulas for the lengths of all the orbits of subspaces under the symplectic, unitary, and orthogonal groups over finite fields, we [34, 12, 35, 36, 14] constructed many asssociation schemes and PBIB designs by taking the 1-dimensional, 2-dimensional, or ν-dimensional totally isotropic subspaces as points and subspaces of any given type as blocks and computed their parameters. Most of the parameters, such as v, b, k, r, and λ_i, follows from our Anzahl theorems, but the parameters n_i and p^i_{jk} need further computation. The computation of the intersection numbers p^i_{jk} is usually not so immediate. It is worth to mention that by taking the ν-dimensional (i.e., maximal) totally isotropic subspaces in the 2ν-dimensional symplectic, the $(2\nu+\delta)$-

dimensional unitary ($\delta = 0$ or 1), or the ($2\nu + \delta$)-dimensional orthogonal ($\delta = 0, 1$, or 2) geometry as points and defining two points P and Q to be the i-th associates ($1 \leq i \leq \nu$) if $\dim(P + Q) = \nu + i$, association schemes with ν classes were obtained and their parameters were computed. In particular, the closed formulas for the intersection numbers p_{jk}^i were obtained. All these results were also compiled in the monograph [14] and were sketched in [37].

However, we would like to mention that when we took the 2-dimensional totally isotropic (or singular) subspaces as points to construct association schemes and PBIB designs, we consider only the symplectic, unitary, and orthogonal spaces of low dimensions, because it gave us PBIB (2) designs.

Of course, we can take any orbit of subspaces under the symplectic, unitary, or orthogonal group over finite fields as the set of points and define the associate relation according to the orbit of pairs of points under that group, then an association scheme is obtained. Moreover, if we take any orbit of subspaces as the set of blocks and define a point to be lie in a block in a certain way, then a PBIB design is obtained. In a short note [38] published in 1965 some association schemes and PBIB designs were constructed by taking the 1-dimensionsl non-isotropic (or non-singular) subspaces in the unitary or orthogonal geometry over some small fields as points and their parameters were computed. In the eighties the idea of taking the 1-dimensional non-isotropic (or non-singular) subspaces or taking the 2-dimensional totally isotropic (or singular) subspaces as points was carried out and generalized by several Chinese mathematicians and their works were sketched in [37] and will not be repeated here.

6. Application to Projective Codes

Let C be a linear $[n, k]$-code over $\mathbb{F}_q$, where q is a prime power. C is called a *projective code*, if the columns of a generator matrix of C are not proportional. Let G be a generator matrix of C, then each column of G can be regarded as a point of $PG(k - 1, \mathbb{F}_q)$, the ($k - 1$)-dimensional projective space over $\mathbb{F}_q$. Different columns of G correspond to different points of $PG(k - 1, \mathbb{F}_q)$. Thus we obtain a point set in $PG(k - 1, \mathbb{F}_q)$, which will be denoted by $\mathcal{S}_{C,G}$ and called the point set arising from C via G. Clearly, different encoding matrices of C give rise to point sets which are projectively equivalent, i.e., there is a tranformation of $PG(k - 1, \mathbb{F}_q)$

of the form

$$^t(x_1, x_2, \cdots, x_k) \rightarrow A \ ^t(x_1, x_2, \cdots, x_k), \tag{34}$$

called a *projective transformation* of $PG(k-1, \mathbb{F}_q)$, which carries one of the point set to the other.

Two projective $[n, k]$-codes over $\mathbb{F}_q$ are said to be *equivalent*, if one can be obtained from the other by permuting the coordinates of the codewords and multiplying them by non-zero elements of $\mathbb{F}_q$.

Let G be a generator matrix of a projective $[n, k]$-code C and C' be a projective code equivalent to C. Then the same transformation which transforms C to C' will transform an encoding matrix G of C to an encoding matrix G' of C'. Clearly, $S_{C,G} = S_{C',G'}$

Let Q be a $k \times k$ matrix over $\mathbb{F}_q$. The set of points $^t(x_1, x_2, \cdots, x_k)$ in $PG(k-1, \mathbb{F}_q)$ satisfying

$$(x_1, x_2, \cdots, x_k)Q \ ^t(x_1, x_2, \cdots, x_k) = 0 \tag{35}$$

is called a *quadric* in $PG(k-1, \mathbb{F}_q)$ and will also be denoted by Q. The number of points of Q will be denoted by $|Q|$ and can be derived from the Anzahl theorems of the singular orthogonal geometry over $\mathbb{F}_q$. In particular, if Q is nondegenerate, i.e., (35) can not be carried under projective transformations into a quadratic homogeneous equation with less than k indeterminates, then $|Q|$ was obtained by Primrose [39] and is also contained in Segre's formula (13). Moreover, $|Q| = |^tAQA|$ for any $k \times k$ nonsingular matrix A.

Let Q be a quadric in $PG(k-1, \mathbb{F}_q)$ and assume that $|Q| = n$. For each point Q choose a system of coordinates and regard it as a k-dimensional column vector. Arrange these n column vectors in any order into a $k \times n$ matrix, denoted by G_Q. It can be proved that G_Q is of rank k. Hence G_Q can be regarded as a generator matrix of a projective $[n, k]$-code, which is denoted by C_Q and called the projective code from the quadric Q in $PG(k-1, \mathbb{F}_q)$. Clearly

$$S_{C_Q, G_Q} = Q. \tag{36}$$

We would like to compute the nonzero weights of C_Q and their multiplicities. Let $(c_1, c_2, \cdots, c_n)$ be a nonzero code word of C_Q. Then it is a nonzero linear combination of the rows of G_Q, say

$$(c_1, c_2, \cdots, c_n) = (a_1, a_2, \cdots, a_k)G_Q, \tag{37}$$

where $a_1, a_2, \cdots, a_k \in \mathbb{F}_q$. $(a_1, a_2, \cdots, a_k)$ defines a hyperplane H of $PG(k-1, \mathbb{F}_q)$, whose equation is

$$a_1 x_1 + a_2 x_2 + \cdots + a_k x_k = 0. \tag{38}$$

Denote by $H \cap Q$ the set of points of Q lying on the hyperplane (38), and by $w_H(c_1, c_2, \cdots, c_n)$ the Hamming weight of the code word $(c_1, c_2, \cdots, c_n)$. Then

$$w_H(c_1, c_2, \cdots, c_n) = n - |H \cap Q|. \tag{39}$$

Thus to compute the nonzero weights of C_Q it is sufficient to compute $|H \cap Q|$ for each hyperplane H. The hyperplanes of $PG(k-1, \mathbb{F}_q)$ are partitioned into orbits under the orthogonal group or the singular orthogonal group defined by Q, when Q is nondegenerate or not, respectively. Hyperplanes H belong to the same orbit have the same $|H \cap Q|$ and $|H \cap Q|$ is given by the Anzahl theorems of the orthogonal geometry or the singular orthogonal geometry. Therefore the values of the nonzero weights of C_Q can be deduced from our results of the geometry of orthogonal or singular orthogonal groups. Moreover, to compute the weight multiplicities it is sufficient to enumerate the number of hyperplanes H having the same value $|H \cap Q|$, which also follows from our results of the geometry of orthogonal or singular orthogonal groups.

In 1975 J.Wolfmann [40] proved that if Q is non-degenerate, then C_Q has only two or three distinct nonzero weights when k is even or odd, respectively, by computing the values of the nonzero weights of C_Q, and he computed also their multiplicities.

In his studies of the wire-tap channel of type II, V.K.Wei [41] introduced a generalization of the minimum Hamming weight of a linear code. Let C be a linear $[n, k]$-code over $\mathbb{F}_q$, where q is a prime power. For a linear subcode D of C, let $\chi(D)$ be the *support* of D, namely,

$$\chi(D) = \{i \,|\, x_i \neq 0 \text{ for some } (x_1, x_2, \cdots, x_n) \in D\}. \tag{40}$$

The r-th generalized Hamming weight of C, denoted by $d_r(C)$, is then defined as

$$d_r(C) = \min\{|\chi(D)| \,|\, D \text{ is an } r\text{-dimensional subcode of } C\}. \tag{41}$$

Obviously, $d_1(C)$ is just the minimum Hamming weight of the code C. The *weight hierachy* of C is then defined to be the set of generalized Hamming weights

$$\{d_1(C), d_2(C), \cdots, d_k(C)\}. \tag{42}$$

It has been shown in [41] that the weight hierachy of a linear code completely characterizes the performance of the code on the type II wire-tap channel. In addition, it has been pointed out that, for a related cryptographical application, the generalized Hamming weights also characterize a linear code's performance as a t-resilient function [42] in every detail.

Using our results of the geometry of orthogonal groups over finite field, the author [43] computed the weight hierachies of the q-ary projective codes from non-degenerate quadrics in projective spaces and their duals.

The corresponding problems for the projective codes from nonsingular Hermitian varieties in projective spaces over $\mathbb{F}_{q^2}$ was studied by I. M. Chakravarti [45], and J. W. P. Hirschfeld, M.A. Tsfasman and S.G. Vladut [46]. More precisely, the former proved that the code has only two distinct nonzero weights and computed the values and multiplicities of them, and the latter three computed the weight hierachies of the code and its dual code. Moreover, Chakravarti [45] also determined the number of distinct nonzero weights of codes from some singular Hermitian varieties or quadrics and computed their values and multiplicities.

In the terminology of V.K.Wei and K.Yang [44], a linear $[n, k]$-code C is said to satisfy the *chain condition*, if there exists r-dimensioual subcode D_r of C for $1 \leq r \leq k$ such that

$$|\chi(D_r)| = d_r(C), \quad r = 1, 2, \cdots, k, \tag{43}$$

and

$$D_1 \subseteq D_2 \subseteq \cdots \subseteq D_k. \tag{44}$$

In their analysis of the weight hierachy of product codes, they showed that it is important to know if a code satisfies the chain condition or not. They also showed that the Hamming codes, Reed-Muller codes, MDS codes, and the extended Golay codes satisfy the chain condition. Moreover, they showed that if C satisfies the chain condition so does its dual code $C^{\perp}$.

Using the geometry of orthogonal groups over finite fields, the author [43] proved that the projective codes C_Q from nondegenerate quadrics Q in $PG(k - 1, \mathbb{F}_q)$ satisfy the chain condition.

7. Application to Authentication Codes

Let $\mathcal{S}, \mathcal{E}$, and $\mathcal{M}$ be three nonempty finite sets and let $f : \mathcal{S} \times \mathcal{E} \rightarrow \mathcal{M}$ be a map. The four tuple $(\mathcal{S}, \mathcal{E}, \mathcal{M}; f)$ is called an *authentication code* [47], if

1) The map $f : \mathcal{S} \times \mathcal{E} \to \mathcal{M}$ is surjective and

2) For any $m \in \mathcal{M}$ and $e \in \mathcal{E}$, if there is an $s \in \mathcal{S}$ satisfying $f(s, e) = m$, then such an s is uniquely determined by the given m and e.

Suppose that $(\mathcal{S}, \mathcal{E}, \mathcal{M}; f)$ is an authentication code. Then $\mathcal{S}, \mathcal{E}$, and $\mathcal{M}$ are called the set of *source states*, the set of *encoding rules*, and the set of *messages*, respectively, and f is called the *encoding map*. Let $s \in \mathcal{S}, e \in \mathcal{E}$, and $m \in \mathcal{M}$ be such that $m = f(s, e)$. Then we say that the source state s is encoded into the message m under the encoding rule e, and for convenience we say that the message m contains the encoding rule e. The cardinals $|\mathcal{S}|, |\mathcal{E}|, |\mathcal{M}|$ are called the *size parameters* of the code . Moreover if the authentication code satisfies the further requirement that given any message m there is a unique source state s such that $m = f(s, e)$ for every encoding rule e contained in m, then the code is called a *Cartesian* authentication code.

Authentication codes are used in communication channels where besides the transmitter and the receiver there is an opponent who may play either the impersonation attack or the substitution attack. By an *impersonation attack* we mean that the opponent sends a message through the channel to the receiver and hopes the receiver will accept it as authentic. i.e., as a message sent by the transmitter. By a *substitution attack* we mean that after the opponent intercepts a message sent by the transmitter to the receiver, he sends another message instead and hopes the receiver will accept it as authentic. To protect against these attacks the transmitter-receiver may use an authentication code which is publicly known and choose a fixed encoding rule e in secret. The set of information which the transmitter would like to be able to transmit to the receiver should be identified with the set of source states of the code. Suppose that the transmitter wants to send a source state s to the receiver, he first encodes s into a message m under the encoding rule e, i.e., $m = f(s, e)$, and then sends m to the receiver. Once the receiver receives a message m', he first has to judge whether m' is authentic, i.e., whether the encoding rule e is contained in m'. If $e \in m'$. then he regards m' as authentic and decodes m' by e to get a source state s', where $m' = f(s', e)$. If $e \notin m'$ then he regards m' as a false message. The object of the opponent is to choose a message and send it to the receiver so that the probability of deceiving the receiver, i.e., of causing him to accept as authentic a message not sent by the transmitter is as large as possible. We denote by P_I and P_S, respectively, the largest probabilities that he could deceive the

receiver when he plays an impersonation attack and a substitution attack
and call them the probabilities of a successful impersonation attack and
of a successful substitution attack, respectively.

In [48] some authentication codes based on projective geometry over
finite fields were constructed. Projective geometry, according to Klein's
Erlangen Program, is the geometry of the projective general linear group.
Then it is natural to propose the problem whether it is possible to construct authentication codes from the geometry of symplectic, unitary, or
orthogonal groups over finite fields. The answer is of course positive and
some authentication codes have been so constructed [49–52]. To illustrate
we give a construction [52] below.

Consider the 2ν-dimensional symplectic space over $\mathbb{F}_q$, i.e., the 2ν-
dimensional row vector space $\mathbb{F}_q^{(2\nu)}$ on which the symplectic group $Sp_{2\nu}(\mathbb{F}_q)$
acts. Assume that $\nu \geq 2$ and let s be an integer such that $1 \leq s < \nu$.
Let P_0 be a fixed subspace of type $(s,0)$. Take the set of subspces of
type $(2s,s)$ containing P_0 to be the set S of source states, the set of s-
dimensional subspaces whose joins with P_0 are subspaces of type $(2s,s)$
to be the set $\mathcal{E}$ of encoding rules and also the set $\mathcal{M}$ of messages. For any
source state s and encoding rule e, let $f(s,e) = s \cap e^\perp$, where

$$e^\perp = \{x \in \mathbb{F}_q^{(2\nu)} \mid xK\,^te = 0\}.$$

It can be proved that $s \cap e^\perp$ is an s-dimensional subspace whose join with
P_0 is of type $(2s,s)$. Thus we may define $f(s,e) = s \cap e^\perp$ to be the message
into which the source state s is encoded using the encoding rule e. Then
a Cartesian authentication code is obtained and its size parameters are

$$|S| = q^{2s(\nu-s)}, \quad |\mathcal{E}| = |\mathcal{M}| = q^{s(2\nu-s)}.$$

Now assume that the encoding rules are chosen according to a uniform
probability distribution. Then the probabilities of a successful impersonation attack and a successful substitution attack are, respectively,

$$P_I = \frac{1}{q^{s^2}}, \quad P_S = \frac{1}{q^s}.$$

In virtue of the combinatorial lower bounds $P_I \geq |S|/|\mathcal{M}|$ and $P_S \geq$
$(|S| - 1)/(|\mathcal{M}| - 1)$, for the authentication code constructed above P_I is
optimal. If we require the order of magnitude of P_S as a function of q to
be optimal, then for the code, P_S is nearly optimal when and only when
$s = 1$.

Similar constructions can be done for the unitary and orthogonal cases. Finally, it should be added that the geometry of classical groups was also used in the study of correlation properties of binary m-sequences [53–56, 23].

References

[1] E. Witt, Theorie der quadratischen Formen in beliebigen Körpern, *J. Reine Angew. Math.*, **176**(1937), 31–44.

[2] C. Arf, Untersuchungen über quadratischen Formen in Körpern der Charakteristik 2, Teil I, *J. Reine Angew. Math.*, **183**(1941), 148–167.

[3] J. Dieudonné, *Sur les groupes classiques*, Hermann, Paris, 1948.

[4] J. Dieudonné, On the structure of unitary groups, *Trans. Amer. Math. Soc.*, **72**(1952), 367–385.

[5] L. K. Hua, A generalization of Hermitian matrices, *Acta Scientia Sinica*, **2**(1953), 1–58.

[6] V. Pless, On Witt's theorem for nonalternating symmetric bilinear forms over a field of characteristic 2, *Proc. Amer. Math. Sco.*, **15**(1964), 979–983.

[7] X. Feng and Z.Dai, Notes on finite geometries and the construction of PBIB designs V, Some " Anzahl " theorem in orthogonal geometry over finite fields of characteristic 2, *Acta Scientia*, **13**(1964), 2005–2008, Also, Studies in finite geometries and the construction of incomplete block designs V, Some "Anzahl" theorems in orthogonal geometry over finite fields of characteristic 2, *Acta Mathematica Sinica*, **15**(1965), 664–682 (in Chinese); English translation: *Chinese Mathematics*, **7**(1965), 392–410.

[8] Z. Dai, On transitivity of subspaces in orthogonal geometry over fields of characteristic 2, *Acta Mathematica Sinica*, **16**(1966), 545–560 (in Chinese). English translation: *Chinese Mathematics*, **8**(1966), 569–584.

[9] B. Segre, On Galois geometries, *Proc. Intern. Congress Math.* (Edinburgh, 1958), Cambridge, 1960, 488–499. Also, Le geometrie di Galois, *Annali di Mathematica Pura ed Applicata*, **48**(1959), 1–97.

[10] D. K. Ray-Chaudhuri, Some results on quadrics in finite projective geometry based on Galois fields, *Canadian Journal of Mathematics*, **14** (1962), 129–138.

[11] Z, Wan, Notes on finite geometries and the construction of PBIB designs I, Some "Anzahl" theorems in symplectic geometry over finite fields, *Acta Scientia Sinica*, **13**(1964), 515–516. Also, Studies in finite geometries and the construction of incomplete block designs I, Some " Anzahl " theorems in symplectic geometry over finite fields, *Acta Mathematica Sinica*, **15**(1965), 354–361 (in Chinese); English translation: *Chinese Mathematics*, **7**(1965), 55–62.

[12] Z. Wan and B. Yang, Notes on finite geometries and the construction of PBIB designs III, Some "Anzahl" theorems in unitary geometry over finite fields and their applications, *Acta Scientia Sinica*, **13**(1964), 1006–1007. Also, Studies in finite geometries and the construction of incomplete block designs III, Some "Anzahl" theorems in unitary geometry over finite fields and their applications, *Acta*

Mathematica Sinica, **15**(1965), 533–544 (in Chinese); English translation: *Chinese Mathematics*, **7**(1965), 252–264.

[13] Z. Dai and X. Feng, Notes on finite geometries and the construction of PBIB designs IV, Some "Anzahl" theorems in orthogonal geometry over finite fields of characteristic not 2, *Acta Scientia Sinica*, **13**(1964), 2001–2004. Also, Studies in finite geometries and the construction of incomplete block designs IV, Some "Anzahl" theorems in orthogonal geometry over finite fields of characteristic $\neq 2$. *Acta Mathematica Sinica*, **15**(1965), 545–558 (in Chinse); English translation: *Chinese Mathematics*, **7**(1965), 265–280.

[14] Z. Wan, Z. Dai, X. Feng, and B. Yang, *Studies on Finite Geometry and the Construction of Incomplete Block Designs*, Science Press, Beijing, 1966. (In Chinese.)

[15] V. Pless, The number of isotropic subspaces in a finite geometry, *Atti Accad. Naz. Lincei. Rend.*, (8) **39**(1965), 418–421.

[16] B. Segre, Forme e geometrie hermitiane con particolare riguardo al caso finito, *Ann. Mat. Pura Appl.*, (4) **70** (1965), 1–201.

[17] R. C. Bose and I. M. Chakravati, Hermitian varieties in a finite projective space $PG(N, q^2)$, *Canad. J. Math.*, **18**(1966), 1161–1182.

[18] Z. Wan, On the symplectic invariants of a subspace of a vector space, *Acta Mathematica Scientia*, **11**(1991), 251–253.

[19] Z. Wan, On the unitary invariants of a subspace of a vector space over a finite field, *Chinese Science Bulletin*, **37**(1992), 705–707.

[20] Z. Wan, On the orthogonal invariants of a subspace of a vector space over a finite field of odd characteristic, *Linear Algebra and Its Applications*, **184**(1993), 123–133.

[21] Z. Wan, On the orthogonal invariants of a subspace of a vector space over a finite field of even characteristic, *Linear Algebra and Its Applications*, **184**(1993), 135–143.

[22] Z. Wan, Some Anzahl theorems in finite singular symplectic, unitary and orthogonal geometries, accepted for publication in *Discrete Mathematics*.

[23] Z. Wan, Further studies on singular symplectic, unitary, and orthogonal geometries over finite fields, accepted for publication in *Southeast Asian Bulletin of Mathematics*.

[24] Y. Liu and Z. Wan, Pseudo symplectic geometries over finite fields of characteristic two, *Recent Advances on Finite Geometries and Designs*, ed. by J. Hirschfeld et al., Oxford University Press, 1991, 265–288.

[25] Z. Wan, Singular pseudo symplectic geometry over finite fields of characteristic 2, *Northeastern Mathematical Journal*, **8**(1992), 391–416.

[26] Z. Wan, Quadrics in $AG(n, \mathbb{F}_q)$ for q odd, *Chinese Science Bulletin*, **36**(1991), 2014–2015.

[27] Z. Wan, Quadrics in $AG(n, \mathbb{F}_q)$ for q even, *Chinese Science Bulletin*, **36**(1991), 2016–2017.

[28] Z. Wan, *Geometry of Classical Groups over Finite Fields*, studentlitteratur, Lund, 1993.

[29] R. C. Bose and T. Shimamato, Classification and analysis of partially balanced incomplete block designs with two associate classes, *Journal of American Statistical Association*, **47** (1952), 151–184.

[30] R. C. Bose and K. R. Nair, Partially balanced incomplete block designs, *Sankhya*, **4**(1939), 337–372.

[31] F. Yates, Incomplete randomized blocks, *Annals of Eugenics*, **7**(1936), 121–140.

[32] W. H. Clatworthy, A geometrical configuration which is a partially balanced incomplete block design, *Proc. Amer. Math. Soc*, **5**(1954), 47–55.

[33] D. K. Ray-Chaudhuri, Application of the geometry of quadrics for constructing PBIB designs, *Annals of Mathematical Statistics*, **33**(1962), 1175–1186.

[34] Z. Wan, Notes on finite geometries and the construction of PBIB designs II, Some PBIB designs with two associate classes based on the symplectic geometry over finite fields, *Scientia Sinica*, **13**(1964), 516–517. Also, Studies in finite geometries and the construction of incomplete block designs II, Some PBIB designs with two associate classes based on symplectic geometry over finite fields, *Acta Mathematica Sinica*, **15**(1965), 362–371 (in Chinese); English translation: *Chinese Mathematics*, **7**(1965), 63–72.

[35] B. Yang, Studies in finite geometries and the construction of incomplete block designs VII, Association schemes with several associate classes by taking the maximal totally isotropic subspaces in the symplectic geometry over finite fields as treatments, *Acta Mathematica Sinica*, **15**(1965), 812–825. (in Chinese); English translation: *Chinese Mathematics,* **7**(1965), 547–560.

[36] B. Yang, Studies in finite geometries and the construction of incomplete block designs VIII, Association schemes with several associate classes by taking the maximal totally isotropic subspaces in the unitary geometry over finite fields as treatments, *Acta Mathematica Sinica*, **15**(1965), 826–841. (in Chinese); English translation: *Chinese Mathematics*, **7**(1965), 561–576.

[37] Z. Wan, Finite geometries and block designs, *Sankhyā: The Indian Journal of Statistics, Special volume*, **54**(1992), 531–543.

[38] Z. Wan, Notes on finite geometries and the construction of PBIB designs VI, Some association schemes and PBIB designs based on finite geometries, *Acta Scientia Sinica*, **14**(1965), 1872–1876.

[39] Primrose, Quadrics in finite geometries, *Proceedings of Cambridge Philosophical Society*, **47**(1951), 299–304.

[40] J. Wolfman, Codes projectifs à deux ou trois poids associés aux hyperquadriques d'une géométrie finie, *Discrete Math.*, **13**(1975), 185–211.

[41] V. K. Wei, Generalized Hamming weights for linear codes, *IEEE Transations on Information Theory*, **IT–37**(1991), 1412–1418.

[42] B. Chor, O. Goldreich, J. Hastad, J. Friedmann, S. Rudish, and R. Smolesky, The bit extraction problem of t-resilient functions, *Proc 26th Symposium on Foundations of Computer Science*, 1985, 396–407.

[43] Z. Wan, The weight hierachies of the projective codes from nondegenerate quadrics, accepted for publication in *Designs, Codes and Cryptography*.

[44] V. K. Wei and K. Yang, On the generalized Hamming weights of squaring construction and product codes, preprint.

[45] I. M. Chakravarti, Families of codes with few distinct weights from singular and non-singular Hermitian varieties and quadrics in projective geometries and Hadamard difference sets and designs associated with two-weight codes, *IMA Volumes in Math. and Its Applications.*, **20** Springer, 1990, 35–50.

[46] J. W. P. Hirschfeld, M. A. Tafasman, and S. G. Vladut, The weight hierachy of higher-dimensional Hermitian codes, to appear in *IEEE Transactions on Information Theory*.

[47] G. Simmons, Authentication theory/coding theory, *Advances in Cryptology, Proc. of Crypto 84, Lecture Notes in Computer Science*, No. 196, Springer, 1985, 411–431.

[48] E. N. Gilbert, F. J. MacWilliams, and N. J. A. Sloane, Codes which detect deception, *Bell System Technical Journal* 53(1974), 405–424.

[49] Z. Wan, B. Smeets and P. Vanroose, On the construction of authentication codes over symplectic spaces, to appear in *IEEE Transactions on Information Theory*.

[50] Z. Wan, Further constructions of Cartesian authentication codes from symplectic geometry, *Northeastern Mathematical Journal*, 8(1992), 4–20.

[51] Z. Wan, Construction of Cartesian authentication codes from unitary geometry, *Designs, Codes and Cryptography*, 2(1992), 333–356.

[52] R. Feng and Z. Wan, A construction of Cartesian authentication codes from geometry of classical groups, accepted for publication in *Journal of Combinatorics, Information and System Sciences*.

[53] T. Høholdt and J. Justeen, Tenary sequences with perfect periodic autocorrelation, *IEEE Transactions on Information Theory*, IT–19(1983) 597–600.

[54] R. A. Games, The geometry quadrics and correlations of sequences, *IEEE Transactions on Information Theory*, IT–32(1986), 423–426.

[55] R. A. Games, The geometry of m-sequences: three-valued crosscorrelations and quadrics in finite projective geometry, *SIAM J. Alg. Disc. Math.*, 7(1986), 43–52.

[56] W. A. Jackson and P. R. Wild, Relations between two perfect sequence constructions, *Designs. Codes and Cryptography*, 2(1992), 325–332.

Representation Theory of Algebraic and Quantum Groups in China

Jianpan Wang*

Department of Mathematics, East China Normal University
Shanghai 200062, The People's Republic of China

Jiachen Ye

Department of Applied Mathematics, Tongji University
Shanghai 200092, The People's Republic of China

§1. Achievements in Modular Representations of Algebraic Groups

1.1 Let G be a connected simply-connected simple algebraic group over an algebraically closed field K of characteristic $p > 0$. Let $B \subset G$ be a Borel subgroup, and $T \subset G$ a maximal torus. Denote by $X(T) = X(B)$ the character (or weight) group, which, as is well-known, is the (integral) weight lattice in the real space $\mathbf{E}$ spanned by the root system R associated to (G, T). Choose a set of positive roots R_+ in such a way that $-R_+$ corresponds to B. We call

$$X(T)_+ = \{\lambda \in X(T) \mid \langle \lambda, \alpha^\vee \rangle \geq 0, \quad \text{for all } \alpha \in R_+\}$$

the set of dominant weights, where $\alpha^\vee = 2\alpha/\langle \alpha, \alpha \rangle$ is the coroot of α and $\langle \cdot, \cdot \rangle$ is the Euclidean inner product on $\mathbf{E}$ invariant under the natural action of the Weyl group $W = N_G(T)/T$.

Any finite dimensional T-module V has an isotypical decomposition

$$V = \bigoplus_{\lambda \in X(T)} V_\lambda,$$

* Both the authors are supported by the National Natural Science Foundation of China, and the first author is also supported by the Science Foundation of the University Doctoral Program of CNEC

where $V_\lambda = \{v \in V \mid tv = \lambda(t)v, t \in T\}$. The formal character of V, $\mathrm{ch}(V)$, by definition, is

$$\mathrm{ch}(V) = \sum_{\lambda \in X(T)} (\dim V_\lambda)\, e(\lambda) \in Z[X(T)].$$

The formal character of a G-module or a B-module is its formal character as a T-module obtained by restriction.

Let E be a finite dimensional B-module. We set

$$H^0(E) = H^0(G/B, E)$$
$$= \{f\colon G \to E \mid f \text{ is polynomial and } f(gb) = b^{-1}f(g),\ g \in G,\ b \in B\},$$

which is the G-module induced by E. The G-action on $H^0(E)$ is given by

$$(gf)(x) = f(g^{-1}x), \quad f \in H^0(E),\ x, g \in G.$$

The assignment $E \mapsto H^0(E)$ defines a left exact functor from the category of rational B-module to the category of rational G-modules. Moreover, we have its right derived functors $H^i(*) = H^i(G/B, *)$.

It is well-known that for $\lambda \in X(T)_+$ the G-module $H^0(\lambda)$, which is induced from the 1-dimensional B-module K_λ, contains a unique simple G-submodule $L(\lambda)$, and in this way $X(T)_+$ parameterizes the finite dimensional simple G-modules. Moreover, $\mathrm{ch}(H^0(\lambda))$ is given by Weyl's character formula, i.e.,

$$\mathrm{ch}(H^0(\lambda)) = \chi(\lambda) = \frac{\sum_{w \in W} \det(w)e(w(\lambda + \rho))}{\sum_{w \in W} \det(w)e(w\rho)},$$

where ρ is half the sum of positive roots. The most important problem in the representation theory of algebraic groups is to determine the formal characters of simple modules. However, this is also the biggest open problem in this area — $\mathrm{ch}(L(\lambda))$'s are unknown in general except that G is of type A_1, A_2, A_3, B_2 or G_2. Obviously, the problem of calculating $\mathrm{ch}(L(\lambda))$ for $\lambda \in X(T)_+$ is equivalent to that of calculating the composition factor multiplicity $[H^0(\lambda) : L(\mu)]$ of $L(\mu)$ in $H^0(\lambda)$ for $\lambda, \mu \in X(T)_+$.

One may define a new action, the dot action, of W on $X(T)$ as follows:

$$w \cdot \lambda = w(\lambda + \rho) - \rho, \quad \text{for all } \lambda \in X(T),\ w \in W.$$

The affine Weyl group W_p of G is the group generated by $s_{\alpha,np}$ for $\alpha \in R_+$ and $n \in Z$, where $s_{\alpha,np}$ is the affine reflection which takes $\lambda \in X(T)$

into $s_\alpha \cdot \lambda + np\alpha$. With this notation, we have the following conjecture, known as Lusztig's conjecture [L1], [A1], on the direction of calculating $[H^0(\lambda) : L(\mu)]$:

Conjecture *Assume $p \geq h$ and fix $\lambda \in X(T)_+$ such that $\langle \lambda + \rho, \alpha^\vee \rangle < p$ for all $\alpha \in R_+$. Then for all $y, w \in W_p$ with $yw_0 \cdot \lambda, ww_0 \cdot \lambda \in X(T)_+$ and $\langle ww_0 \cdot \lambda + \rho, \alpha^\vee \rangle \leq p(p - h + 2)$ for all $\alpha \in R_+$ one has*

$$[H^0(ww_0 \cdot \lambda) : L(yw_0 \cdot \lambda)] = Q_{y,w}(1).$$

Here h is the Coxeter number, w_0 is the longest element in W, and $Q_{y,w}$'s are the "inverse" Kazhdan-Lusztig polynomials, i. e., the polynomials determined by the inversion formula

$$\sum_{z \in W_p} (-1)^{l(z)-l(y)} P_{y,z} Q_{z,w} = \delta_{y,w},$$

where $P_{y,z}$ denotes the Kazhdan-Lusztig polynomial associated to $y, z \in W_p$, and l is the length function on W_p. Although the conjecture only deal with the problem for λ p-regular, Jantzen's translation principle [J4] allows us to deduce the general case from the p-regular case. It should be mentioned here that this conjecture is verified only for groups of rank ≤ 2 [L2] and for the group of type A_3 [Y12].

1.2 We define a good filtration of a rational G-module V to be an ascending chain of submodules of V

$$0 = V_0 \subset V_1 \subset V_2 \subset \cdots$$

such that $V = \bigcup_{i=1}^\infty V_i$ and for each $i > 0$, and V_i/V_{i-1} is either 0 or isomorphic to $H^0(\lambda_i)$ for some $\lambda_i \in X(T)_+$. One has the following deepest result on good filtrations.

Theorem *If both G-modules V and V' admit good filtrations, then so does $V \otimes V'$.*

This result was first obtained by Wang Jianpan [W1] in the case that G is of type A_n or that char K is large compared to the rank of G; then S. Donkin [D1] proved it, with a case-by-case analysis, under the assumption that G has no component of type E_7 and E_8 or char $K > 2$. Finally, O. Mathieu [M1] completed the proof of the theorem by using Ramanathan's "Frobenius splitting".

1.3 We call a finite dimensional rational G-module V cyclic if it is generated by a single element, and cocyclic if its dual V^* is cyclic. Then we call G a c. c. group if the conditions of being cyclic and cocyclic are equivalent for G, and a c. r. group if every rational G-module is completely reducible. Wang Jianpan [W7], [W8] proved that G is a c. c. group if and only if G is a c. r. group, i. e., G is a torus. (Recall that we assume that char $K = p > 0$. In the case char $K = 0$, the result should be modified as follows: If G has no non-trivial unipotent quotient, then G is a c. c. group if and only if G is a c. r. group, i. e., a reductive group.)

1.4   The pth power map on K induces an endomorphism on $GL(n, K)$, namely the map which raise the entries to their pth power. In the same way, we get the Frobenius morphism F on G. Let $G_n \subset G$ denote the scheme-theoretic kernel of F^n. This is a normal subgroup scheme of G with just a single point over K. Let $V^{[n]}$ be the Frobenius twist for any G-module V, that is, the representation of G on V given by the composition $\cdot G \xrightarrow{F^n} G \longrightarrow GL(V)$. It is well-known that $V^{[n]}$ is trivial when considered as a G_n-module and that any G-module that becomes trivial upon restriction to G_n is of this form. Set

$$X_n(T) = \{\lambda \in X(T)_+ \mid \langle \lambda, \alpha^\vee \rangle < p^n \text{ for all simple roots } \alpha\}.$$

Then the $L(\lambda)$'s with $\lambda \in X_n(T)$ remain simple upon restriction to G_n and any simple G_n-module is isomorphic to exactly one of these. For $\lambda \in X(T)_+$ one has the "p^n-adic" decomposition

$$\lambda = \lambda_0 + p^n \lambda_1, \qquad \lambda_0 \in X_n(T), \ \lambda_1 \in X(T)_+.$$

Then Steinberg's tensor product theorem [S] says that

$$L(\lambda) \cong L(\lambda_0) \otimes L(\lambda_1)^{[n]}.$$

Let us assume that both T and B are stable under F^n so that we also have Frobenius kernels T_n and B_n, and let us consider the subgroup schemes $G_n T$ and $G_n B$ of G. For any B-module E, we define

$$\widehat{Z}_n(E) = H^0(G_n B/B, E).$$

It is also known that $\widehat{Z}_n(\lambda)$ with $\lambda \in X(T)$ contains a unique simple submodule $\widehat{L}_n(\lambda)$ and in this way $X(T)$ parameterizes the simple G_n-modules. Moreover, one has for $\lambda \in X_n(T)$ and $\lambda' \in X(T)$ that

$$\widehat{L}_n(\lambda) \cong L(\lambda)|_{G_n B},$$
$$\widehat{L}_n(\lambda + p^n \lambda') \cong \widehat{L}_n(\lambda) \otimes p^n \lambda',$$
$$\widehat{Z}_n(\lambda + p^n \lambda') \cong \widehat{Z}_n(\lambda) \otimes p^n \lambda'.$$

For $\lambda \in X(T)$ we let $\widehat{Q}_n(\lambda)$ denote the injective G_nT-hull of $\widehat{L}_n(\lambda)$. Then

$$\widehat{Q}_n(\lambda + p^n\lambda') \cong \widehat{Q}_n(\lambda) \otimes p^n\lambda'.$$

We call a G-module p^n-bounded if all its weights μ satisfy $\langle \mu, \alpha^\vee \rangle < 2p^n(h-1)$ for all $\alpha \in R_+$. When $p \geq 2(h-1)$, $\widehat{Q}_n(\lambda)$ with $\lambda \in X_n(T)$ is a G-summand of $St_n \otimes H^0((p^n-1)\rho + w_0(\lambda))$, where St_n is the Steinberg module. So $\widehat{Q}_n(\lambda)$ has a G-module structure and admits a good filtration. Moreover, it can be characterized as the injective hull of $L(\lambda)$ in the category of p^n-bounded G-modules.

1.5   Let us now consider the case $n = 1$. Set

$$C_0 = \{\lambda \in \mathbf{E} \mid 0 < \langle \lambda + \rho, \alpha^\vee \rangle < p \quad \text{for all } \alpha \in R_+\},$$

the lowest alcove. J. C. Jantzen proved the following results:

(1) Let $\lambda, \mu \in X(T)_+$. Then

$$[H^0(\lambda) : L(\mu)] = \sum_\nu [\widehat{Z}_1(\lambda) : \widehat{L}_1(\nu)]\, [L(\nu_0) \otimes \chi(\nu_1)^{[1]} : L(\mu)],$$

where $\nu = \nu_0 + p\nu_1$ with $\nu_0 \in X_1(T)$.

(2) If $[\widehat{Z}_1(\lambda) : \widehat{L}_1(\nu)] \neq 0$ for $\nu_1 \notin \overline{C}_0$, then one has

$$[H^0(\lambda) : L(\mu)] = [\widehat{Z}_1(\lambda) : \widehat{L}_1(\mu)].$$

(3) Let $\mu \in X_1(T)$, $\lambda, \nu \in X(T)$ and $w \in W$. Then

$$[\widehat{Z}_1(\lambda + p\rho) : \widehat{L}_1(\mu)] = [\widehat{Z}_1(w \cdot \lambda + p\rho) : \widehat{L}_1(\mu)]$$
$$= [\widehat{Z}_1(w \cdot \lambda + p(\nu + \rho)) : \widehat{L}_1(\mu + p\nu)].$$

(4) $\widehat{Q}_1(\lambda)$ with $\lambda \in X(T)_+$ has a G_1B-filtration in which the successive quotients have the form $\widehat{Z}_1(\mu)$ for $\mu \in X(T)$. Let $(\widehat{Q}_1(\lambda) : \widehat{Z}_1(\mu))$ denote the number of times that $\widehat{Z}_1(\mu)$ occurs in this filtration. Then one has the reciprocity

$$(\widehat{Q}_1(\lambda) : \widehat{Z}_1(\mu)) = [\widehat{Z}_1(\mu) : \widehat{L}_1(\lambda)].$$

1.6   For $\lambda \in X_1(T)$ Ye Jiachen [Y3] defined a set

$$S_{\lambda,1} = \{\xi \in X(T) \mid \xi + \rho \in X(T)_+, \ \xi \text{ is strongly linked to } w_0 \cdot \lambda + p\rho\},$$

and proved the following result, which can be viewed as a converse to the linkage principle [J4].

Theorem *Suppose that $p \geq 2(h-1)$. Let $\lambda = \lambda_0 + p\lambda_1 \in X(T)$ with $\lambda_0 \in X_1(T)$ p-regular. Then*

$$\{\mu \in X(T) \mid [\widehat{Z}_1(\mu) : \widehat{L}_1(\lambda)] \neq 0\} = \{w \cdot \xi + p(\rho + \lambda_1) \mid \xi \in \mathcal{S}_{\lambda_0,1},\ w \in W\}.$$

In particular, we know the composition factors of $H^0(\mu)$ for $\mu \in X(T)_+$, but we still do not know the multiplicities with which they occur.

Bai Yuanhuai [B1], [B2], [B3], [B5] described the composition factor multiplicities of simple G-modules in cohomology groups of line bundle on G/B. Also, Bai Yuanhuai, Wang Jianpan and Wen Kexin [BWW] considered the socle series of these cohomology groups for G.

1.7 Wang Jianpan [W4], [W5], [W9] considered the category of quasirational G-modules. A G-module V is called quasirational, if it is locally finite and each finite dimensional submodule of V can be rationalized by a Frobenius twist. This category is an abelian category, and it admits enough injectives. Hence one may define right derived functors $\widetilde{R}^i \mathcal{F}_G$ in this category for the fixed point functor $\mathcal{F}_G$. He proved the following theorem.

Theorem *Let V be a rational G-module. Then*

(1) *The generic cohomology $H^*_{\mathrm{gen}}(G, V)$ coincides with $\widetilde{R}^* \mathcal{F}_G(V)$.*

(2) $\varprojlim_r H^*(G, V^{[r]})$ *coincides with $\widetilde{R}^* \mathcal{F}_G(V)$.*

(3) *Consider the Z_+-inverse system*

$$G_0 \xleftarrow{\mu_{10}} G_1 \xleftarrow{\mu_{21}} G_2 \xleftarrow{\mu_{32}} G_3 \longleftarrow \cdots$$

with $G_r = G$ and $\mu_{r+1,r} = F$ for all $r \in Z_+$. Let $\widetilde{G} = \varprojlim G_r$, the inverse limit of the inverse system. Then

$$H^*_{\mathrm{gen}}(G, V) \cong H^*(\widetilde{G}, V).$$

1.8 From the following spectral sequence

$$\mathrm{Ext}^i_{G/G_1}(L(\mu_1)^{[1]}, \mathrm{Ext}^j_{G_1}(L(\mu_0), L(\lambda))) \Longrightarrow \mathrm{Ext}^{i+j}_G(L(\mu), L(\lambda)),$$

we get the Lyndon-Hochschild-Serre five-term exact sequence

$$
\begin{aligned}
0 &\longrightarrow \mathrm{Ext}^1_G(L(\mu_1), \mathrm{Hom}_{G_1}(L(\mu_0), L(\lambda))^{[-1]}) \\
&\longrightarrow \mathrm{Ext}^1_G(L(\mu), L(\lambda)) \\
&\longrightarrow \mathrm{Hom}_G(L(\mu_1), \mathrm{Ext}^1_{G_1}(L(\mu_0), L(\lambda))^{[-1]}) \\
&\longrightarrow \mathrm{Ext}^2_G(L(\mu_1), \mathrm{Hom}_{G_1}(L(\mu_0), L(\lambda))^{[-1]}) \\
&\longrightarrow \mathrm{Ext}^2_G(L(\mu), L(\dot{\lambda})).
\end{aligned}
$$

Andersen [A2] showed for $p \geq 3h - 3$ and for $\lambda, \mu \in X_1(T)$ with $\lambda \neq \mu$ that $\mathrm{Ext}^1_{G_1}(L(\mu), L(\lambda))$ is a completely reducible G-module, and its simple G-summands are of the form $L(\nu)^{[1]}$ with $\mu + p\nu \leq w_0 \cdot \lambda + 2p\rho$. Here the partial ordering $\leq$ in $X(T)$ is defined by $\mu \leq \lambda$ if and only if $\lambda - \mu$ can be written as a linear combination of positive roots with non-negative integer coefficients. Moreover, the number of times $L(\nu)^{[1]}$ occurring in $\mathrm{Ext}^1_{G_1}(L(\mu), L(\lambda))$ is equal to the dimension of $\mathrm{Ext}^1_G(L(\mu + p\nu), L(\lambda))$; that is

$$[\mathrm{Ext}^1_{G_1}(L(\mu), L(\lambda)) : L(\nu)^{[1]}] = \dim \mathrm{Ext}^1_G(L(\mu + p\nu), L(\lambda)).$$

It is known that $\mathrm{Ext}^1_G(L(\mu), L(\lambda)) \neq 0$ implies that $\mu \in W_p \cdot \lambda$. Therefore, $\mathrm{Ext}^1_G(L(\mu + p\nu), L(\lambda)) \neq 0$ implies that (1) $\mu + p\nu \in X(T)_+$, and (2) $\mu + p(\nu - \rho) \in W \cdot S_{\lambda,1}$ or $\mu + p\nu$ is strongly linked to λ.

Let $\lambda = \lambda_0 + p\lambda_1$, $\mu = \mu_0 + p\mu_1$ be dominant weights with $\lambda_0, \mu_0 \in X_1(T)$. Once all $\mathrm{Ext}^1_G(L(\mu_0 + p\nu), L(\lambda_0))$ are known for all weights ν with $\mu_0 + p(\nu - \rho) \in X(T)_+ \cap W \cdot S_{\lambda_0,1}$ or $\mu_0 + p\nu$ strongly linked to λ_0, we can in turn calculate $\mathrm{Ext}^1_{G_1}(L(\mu_0), L(\lambda_0))$. Moreover, we can completely determine all $\mathrm{Ext}^1_G(L(\mu), L(\lambda))$ as soon as all $\mathrm{Soc}_G L(\lambda_1) \otimes L(\nu)$ with ν small are known. In this way, Ye Jiachen [Y13], [Y14] completely determined Ext^1_G-groups of simple G-modules for G being of type B_2 and for all prime p. Also Liu Jiachun and Ye Jiachen [LY] computed all Ext^1_G-groups of simple G-modules for G being of type G_2 with $p \geq 13$. As a byproduct, they also determined all $\mathrm{Ext}^1_{G_1}$-groups of simple G_1-modules for these groups.

1.9   Let $H_{\beta,m}$ $(\beta \in R_+, m \in Z)$ denote the reflecting hyperplane with respect to $S_{\beta,mp}$ under the dot action. Then the connected components of $\mathbf{E} \setminus \bigcup_{\beta \in R_+, m \in Z} H_{\beta,m}$ are called alcoves. We denote by $\mathcal{C}$ the set of alcoves, and let $\mathcal{C}^+$ be the set of dominant alcoves, i. e.,

$$\mathcal{C}^+ = \{C \in \mathcal{C} \mid \langle \chi + \rho, \alpha^\vee \rangle > 0 \text{ for all } \alpha \in R_+ \text{ and } \chi \in C\}.$$

By sending w to $w \cdot C_0$, we get a 1–1 correspondence between W_p and $\mathcal{C}$. If $C = w \cdot C_0$ for some $w \in W_p$ and $\lambda \in X(T) \cap C_0$, then we let $\lambda_C = w \cdot \lambda$.

It is known that there are three partial orderings defined on $\mathcal{C}$, and they are equivalent when restricted to $\mathcal{C}^+$ (See [W6] and [Y5]).

Let $A^+_{p\rho} = \{v \in \mathbf{E} \mid \langle \lambda + \rho, \alpha^\vee \rangle > p \quad \text{for all } \alpha \in R_+\}$, and define $\mathcal{C}(v)$ for $v \in A^+_{p\rho}$ to be the collection of alcoves containing v in their closure.

Wen Kexin [Wen] investigated the homomorphism between the cohomology groups of the line bundle on G/B. He proved the following result.

Theorem *Let A, B, C be alcoves in $C(v)$ for some $v \in A_{p\rho}^+$, and let $\lambda \in X(T) \cap C_0$ be a dominant weight. Then*

(1) *If $A > B$ and if $A \subset A_{p\rho}^+$, then $\mathrm{Hom}_G(H^0(\lambda_A), H^0(\lambda_B)) = K$.*

(2) *For such A and B, $\mathrm{Ext}_G^1(H^0(\lambda_A), H^0(\lambda_B)) \neq 0$.*

(3) *If $B \subset A_{p\rho}^+$ also, then the composition of nonzero homomorphisms from $H^0(\lambda_A)$ (resp. $H^0(\lambda_B)$) to $H^0(\lambda_B)$ (resp. $H^0(\lambda_C)$) remains non-zero.*

Moreover, he showed that infinitesimal versions of (1) and (3) for $G_n B$ and G_n are available. In fact, for A and B as in (1), one has

$$\mathrm{Hom}_G(H^0(\lambda_A), H^0(\lambda_B)) \cong \mathrm{Hom}_{G_n B}(\widehat{Z}_n(\lambda_A), \widehat{Z}_n(\lambda_B)).$$

§2. Applications to Modular Representations of Finite Groups of Lie Type

2.1 Let π be the automorphism of G coming from a Dynkin diagram automorphism of G. We denote by $G_\pi(n)$ the finite group consisting of fixed points under $\pi \circ F^n$ in G, which is called a finite group of Lie type. It is reasonable that we study modular representations of finite groups of Lie type via those for the ambient algebraic groups.

It is well-known that the restriction of the simple G-module $L(\lambda)$ to $G_\pi(n)$, denoted by $L_n(\lambda)$, remains simple and that any simple $G_\pi(n)$-module is isomorphic to exactly one of the $L_n(\lambda)$'s with $\lambda \in X_n(T)$. We denote by $U_n(\lambda)$ the projective $G_\pi(n)$-cover of $L_n(\lambda)$. Recall that $L_n(\lambda)$ is also the unique bottom composition factor of $U_n(\lambda)$. When $p \geq 2(h-1)$, the restriction $\widehat{Q}_n(\lambda)$ is also a projective $G_\pi(n)$-module and is decomposed into a direct sum of $G_\pi(n)$-PIM's such that $U_n(\lambda)$ occurs exactly once. We denote by $[\widehat{Q}_n(\lambda) : U_n(\mu)]_{G_\pi(n)}$ the number of times of $U_n(\mu)$ occurring as a $G_\pi(n)$-summand in $\widehat{Q}_n(\lambda)$ for $\lambda, \mu \in X_n(T)$.

2.2 For a finite dimensional G-module V and any $\lambda \in X_n(T)$, we define an $X(T)_+ \times X(T)_+$-matrix $M_n^\lambda(V)$, whose (ν, ν')-entry is

$$M_n^\lambda(V)_{\nu,\nu'} = [V \otimes L(\nu): L(\lambda + p^n \nu')].$$

One has obviously

$$M_n^\lambda(L(\mu))_{0,\nu} = \begin{cases} 1, & \text{if } \lambda = \mu \text{ and } \nu = 0; \\ 0, & \text{otherwise.} \end{cases}$$

Removing the row and the column corresponding to $\nu = 0$ from the matrix $M_n^\lambda(L(\mu))$, one obtains a new matrix, denoted still by $M_n^\lambda(L(\mu))$. Then one has for all $\lambda, \mu \in X_n(T)$ (cf. [Cha1])

$$\text{Tr} M_n^\lambda(L(\mu)) = \begin{cases} [\widehat{Q}_n(\lambda) : U_n(\mu)]_{G_\pi(n)}, & \text{if } \lambda \neq \mu; \\ 0, & \text{otherwise.} \end{cases}$$

Also, we define an $X_n(T)_+ \times X_n(T)_+$-permutation matrix P_π, whose (ν, ν')-entry with $\nu, \nu' \in X(T)_+$ is equal to 1 if $\pi(\nu') = \nu$, or to 0 otherwise. Moreover, Let $\widetilde{S}_{n,k}(\lambda, \mu)$ for $\lambda, \mu \in X_n(T)$ and $k > 0$ be the set of sequences Θ: $\lambda = \theta_0, \theta_1, \cdots, \theta_{k-1}, \theta_k = \mu$ with the property that all $\theta_i \in X_n(T)$. Then, we define, for $\lambda, \mu \in X_n(T)$ and $k > 0$, a new matrix

$$N_{n,k}^\lambda(\mu) = \sum_{\Theta \in \widetilde{S}_{n,k}(\lambda, \mu)} M_n^\lambda(L(\theta_1)) \otimes M_n^{\theta_1}(L(\theta_2)) \otimes \cdots \otimes M_n^{\theta_{k-1}}(L(\mu));$$

define also $N_{n,0}^\lambda(\mu) = (\delta_{\lambda\mu})$.

For $\lambda, \mu \in X_n(T)$, the integer $C_{\lambda\mu} = [U_n(\lambda) : L_n(\mu)]_{G_\pi(n)}$, the multiplicity of $L_n(\mu)$ occurring as composition factors of $U_n(\lambda)$, is called a Cartan invariant of $G_\pi(n)$. In particular, $C_{00} = [U_n(0) : L_n(0)]_{G_\pi(n)}$ is called the first Cartan invariant.

Hu Yuwang and Ye Jiachen [HY] proved the following result (also see [J1] and [J3]).

Theorem   *Let $\lambda, \lambda' \in X_n(T)$. Then*

$$C_{\lambda, \lambda'} = \sum_{\substack{\mu, \mu' \in X_n(T) \\ k, k' \geq 0}} (-1)^{k+k'} \text{Tr}\left(N_{n,k}^\lambda(\mu) \otimes N_{n,k'}^{\lambda'}(\mu') \otimes M_n^{\mu'}(\widehat{Q}_n(\mu)))(P_\pi^{\otimes k} \otimes P_\pi^{\otimes k'} \otimes P_\pi \right),$$

and

$$\dim U_n(\lambda) = \sum_{\substack{\mu \in X_n(T) \\ k \geq 0}} (-1)^k \text{Tr}\left(N_{n,k}^\lambda(\mu) P_\pi^{\otimes k} \right) \dim \widehat{Q}_n(\mu).$$

In particular, if $\lambda = \lambda' = 0$ in the above Theorem, by using the same calculation as in [J3, p.138, Appendix 2], one has the following formulas to calculate the first Cartan invariant $C_{00}^{(n)}$ and $\dim U_n(0)$:

$$C_{00}^{(n)} = \sum_{k, k' \geq 0} (-1)^{k+k'} \text{Tr}\left(\left(\sum_{\mu, \mu' \in X_1(T)} N_{1,k}^0(\mu) \otimes N_{1,k'}^0(\mu') \otimes M_1^{\mu'}(\widehat{Q}_1(\mu)) \right)^n \cdot \right.$$

$$\left. \cdot \left(P_\pi^{\otimes k} \otimes P_\pi^{\otimes k'} \otimes P_\pi \right) \right),$$

and

$$\dim U_n(0) = \sum_{k \geq 0} (-1)^k \text{Tr} \left(\left(\sum_{\mu \in X_1(T)} N_{1,k}^0(\mu) \dim \widehat{Q}_1(\mu) \right)^n P_\pi^{\otimes k} \right).$$

Note that $\text{Tr} N_{n,k}^\lambda(\mu) = 0$ for sufficiently large k. Thus all the above formulas are reasonable.

As an example, Hu Yuwang and Ye Jiachen [HY] calculated the first Cartan invariant for group $G_\pi(n)$, where G is of type B_2, $\pi = 1$ and $p = 7$. The result is

$$C_{00} = a^n + b^n + c^n + d^n + e^n + f^n + g^n + 2^n - 2(\alpha^n + \beta^n + \gamma^n),$$

where a, b, c, d are the roots of $x^4 - 64x^3 + 804x^2 - 2672x + 2048$, e, f, g are the roots of $x^3 - 36x^2 + 256x - 512$, and α, β, γ are the roots of $x^3 - 43x^2 + 312x - 512$.

Earlier, Ye Jiachen [Y6] and [Y7] calculated C_{00} for groups $SL(3, p^n)$, $SU(3, p^n)$ with $p \geq 7$ and $Sp(4, p^n)$ with $p \geq 11$. Also, he wrote down the full Cartan matrices for $SL(3, 7)$, $Sp(4, 7)$ in [Y1], [Y4]. In addition, Chen Chengdong [Chen1] and Du Jie [Du1] calculated the Cartan matrices for $SL(3, 9)$ and for $SL(4, 2)$, $SL(4, 4)$, separately. (Du used computer.) However, the methods used in these papers are different from what mentioned above.

Furthermore, L. Chastkofsky [Cha3] introduced a notion of generic Cartan invariants and showed that it coincides with Cartan invariants for weights in generic position. Based on §1.5 (also see [Y3]), Ye Jiachen [Y10], [Y11] gave another proof of Chastkofsky's theorem. He also showed how the Steinberg's tensor product theorem is used to reduce, through a nice recursion formula, the study of Cartan invariants of arbitrary finite groups of Lie type to the case where the groups are over the prime field and the weights are in generic position. Some new symmetries were also obtained.

2.3 The extensions between simple $G_\pi(n)$-modules are known only in very few cases. It is a reasonable method to determine these Ext-groups for the finite groups of Lie type via those for the ambient algebraic groups. First we should introduce Andersen's theorem [A2].

Theorem (1) Let $\lambda, \mu \in X_n(T)$ with $\langle \lambda + \mu, \alpha_0^\vee \rangle < p^n - p^{n-1} - 1$. Then the restriction map

$$\text{Ext}_G^1(L(\mu), L(\lambda)) \to \text{Ext}_{G_\pi(n)}^1(L_n(\mu), L_n(\lambda))$$

is an isomorphism.

(2) *Let* $\lambda, \mu \in X_n(T)$ *such that the distances from* λ *(resp.* μ*) to walls are at least* $2h - 2$ *(resp.* $h - 1$*). That is, if* c *is an integer such that* p *divides* $\langle \lambda + \rho, \alpha^\vee \rangle + c$ *(resp.* $\langle \mu + \rho, \alpha^\vee \rangle + c$*) for all* $\alpha \in R$*, then* $|c| \geq 2h - 2$ *(resp.* $h - 1$*). Then*

$$\bigoplus \mathrm{Hom}_G(L(\mu), L(\nu_0) \otimes L(\pi\nu_1)) \otimes \mathrm{Ext}_G^1(L(\nu), L(\lambda))$$

$$\cong \mathrm{Ext}_{G_\pi(n)}^1(L_n(\mu), L_n(\lambda)),$$

where the direct summation is over all p^n*-bounded weights* $\nu = \nu_0 + p^n \nu_1$ *with* $\nu_0 \in X_n(T)$ *and* $\nu_1 \in X(T)_+$.

Recently, Ye Jiachen [Y16] has determined all $\mathrm{Ext}_{G_2(p)}^1$-groups between simple $G_2(p)$-modules for the finite group of type G_2 with $p \geq 13$. In addition, Chen Chengdong and Pan Jiezheng [CP] determined all indecomposable modules for the group $SL(2,p)$, and computed all $\mathrm{Ext}_{SL(2,p)}^n$-groups of simple modules through determining submodule structures of all projective indecomposable $SL(2,p)$-modules.

We should mention here that the above methods used to calculate Cartan invariants and to determine Ext-groups are actually effective in low rank cases, but they heavily depend on the solution of the Lusztig conjecture [L1], the biggest open problem in the representation theory of G.

§3. Achievements in Representations of Quantum Groups

3.1 The concept of quantum groups (see, for example, [Dr1], [Dr2], [FRT], [Jim], [M]) is introduced in 1980s by some Russian mathematical physicists in their study of statistical quantum mechanics. They found a class of algebras that are quantizations of the functional algebras of Lie (or algebraic) groups or quantizations of the universal enveloping algebras of Lie algebras, preserving some "group-like" or "Lie-algebra-like" properties. In fact, an algebra with the above-mentioned "group-like" or "Lie-algebra-like" properties is a Hopf algebra. Thus, a quantum group can be thought as an object whose "functional algebra" is a Hopf algebra quantizing the functional algebra of a Lie (or an algebraic) group, or an object whose "Lie algebra" is a Hopf algebra quantizing the universal enveloping algebra of a Lie algebra. The representation theory of a quantum group, therefore, is the theory of comodules over the quantized functional algebra, or the theory of modules over the quantized universal enveloping algebra.

Instead of Drinfel'd and Jimbo's quantized universal enveloping algebra, G. Lusztig [L3], [L4] considered an "arithmetic version", which is constructed via a lattice (over a suitable ring) in Drinfel'd and Jimbo's algebra, and has no difference from Drinfel'd and Jimbo's algebra when the parameter is not a root of unity. Lusztig discovered a parallel between the representation theory of algebraic groups and that of the arithmetic quantized universal enveloping algebra, and raised a conjecture for quantum groups similar to (but simpler than) the Lusztig conjecture in the representation theory of algebraic groups (see §1.1). It is hopeful that the quantum Lusztig conjecture will imply the old Lusztig conjecture, and then a thread of sunshine will light up the representation theory of algebraic groups. However, up to now, neither the quantum Lusztig conjecture nor the implication has been completely proved.

Wen Kexin and his collaborators, H. H. Andersen and P. Polo, made a deep investigation to the representations of (arithmetic) quantized universal enveloping algebras, paying attention on the parallel between the theory of algebraic groups and quantum groups [APW1], [APW2]. As the base ring for Lusztig's lattice, they considered the local ring $\mathcal{A} = Z[\nu]_{\mathfrak{m}}$, where ν is an indeterminate and $\mathfrak{m}$ is the maximal ideal generated by $\nu - 1$ and a fixed odd prime p. Thus, $\mathcal{A}$ has $Q(\nu)$ as its quotient field and has F_p as its residue field. In this way, the Lusztig's lattice U over $\mathcal{A}$ becomes an intermediary between Lusztig's arithmetic quantized universal enveloping algebra and (the hyperalgebra of) the corresponding algebraic group. With this set-up, they obtained the following results:

(1) Introduced the coordinate algebra $\mathcal{A}[U]$ as a dual of U in some sense, and proved that $\mathcal{A}[U]$ is free as an $\mathcal{A}$-module.

(2) Using the coordinate algebra, defined induction functors between certain subalgebras of U, and proved the Frobenius reciprocity, the tensor product identity for induction functors and their derived functors. Studied the behavior of induction functors under base change, getting explicit connections to induction functors in the representation theory of algebraic groups.

(3) Generalized many important results on the sheaf cohomology on the flag variety to the quantum case, including Grothendieck's vanishing theorem, Serre's duality theorem, Borel-Weil-Bott's theory, Kempf's vanishing theorem, linkage and translation principles, as well as the Demazure

character formula and Jantzen's sum formula.

(4) Investigated injective modules for arithmetic quantized universal enveloping algebras $U_K = U \otimes_A K$ for any ring homomorphism $\sigma: A \to K$ with K a field of characteristic 0 and $\sigma(\nu)$ is a primitive pth root of unity, p being an odd prime. They proved that the injective hull of any irreducible module is finite dimensional and its U_K-socle equals its $\mathbf{u}_K$-socle, where $\mathbf{u}_K$ is a finite dimensional sub-Hopf algebra of U_K, similar to the restricted enveloping algebra of the Lie algebra of a semisimple algebraic group in prime characteristic.

Some other authors also made investigations into the parallel of the representation theory of arithmetic quantized universal enveloping algebras and that of semisimple algebraic groups. See, for example, [B7], [B8], [B9], [X1], [X2].

3.2 In [PW1] Wang Jianpan and his collaborator B. Parshall took the other approach to the representations of quantum groups — the approach of comodules over the functional algebras of quantum groups. They consider only the quantum linear groups, the quantizations of general linear groups $GL_q(n)$ and special linear groups $SL_q(n)$. They systematically developed a satisfactory theory for the representations of these quantum groups. Their results include:

(1) Completed the theory of restriction and induction in the general context of comodules over Hopf algebras. The definition of induction in the context of comodules over coalgebras can be traced back to S. Donkin [D2]. The main results of [PW1] in this direction are the Frobenius reciprocity, the tensor product identity for induction and its derived functors. Discussed exact sequences and fiber products of quantum groups, and prove the Hochschild-Serre spectral sequence theorem under some assumption (examples show that the assumption is necessary for such a theorem).

(2) Defined and studied some important closed subgroups of the quantum general and special linear groups, including maximal tori, Borel subgroups, parabolic subgroups and Levi subgroups. Established the Levi decomposition for parabolic subgroups and the density of the "big cell". Gave a natural description of the root systems for quantum linear groups.

(3) Developed the representation theory of Borel subgroups — proved that all irreducible representations are 1-dimensional (thus they correspond to characters) and gave a description of the injective modules.

(4) Established the basic theory of the representations of $GL_q(n)$ and $SL_q(n)$, including the highest weight theory, the definitions and the im-

portant properties of Weyl modules and induced modules, the correspondence between irreducible modules and dominant weights. For $SL_q(2)$, modules were constructed very concretely. Based on some known facts about Hecke algebras, proved the complete reducibility of representations of $GL_q(n)$ and $SL_q(n)$ in the case q is not a root of unity,

(5) In the case q is a primitive odd degree root of unity, defined the Frobenius morphism $GL_q(n) \to GL(n)$ and $SL_q(n) \to SL(n)$, regarding the quantum linear groups as "coverings" of the corresponding algebraic groups. Using the Hochschild-Serre spectral sequence, reduced, in some sense, the representation theory of quantum linear groups to that of the corresponding algebraic groups and that of a finite dimensional Hopf algebras (called the infinitesimal quantum groups or Frobenius kernels).

(6) Established the representation theory of infinitesimal quantum groups (Frobenius kernel) and that of the fiber products of the infinitesimal quantum groups with tori. The theory is similar to the corresponding theory for algebraic groups in characteristic p (see §§1.4–1.6 for some results in this theory). Using the theory, stated and proved the twisted tensor product theorem, the q-analogue of the Steinberg's tensor product theorem (see §1.4).

(7) Generalized many important results on the sheaf cohomology on the flag variety to the quantum case, including Grothendieck's vanishing theorem, Serre's finiteness theorem and duality theorem, Kempf's vanishing theorem, strong linkage theorem, as well as the Weyl character formula for the Euler character.

3.3 Lin Zongzhu [Lin1], [Lin2] discussed duality theory and induction theory of Hopf algebras, obtaining a connection between the above-mentioned Andersen-Polo-Wen's theory and Parshall-Wang's theory.

The quantization of a Hopf algebra depends on its bialgebra cohomology. In [PW2], B. Parshall and Wang Jianpan gave a detailed investigation, obtaining some useful results and calculations.

3.4 There is a class of finite dimensional algebras closely related with the coordinate algebras of quantum linear groups — the q-Schur algebras. In [PW1] B. Parshall and Wang Jianpan showed how these algebras appear in the quantum linear group theory, and established a theory on their representations, i. e., a theory of polynomial representations of general quantum groups. Moreover, they proved that q-Schur algebras are quasi-hereditary in the sense of [CPS].

The q-Schur algebras are first introduced by R. Dipper and G. James as the centralizers of certain representations of Hecke algebras. In a series of papers (see [Du3] – [Du8] and [DD1]) Du Jie made deep investigation into the representations of q-Schur algebras, focusing on the connection of q-Schur algebras with quantum groups. He developed a "modular theory of characteristic 0" for q-Schur algebras, generalizing some interesting ideas and results in prime characteristic situation, e. g., the defect group of a primitive idempotent, the vertex of an indecomposable module, the Brauer homomorphism (which has been shown to correspond to the Frobenius morphism of quantum linear groups in the sense of [PW1]) and the Green correspondence. He also found Kazhdan-Lusztig style canonical bases for q-Schur algebras, and, based on the bases, gave an isomorphism of certain q-Schur algebras with classical Schur algebras (over an extension of the ground field). The canonical bases for q-Schur algebras were also used to obtain canonical bases for representations of quantum linear groups.

There are also multi-parameter quantum general linear groups and q-Schur algebras. In [DPW] Du Jie, Wang Jianpan and their collaborator B. Parshall made an investigation into two-parameter quantum general linear groups $GL_{\alpha,\beta}(n)$, α, β being parameters, and q-Schur algebras (more precisely, (α, β)-Schur algebras). Most significant result in this paper is a hyperbolic invariance of these quantum groups and q-Schur algebras. That is, if α, β vary with their product invariant, the coalgebra structure of the functional algebras of $G_{\alpha,\beta}(n)$'s and the algebra structure of (α, β)-Schur algebras remain unchanged.

The induction from the naturally defined Borel subalgebra to a q-Schur algebra is not exact, even in the case $q = 1$ (the case of classical Schur algebras). Can we find an exact Borel subalgebra (by this we mean a subalgebra with some properties Borel subalgebras have and the induction from this subalgebra to the whole algebra is exact) for a q-Schur algebra (or more generally a finite dimensional quasi-hereditary algebra) or its Morita equivalence? This question is raised by S. König, and has been positively answered by L. Scott for classical Schur algebras. In [PW5] B. Parshall and Wang Jianpan gave a categorical approach to this question, obtaining some useful sufficient conditions for a given quasi-hereditary algebra to have a Morita equivalence having an exact Borel subalgebra. As a result, they proved that (suitable Morita equivalences of) q-Schur algebras (and some other important finite dimensional quasi-hereditary algebras arising naturally in the representation theory of algebraic and quantum groups) have exact Borel subalgebras.

3.5 In discussing representations and cohomology of quantum groups, the categorical language gives a uniform treatment for the viewpoint of quantized enveloping algebras and the viewpoint of quantized functional algebras. In [PW3] and [PW4], B. Parshall and Wang Jianpan took the categorical approach, hence the results in these papers apply equally to both quantized enveloping algebra and quantized functional algebra viewpoints.

In [PW3], some calculations for infinitesimal quantum groups were carried out. (Some of them were obtained by assuming Lusztig's conjecture.) In particular, the authors expressed the "infinitesimal extensions" in terms of "global extensions", obtained an "even-odd vanishing behavior" of the "infinitesimal extensions" and proved that the infinitesimal Hochschild cohomology of the trivial module has the same graded formal character as the coordinate algebra of the nullcone.

In [PW4], the authors paid their attention on the investigation of properties of the "quantum dimension", and based on them to obtained some interesting results on quantum support varieties. They proved that a Weyl module (induced module) has nonzero quantum dimension if and only if it has regular highest weight. As a result, if a Weyl module (induced module) has regular highest weight, it has the biggest support variety (the nullcone). The above results hold for irreducible modules if Lusztig's conjecture is true. They proved also that if a Weyl module (induced module, irreducible module) has singular highest weight, its support variety is a proper subvariety of the nullcone. By generalizing some calculation for algebraic groups of type A, they gave an explicit description of the support variety of Weyl module when the highest weight is in a facet of codimension 1 for quantum groups of type A.

*   *   *

The research of representations of algebraic and quantum groups is still going on in China. We hope that in the near future we shall have some new breakthroughs, making our contributions to the worldwide development of these theories and related fields.

References

[A1] H. H. Andersen, *Modular representations of algebraic groups*, Proc. Symp. Pure Math., AMS, **47–I** (1987), 23–36.

[A2] H. H. Andersen, *Extensions of modules for algebraic groups*, Amer. J. Math., **206** (1984), 489–504.

[APW1] H. H. Andersen, P. Polo and Wen Kexin, *Representations of quantum algebras*, Invent. Math., **104** (1991), 1–59.

[APW2] H. H. Andersen, P. Polo and Wen Kexin, *Injective modules for quantum algebras*, Amer. J. Math., **114** (1992), 571–604.

[AW] H. H. Andersen and Wen Kexin, *Representations of quantum algebras: the mixed case*, preprint.

[B1] Bai Yuanhuai, *Composition factors of sheaf cohomology groups on G/B*, Northeastern Math. J., **4** (1988), 222–232. (in Chinese)

[B2] Bai Yuanhuai, *Composition factors with multiplicity 1 of sheaf cohomology groups on G/B*, Chin. Ann. of Math., **10A** (1989), 109–118 (in Chinese).

[B3] Bai Yuanhuai, *The decomposition pattern and socle series of $H^0(\lambda)$ in non-generic position for type A_2*, Chin. Ann. of Math., **10A** (1989), 398–406 (in Chinese).

[B4] Bai Yuanhuai, *On the intertwining homomorphism of Weyl modules*, Acta Math. Sinica, **32** (1989), 639–646. (in Chinese)

[B5] Bai Yuanhuai, *Simple G-modules $H^1(\chi)$ for type B_2*, Quarterly J. of Math., **4** No. 4 (1989), 15–23 (in Chinese).

[B6] Bai Yuanhuai, *A vanishing property for the higher G/B-cohomology of type E*, Acta Math. Sinica, **33** (1990), 472–481 (in Chinese).

[B7] Bai Yuanhuai, *Extensions of representations and vanishing property for quantum groups*, preprint (in Chinese).

[B8] Bai Yuanhuai, *The first cohomology group of quantum algebras*, Algebraic Geometry and Algebraic Number Theory World Scientific, Singapore (1992), 1–11.

[B9] Bai Yuanhuai, *Extensions and the vanishing behavior of induction for finite dimensional representations of quantum groups of rank 1*, preprint (in Chinese).

[BWW] Bai Yuanhuai, Wang Jianpan and Wen Kexin, *Translation and cancellation of socle series patterns*, Tôhoku Math. J., **40** (1988), 663–643.

[CW1] Cao Xihua   and Wang Jianpan, *A survey of Chevalley groups and their representations*, Adv. in Math., **14** (1985), 1–22 (In Chinese)

[CW2] Cao Xihua and Wang Jianpan, "*Introduction to the Representation Theory of Linear Algebraic Groups*, Science Press, Beijing (1987 (in Chinese).

[CW3] Cao Xihua and Wang Jianpan, *The research of the modular representations of algebraic groups in China*, Adv. in Sci. of China, Math., **3** (1990), 1–28.

[Car] R. W. Carter, *The relation between characteristic 0 representations and characteristic p representations of finite groups of Lie type*, Proc. Symp. Pure Math., AMS, **37** (1980), 301–311.

[Cha1] L. Chastkofsky, *Projective character for finite Chevalley groups*, J. Algebra, **69** (1981), 347–357.

[Cha2] L. Chastkofsky, *Rationality of certain zeta functions associated with modular representation theory*, Contemp. Math., **45** (1985), 41–50.

[Cha3] L. Chastkofsky, *Generic Cartan invariants for Chevalley groups*, J. Algebra, **103** (1986), 466–478.

[Chen1] Chen Chengdong, *The Cartan matrix for the group $SL(3,9)$*, Master Dissertation, East China Normal University, 1981 (in Chinese).

[Chen2] Chen Chengdong, *The conjugacy classes of SL(n, q)*, J. Tongji University, **14** No. 2 (1986), 193–204 (in Chinese).

[Chen3] Chen Chengdong, *(B, B)-double cosets for classical algebraic groups*, J. Tongji University, **15** No. 3 (1987), 381–390 (in Chinese).

[Chen4] Chen Chengdong, *Dimensions of principal indecomposable modules for classical 0-Hecke algebras*, Chin. Sci. Bull., **36** (1991), 410–413 (in Chinese).

[CP]   Chen Chengdong and Pan Jiezheng, *The indecomposable modules for SL(2, p) and their n-extensions*, Chin. Ann. Math., **6A** (1985), 331–346 (in Chinese).

[CPS]   E. Cline, B. Parshall and L. Scott, *Finite dimensional algebras and highest weight categories*, J. reine angew. Math., **391** (1988), 85–99.

[DD1]   R. Dipper and Du Jie, *Trivial and alternating source modules of Hecke algebras of type A*, Proc. London Math. Soc., **66** (1993), 479–506.

[DD2]   R. Dipper and Du Jie, *Harish-Chandra vertices*, J. reine angew. Math., **437** (1993), 101–130.

[D1]   S. Donkin, "*Rational representations of algebraic groups*, Lecture Notes in Mathematics, 1140, Springer-Verlag (1985).

[D2]   S. Donkin, *Hopf complements and injective comodules for algebraic groups*, Proc. London Math. Soc. (3), **40** (1980), 298–319.

[Dr1]   V. G. Drinfel'd, *Hamiltonian structure on Lie groups, Lie algebras, and the geometric meaning of the Yang-Baxter equation*, Dokl. Akad. Nauk SSSR, **269** (1983), 285–287.

[Dr2]   V. G. Drinfel'd, *Quantum groups*, Proceedings ICM 1986), 798–820.

[Du1]   Du Jie, *The Cartan invariants of SL(4, 2^n)*, J. East China Normal University (Nat. Sci. Ed.) No. 4 (1986), 17–26 (in Chinese).

[Du2]   Du Jie, *Tensor products of Weyl modules and Ext¹-groups for type A₁*, J. Math. Research & Exposition, **7** (1987), 403–407 (in Chinese).

[Du3]   Du Jie, *The modular representation theory of q-Schur algebras*, Trans. Amer. Math. Soc., **329** (1992), 253–271.

[Du4]   Du Jie, *The modular representation theory of q-Schur algebras, II*, Math. Zeit., **208** (1991), 503–536.

[Du5]   Du Jie, *The Green correspondence for the representations of Hecke algebras of type A_{r−1}]*, Trans. Amer. Math. Soc., **329** (1992), 273–287.

[Du6]   Du Jie, *Kazhdan-Lusztig bases and isomorphism theorems for q-Schur algebras*, Contemp. Math. (to appear).

[Du7]   Du Jie, *Canonical bases for irreducible representations of quantun GL_n*, Bull. London Math. Soc., **24** (1992), 325–334.

[Du8]   Du Jie, *Integral Schur algebras for GL₂*, Manuscr. Math., **87** (1992), 411–427.

[DPW]   Du Jie, B. Parshall and Wang Jianpan, *Two-parameter quantum linear groups and the hyperbolic invariance of q-Schur algebras*, J. London Math. Soc. (2), **44** (1991), 420–436.

[DS]   Du Jie and L. Scott, *Lusztig conjectures, old and new, I*, J. reine angew. Math. (to appear).

[FRT]   L. D. Faddeev, N. Yu. Reshetikhin and L. A. Takhtajan, *Quantization of Lie groups and Lie algebras*, Algebraic Analysis, Academic Press (1988), 129–140.

[H1]   J. E. Humphreys, "*Linear Algebraic Groups*," Corrected third printing, Springer-Verlag (1981).

[H2]   J. E. Humphreys. *"Representations of Algebraic Groups,"* Notes of lectures at East China Normal University (1980).

[H3]   J. E. Humphreys, *Cartan invariants*, Bull. London Math. Soc., **17** (1985), 1–14.

[HY]   Hu Yuwang and Ye Jiachen, *On the first Cartan invariant of finite groups of Lie type*, Comm. Algebra, **21** (1993), 935–950.

[J1]   J. C. Jantzen, *"Lectures on representations of algebraic groups at East China Normal University*, (1984).

[J2]   J. C. Jantzen, *Modular representations of reductive groups*, Lecture Notes in Mathematics, 1185 (1986), 118–154.

[J3]   J. C. Jantzen, *Representations of Chevalley groups in their own characteristic*, Proc. Symp. Pure Math. AMS, **47–I** (1987), 127–146.

[J4]   J. C. Jantzen, *"Representations of Algebraic Groups,"* Academic Press, Orlando (1987).

[Jim]   M. Jimbo, *A q-defference analogue of $U(\mathfrak{g})$ and the Yang-Baxter equation*, Letters Math. Phys., **10** (1985), 63–69.

[Lin1]   Lin Zongzhu, *Induced representations of Hopf algebras: application to quantum groups at roots of 1*, J. Algebra, **154** (1993), 152–187.

[Lin2]   Lin Zongzhu, *Rational representations of Hopf algebras and quantum groups*, preprint.

[L1]   G. Lusztig, *Some problems in the representation theory of finite Chevalley groups*, Proc. Symp. Pure Math., AMS, **37** (1980), 313–317.

[L2]   G. Lusztig, *Hecke algebra and Jantzen's generic decomposition patterns*, Adv. in Math., **37** (1980), 121–164.

[L3]   G. Lusztig, *Quantum deformations of certain simple modules over enveloping algebras*, Adv. Math., **70** (1988), 237–249.

[L4]   G. Lusztig, *Modular representations and quantum groups*, Contemp. Math, **82** (1989), 59–77.

[LY]   Liu Jiachun and Ye Jiachen, *Extensions of simple modules for the algebraic group of type G_2*, Comm. Algebra, **21** (1993), 1909–1946.

[M]   Yu. I. Manin, *"Quantum Groups and Non-Commutative Geometry,"* Université de Montréal (1988).

[PW1]   B. Parshall and Wang Jianpan, *"Quantum Linear Groups,"* Memoirs AMS, no. 439, Providence, RI (1991).

[PW2]   B. Parshall and Wang Jianpan, *On bialgebra cohomology*, Bull. Soc. Math. Belg., **42** (1990), 607–642.

[PW3]   B. Parshall and Wang Jianpan, *Cohomology of infinitesimal quantum groups, I*, Tôhoku Math. J., **44** (1992), 1231–1259.

[PW4]   B. Parshall and Wang Jianpan, *Cohomology of quantum groups: the quantum dimension*, J. Canad. Math. Soc. (to appear).

[PW5]   B. Parshall and Wang Jianpan, *Exact Borel subalgebras: Examples from algebraic and quantum groups*, preprint.

[S]   R. Steinberg, *Representations of algebraic groups*, Nagoya Math. J., **22** (1963), 33–56.

[W1]   Wang Jianpan, *Sheaf cohomology on G/B and tensor products of Weyl modules*, J. Algebra, **77** (1982), 162–185.

[W2]   Wang Jianpan, *Coinduced representations and injective modules for hyperalgebra* $\mathfrak{b}]_r$, Chin. Ann. of Math., **4B** (1983), 357–364.

[W3] Wang Jianpan, *The inverse image of an induced sheaf on G/P*, J. East China Normal Univ. (Nat. Sci. Ed.) No. 4 (1984), 1–8 (in Chinese).

[W4] Wang Jianpan, *Quasi-rational modules and generic cohomology*, Northeastern Math. J., **1** (1985), 90–100.

[W5] Wang Jianpan, *On the generic cohomology*, Proc. Symp. Pure Math. AMS, **47–II** (1987), 195–200.

[W6] Wang Jianpan, *Partial orderings on affine Weyl groups*, J. East China Normal Univ. (Nat. Sci. Ed.) No. 4 (1987), 15–25 (in Chinese).

[W7] Wang Jianpan, *On the cyclicity and cocyclicity of G-modules*, Acta Math. Sinica, New Series, **4** (1988), 45–54.

[W8] Wang Jianpan, *On the cyclicity and cocyclicity of G-modules*, Contemp. Math., **82** (1989), 231–233.

[W9] Wang Jianpan, *Inverse limits of affine group schemes*, Chin. Ann. of Math., **9B** (1989), 418–428.

[W10] Wang Jianpan, *Notes on some topics in the representation theory of linear algebraic groups*, Comm. Algebra, **18** (1990), 347–355.

[W11] Wang Jianpan, *A summarized account for the representation theory of quantum linear groups*, Algebraic Geometry and Algebraic Number Theory, World Scientific, Singapore (1992), 102–120.

[W12] Wang Jianpan, *A survey of the theory of quantum groups*, (Chinese) Adv. in Math., **20** (1991), 424–454.

[Wen] Wen Kexin, *The composition of interwining homomorphisms*, Comm. Algebra (17 (1989), 587–630.

[X1] Xi Nanhua, *Finite dimensional modules of some quantum groups over $F_p(\nu)$*, J. reine angew. Math., **410** (1990), 109–115.

[X2] Xi Nanhua, *Representations of finite dimensional Hopf algebras arising from quantum groups*, preprint.

[X3] Xi Nanhua, *Representations of finite dimensional Hopf algebras arising from quantum groups, II*, Algebraic Geometry and Algebraic Number Theory, World Scientific, Singapore, (1992), 143–151.

[Y1] Ye Jiachen, *The Cartan invariants of $SL(3, p^n)$*, J. Mathematical Research & Exposition, **2** No. 4 (1982), 9–19 (in Chinese).

[Y2] Ye Jiachen, *On the Weyl filtrations of principal indecomposable modules for hyperalgebra $u]_n$ with quotients isomorphic to Weyl modules*, J. East China Normal Univ. (Nat. Sci. Ed.) No. 3 (1984), 33–42 (in Chinese).

[Y3] Ye Jiachen, *Filtrations of principal indecomposable modules of Frobenius kernels of reductive groups*, Math. Zeit., **189** (1985), 515–527.

[Y4] Ye Jiachen, *The Carter invariants of $Sp(4, p^n)$*, J. of East China Normal Univ. (Nat. Sci. Ed.) No. 2 (1985), 18–25 (in Chinese).

[Y5] Ye Jiachen, *A theorem on the alcove geometry*, J. Tongji University, **14** No. 1 (1986), 57–64 (in Chinese).

[Y6] Ye Jiachen, *On the first Carter invariant of the groups $SL(3, p^n)$ and $SU(3, p^n)$*, Lecture Notes in Math., 1185 (1986), 388–400.

[Y7] Ye Jiachen, *On the first Carter invariant of the group $Sp(4, p^n)$*, Acta Math. Sinica, New Series, **4** (1988), 18–27.

[Y8] Ye Jiachen , *Some properties on the polynomials $Q_{A,C]}$*, J. Tongji University, **16** No. 1 (1988), 85–90 (in Chinese).

[Y9] Ye Jiachen , *Filtrations of principal indecomposable modules of Frobenius kernels of semisimple algebraic groups*, J. Math., **8** No. 2 (1988), 139–142 (in Chinese).

[Y10] Ye Jiachen , *Cartan invariants of finite groups of Lie type*, Contemp. Math., **82** (1989), 235–241.

[Y11] Ye Jiachen , *Cartan invariants of finite groups of Lie type (I)*, Northeastern Math. J., **10** (1989), 93–101.

[Y12] Ye Jiachen , *On the inverse Kazhdan-Lusztig polynomials $Q_{A,C]}$*, Chinese J. Contemp. Math., **10** No. 3 (1989), 281–289.

[Y13] Ye Jiachen , *Extensions of simple modules for the group $Sp(4, K)$*, J. London Math. Soc. (2), **41** (1990), 51–62.

[Y14] Ye Jiachen , *Extensions of simple modules for the group $Sp(4, K)$ II*, Chinese Sci. Bull., **35** No. 6 (1990), 450–454.

[Y15] Ye Jiachen , *Cartan invariants and extensions for finite groups of Lie type*, Algebraic Geometry and Algebraic Number Theory, Proc. of the Special Program at Nankai Institute of Mathematics, 1989–1990, World Scientific, Singapore (1993), 179–188.

[Y16] Ye Jiachen , *Extensions of simple modules for $G_2(p)$*, Comm. Algebra, **22** (1994 (to appear).

Some Work on Vertex-Transitive Graphs by Chinese Mathematicians

Ming-Yao Xu

Dept. of Math., Peking University, Beijing 100871, People's Republic of China

Abstract   In this paper we survey some recent results on vertex-transitive graphs, especially on Cayley graphs and arc-transitive graphs obtained by a group of Chinese mathematicians including myself, and 17 open problems in this research area are proposed.

Like many other areas in discrete mathematics, the classification of finite simple groups, completed in 1981, effects major breakthrough in several problems concerning vertex-transitive graphs. In this paper I would like to survey some recent results obtained by a group of Chinese mathematicians, including myself, in this field; the proofs of the most of our results need the finite simple group classification.

I assume that the reader is familiar with basic concepts of group theory and graph theory.

The following words are just for fixing my notation: we call a pair of sets $X = (V, E)$ a *digraph* (directed graph) if V is a finite nonempty set, called the vertex set of X, and E a subset of $V \times V - V_0$, where $V_0 = \{(v, v) \mid v \in V\}$, called the edge set of X; a digraph $X = (V, E)$ is called an (undirected) *graph* if $(u, v) \in E$ implies $(v, u) \in E$. From the point of view of graph theorists the digraphs and graphs considered here are finite, simple ones.

A permutation α of V is called an automorphism of X if $(v, u) \in E$ is equivalent to $(v^\alpha, u^\alpha) \in E$ for all $v, u \in V$. The set of all automorphisms of X forms a group, which is called the (full) automorphism group of X and is denoted by $\mathrm{Aut}X$.

---

*Supported in part by NNSFC and DPFC.

We call a (di)graph $X = (V, E)$ *vertex-transitive* if $\operatorname{Aut} X$ acts transitively on V. The class of vertex-transitive graphs is a very special one, but there are still many examples of such graphs, and this is the most interesting class of graphs to group theorists. Some leading group theorists, such as M. Aschbacher, P.J. Cameron, D.G. Higman, P.M. Neumann and C.C. Sims, have done a lot of work in this field. Actually it was them and some algebraic graph theorists, such as L. Babai, A. Gardiner, C.D. Godsil, A.A. Ivanov, C.E. Praeger and R. Weiss, (they were originally group theorists), who laid the foundation in this field. So one can say that this research field belongs to group theory, rather than graph theory in its nature.

Recently, many new results have been obtained in this field, and many long-standing problems have been solved. The reader who is interested in this and wants to know the details is refered to the following very good survey articles [B1], [C2], [G1], [Pr1], [Pr4], [W1] and [Y].

As mentioned above, in this paper I only want to give an exposition of some recent results obtained by myself and in a workshop held at Peking University since March 1988. I would also like to propose some open problems and conjectures which we are trying to solve presently.

This paper is divided into four sections: (1) Preliminaries, (2) Cayley graphs of finite groups, (3) Classifications of some special classes of arc-transitive graphs, and (4) Half-transitive graphs.

§1. Preliminaries

Give a digraph $X = (V, E)$. As mentioned above, the automorphism group $\operatorname{Aut} X$ of X is a permutation group on the vertex set V of X. It also induces a permutation group on the edge set E, and on the arc set A, of X. (An arc is an ordered pair of adjacent vertices; so in a directed graph, an arc is just an edge, but in an undirected graph, an edge $uv \in E$ is viewed as two arcs (u, v) and (v, u).) We have the following

Definition 1.1 *An undirected graph $X = (V, E)$ is said to be edge-transitive, if $\operatorname{Aut} X$ induces a transitive action on the edge set E of X.*

Definition 1.2 *A (directed or undirected) graph $X = (V, E)$ is said to be arc-transitive, if $\operatorname{Aut} X$ induces a transitive action on the arc set A of X.*

A vertex transitive graph is not necessarily edge-transitive, and vice versa. Also, a vertex- and edge-transitive graph is not necessarily arc-transitive, we shall say something about this kind of graphs in §4. Note that an arc-transitive (di)graph must be vertex- and edge-transitive.

In the whole paper we assume that all graphs considered are at least vertex-transitive.

Furthermore, if we assume that $\mathrm{Aut}X$ acts on V doubly-transitive, then it is easy to see that X must be the complete graph K_n or its comlementary graph nK_1, where $n = |V|$ and nX denotes the union of n copies of the graph X. Also, if we assume that $\mathrm{Aut}X$ acts on E doubly-transitively, then it is easy to prove the following

Proposition 1.3 *Assume that a graph $X = (V, E)$ has at least two edges. If $\mathrm{Aut}X$ acts on E doubly-transitive, then one of the following holds.*

(1) $X = nK_2 \cup mK_1$.

(2) $X = K_{1,n} \cup mK_1$.

(3) $X = K_3 \cup mK_1$.

In what follows we may assume that $\mathrm{Aut}X$ acts on V and on E not doubly-transitively.

Let G be a transitive permutation group on V. A binary relation $\sim$ is called G-invariant, if

$$v \sim u \Longrightarrow v^g \sim u^g, \ \forall g \in G.$$

If V has no nontrivial G-invariant relation, we say that G is primitive.

Definition 1.4 *A (di)graph $X = (V, E)$ is called vertex-primitive, if $\mathrm{Aut}X$ acts on V primitively.*

Due to the rapid development in the classification of primitive groups in the last ten years, the study of vertex-primitive graphs has achieved a lot; we shall say something about them in §3.2.

Next we shall define two more transitivity conditions on graphs which are stronger than arc-transitivity. The first is so-called s-arc-transitivity due to Tutte (see [Tut1] and [Tut2]). We have the following

Definition 1.5 *Let $X = (V, E)$ be a directed or undirected graph. We call a sequence of vertices $v_0, v_1, ..., v_s$ an s-arc if $(v_i, v_{i+1}) \in A$ for $i = 0, 1, ..., s - 1$, and $v_i \neq v_{i+2}$ for $i = 0, 1, ..., s - 2$.*

Definition 1.6 *A (di)graph $X = (V, E)$ is said to be s-arc-transitive if for any two s-arcs $v_0, v_1, ..., v_s$ and $u_0, u_1, ..., u_s$, there is an automorphism $\alpha \in \mathrm{Aut}X$ such that $v_i^\alpha = u_i$ for $i = 0, 1, ..., s$.*

Note that 1-arc-transitivity is just arc-transitivity for a (di)graph.

The second concept we want to define is so-called distance transitivity. The following definition is only for undirected graphs.

Definition 1.7 *A graph $X = (V, E)$ is said to be distance transitive if for any two pairs of vertices u, v and u', v' with $d(u, v) = d(u', v')$, where $d(x, y)$ is the distance, that is the length of the shortest path, from x to y, there is an automorphism $\alpha \in \text{Aut}X$ such that $u^\alpha = u'$ and $v^\alpha = v'$.*

The theories of s-arc-transitive and distance transitive graphs are two fruitful branches in algebraic graph theory. However they are not the theme in this survey article; we only need the definitions in later discussion. The reader interested is refered to [G1], [W1], [PSY] and a beautiful book [BCN].

Next we shall give some constructions for vertex-transitive graphs using groups, which show the basic connections between those graphs and groups. the first is the Cayley graphs of groups.

Definition 1.8 *Let G be a finite group and S a subset of G not containing the identity element 1. We define the Cayley digraph $X = \text{Cay}(G, S)$ of G with respect to S by*

$$V(X) = G,$$
$$E(X) = \{(g, sg) \mid g \in G, s \in S\}.$$

The following obvious facts are basic for Cayley graphs.

Proposition 1.9 *Let $X = \text{Cay}(G, S)$ be a Cayley digraph of G with respect to S. Then*

(1) $\text{Aut}X$ *contains the right regular representation $R(G)$ of G, so X is vertex-transitive.*

(2) X *is connected if and only if $G = \langle S \rangle$.*

(3) X *is undirected if and only if $S^{-1} = S$.*

Note that the converse of Proposition 1.9(1) is not true; namely there exist vertex-transitive graphs which are not Cayley ones. The Petersen graph, for instance, is an example. We state the following criterion for Cayley digraphs.

Proposition 1.10 *A digraph $X = (V, E)$ is a Cayley digraph of a group if and only if $\text{Aut}X$ contains a regular subgroup.*

As a generalization of Cayley digraphs, Sabidussi [Sa] gave another construction of vertex-transitive digraphs using groups; it is known as Sabidussi coset graph.

Definition 1.11   *Let G be a finite group and H a subgroup of G. Let D be a union of several double cosets of the form HgH, not containing the subgroup H. We define the Sabidussi coset digraph $X = \mathrm{Sab}(G, H, D)$ of G with respect to H and D by*

$$V(X) = [G : H], \text{ the set of right cosets of } H \text{ in } G,$$
$$E(X) = \{(Hg, Hdg) \mid g \in G, d \in D\}.$$

Note that we do not consider multigraphs, so if $Hdg = Hd_1g$, the edges (Hg, Hdg) and (Hg, Hd_1g) are viewed as equal.

The following obvious facts are basic for Sabidussi coset graph.

Proposition 1.12   *Let $X = \mathrm{Sab}(G, H, D)$ be the Sabidussi coset digraph of G with respect to H and D. Then*

(1) *X is a well-defined digraph with in-degree and out-degree $|D : H|$.*

(2) *$\mathrm{Aut}X$ contains G by right multiplication, so X is vertex-transitive. For a vertex Hg, the stabilizer in G is $g^{-1}Hg$.*

(3) *X is connected if and only if $G = \langle D \rangle$.*

(4) *X is undirected if and only if $D^{-1} = D$.*

(5) *X is G-arc-transitive if and only if $D = Hg_iH$ is a single double coset.*

Note that Cayley graphs are the special case of Sabidussi coset graphs with $H = 1$.

Different from Cayley graphs, any vertex-transitive digraph is a Sabidussi coset digraph. In fact, given a vertex-transitive digraph X and a vertex $v \in V(X)$, take $G = \mathrm{Aut}X$, $H = G_v$, and $D = \{g \in G \mid v^g \in X_1(v)\}$, then D is a union of several double cosets of the form HgH with $D \cap H = \emptyset$ and $X \cong \mathrm{Sab}(G, H, D)$.

So, in theory, if we knew all groups and their subgroup structure, then we would know all vertex-transitive digraphs. In this sense, we would say that the theory of vertex-transitive digraphs is a branch in group theory, rather than one in graph theory.

In the construction of Sabidussi coset graphs we actually used the the permutation representation of G on $[G : H]$. Now we give a permutation group theoretic description for vertex-transitive digraphs.

Let G be a transitive group acting on V. Consider the action of G on $V \times V$. Suppose that the orbits of G on $V \times V$ are $\Delta_0, \Delta_1, \cdots, \Delta_{r-1}$, where

$$\Delta_0 = \{(v, v) \mid v \in V\}.$$

For each $i \neq 0$, we define a digraph, which is still denoted by Δ_i, as follows:

$$V(\Delta_i) = V,$$
$$E(\Delta_i) = \Delta_i.$$

We call these digraphs the *orbital digraphs* of G, and r the *rank* of G. By definition orbital digraphs are arc-transitive. So each transitive group of rank r gives $r - 1$ arc-transitive digraphs in this way. The orbital digraph Δ_i could be undirected, in this case we call the orbital and the orbital digraph *self-paired*. It is easy to see that G has a self-paired orbital if and only if $|G|$ is even.

A union of several orbitals of G is called a generalized orbital of G. The corresponding digraph is called a generalized orbital digraph of G. All such digraphs are vertex-transitive, and any vertex-transitive digraph can be obtained in this way. This idea is also very important in the study of vertex-transitive undirected graphs, especially in that of arc-transitive graphs.

By the way, these concepts have been successfully used for studying the suborbit structure of primitive groups. As the first example of this kind of application, a very famous theorem about s-arc-transitive graphs due to Tutte ([Tut1] and [Tut2]), which claimed that there exist s-arc-transitive undirected graphs of valency 3 only for $s \leq 5$, was used to bound the order of primitive groups with a suborbit of length 3 by Sims [Si]; namely Sims proved that if G is a primitive group with a suborbit of length 3, then the order of the point stabilizer G_v divides 48. Using this result, Wong [Wo] successfully classified all primitive groups with a subdegree 3 in 1967. Later, in 1970s, several authors investigated the suborbit structure of primitive groups using graph theoretic methods, and obtained many results; see [C3] and [Neu] for details.

Now I would like to mention a problem about s-arc transitive graphs, that is the following conjecture.

Conjecture 1.13 *If X is an s-arc-transitive undirected graph, then $s \leq 7$ and $s \neq 6$.*

This conjecture has been proved by Weiss [W2] using the finite simple group classification.

Using orbital digraphs, Higman gave the following characterization for primitive groups, which is very useful both in group theory and in graph theory, see [Hi1, Proposition 2.3].

Proposition 1.14   *G is primitive if and only if every nontrivial orbital digraph of G is connected.*

To end this section, we define the concept of *quotient graphs* or *block graphs*. Assume that G acts imprimitively on a vertex-transitive (di)graph X. Let $\overline{X} = \{B_0, B_1, \cdots, B_{n-1}\}$ be a complete block system of G. We define the *quotient graph* of X corresponding to $\overline{X}$, which is still denoted by $\overline{X}$, by

$$V(\overline{X}) = \overline{X},$$

$$E(\overline{X}) = \{B_i B_j \mid \text{there exist } v_i \in B_i, v_j \in B_j \text{ such that } v_i v_j \in E(X)\}.$$

We shall use this concept later in several places.

§2. Cayley Graphs of Finite Groups

In this section we shall survey some results about three problems concerning Cayley (di)graphs which are the most interesting to group theorists. They are: (1) Isomorphism problems; (2) classification problems; and (3) Hamiltonian properties.

§2.1 Isomorphisms of Cayley Digraphs and Graphs

Let $X = \mathrm{Cay}(G, S)$ be a Cayley digraph of G with respect to S. Take an $\alpha \in \mathrm{Aut}\, G$ where $\mathrm{Aut} G$ is the automorphism group of the group G, and set $S^\alpha = T$. Obviously, we have $\mathrm{Cay}(G, S) \cong \mathrm{Cay}(G, T)$. This kind of isomorphism between Cayley digraphs is called *Cayley isomorphism* or *trivial isomorphism*.

Given a subset S of G, we call S a CI-subset of G, (CI stands for "Cayley isomorphism"), if for any isomorphism $\mathrm{Cay}(G, S) \cong \mathrm{Cay}(G, T)$ of Cayley digraphs there is an $\alpha \in \mathrm{Aut} G$ such that $S^\alpha = T$. In other words, S being a CI-subset means that only Cayley isomorphisms can be found between $\mathrm{Cay}(G, S)$ and other Cayley digraphs of G.

We call G a DCI-group if every subset of G is CI; that means there exist only Cayley isomorphisms between Cayley digraphs of G. We call G a GCI-group, or simply a CI-group, if every subset S with $S^{-1} = S$ is CI; that means there exist only Cayley isomorphisms between Cayley graphs of G. Here DCI and GCI stand for "digraph Cayley isomorphism" and "graph Cayley isomorphism", respectively. It is obvious that DCI-groups are also CI-ones.

The study of DCI- and CI-groups is important because it is a necessary step for classifying Cayley digraphs and graphs of finite groups. It began with a conjecture posed by A. Ádám [A] in 1967, which claimed, using our terminology, that all finite cyclic groups are DCI. Although this conjecture was disproved in 1970 by Elspas and Turner [ET] by showing that Z_8 is not DCI, this counterexample was not the end of investigating the conjecture. Very dramatically, it was the real beginning for searching finite DCI- and CI-groups. Since then many authors have done a lot of work on it; more important results can be found in [B2], [BF1], [BF2], [Go], [P2] and [X1].

In the study of CI-groups, a criterion for CI-subsets due to Babai [B2] plays an important role. We state it as the following lemma.

Lemma 2.1 *Let G be a finite group and S a subset of G not containing the identity element 1. Let $X = \mathrm{Cay}(G, S)$ and let $A = \mathrm{Aut}X$. Then S is a CI-subset of G if and only if for any $\sigma \in \mathrm{Sym}(G)$ with $\sigma R(G)\sigma^{-1} \leq A$, there exists an $\alpha \in A$ such that $\alpha R(G)\alpha^{-1} = \sigma R(G)\sigma^{-1}$.*

Now I want to mention three problems and some results obtained so far about them. The first one is a variation of [BF1, Problem 2.2(b)] posed by Babai and Frankl.

Problem 1. *Is the elementary Abelian p-group Z_p^n of order p^n DCI for any positive integer n?*

The case of $n = 1$ was affirmatively solved by J. Turner [Tu] and D. Z. Djokovič [D] independently. This case is very easy. By Lemma 2.1, Z_p being DCI is equivalent to the conjugacy of Sylow subgroups in the automorphism group of the Cayley digraphs of Z_p. However the case of $n = 2$ is more difficult. Although several authors announced, as early as in 1977, that they had proved the conjecture in this case, the first proof of Z_p^2 being CI was published in 1983 by C. D. Godsil [Go]. In his proof a deep result about primitive groups due to M. O'Nan and the Wielandt dissection theorem [Wi, Theorem 6.5] for transitive permutation groups were used. For $n = 3$, an affirmative answer was given by Xu in [X1, X1p], (in fact the author determined all DCI- and CI-groups of order p^3). The cases $n \geq 4$ are still open. (Here I should mention that a referee for my paper [X1p] told me that Lewis Nowitz has found an example to show that Z_2^6 is not CI, but I haven't seen his proof.)

The main difficulty of the problem for $n \geq 4$ is the analysis of the Sylow p-subgroups of $\mathrm{Aut}X$ which act on G imprimitively. After the finite simple group classification, we know more and more about primitive groups, but we know very little about imprimitive ones. Since the full automorphism

group of a digraph must be 2-closed in the sense of H. Wielandt [Wi2], (also
see §3.2 for a definition), and this important concept has not received much
attention by group theorists, I think, as the first step we should study the
properties of 2-closed groups. For example, which normal subgroups and
factor groups of a 2-closed group are still 2-closed? This would provide a
tool for studying the connections between $\mathrm{Aut}X$ and the automorphism
group of the quotient graph corresponding to a complete block system of
$\mathrm{Aut}X$.

The second problem is about cyclic CI-groups, with which the original
Ádám's conjecture was concerned; the problem is to prove or disprove the
following conjecture.

Conjecture 2. *For square-free n, Z_n is DCI (or CI).*

In this direction the best result I have seen is due to Pàlfy [P2], which
claimed that Z_n is DCI provided $(n, \varphi(n)) = 1$. In 1979 Egorov and
Markov [EM] published a sketchy proof of this conjecture, but there is a
gap in it. Recently Ya.Yu. Gol'fand [Gfd] proved this conjecture for odd
n. But I have not found his paper, and I do not know any work about the
even case.

The third problem is about simple groups.

Problem 3. *Determine nonabelian simple DCI- and CI-groups.*

For this problem, Babai and Frankl [BF2] proved that there are only
two simple groups A_5 and $PSL(2, 13)$ which may be CI. In [Zh] J.P. Zhang
excluded $PSL(2, 13)$. So only one candidate is left. Recently, some people
at Peking University are trying to verify (or to disprove) that A_5 is DCI
or CI using GAP on a PC. However, no result is known.

Finally I would like to suggest the following problem; a solution of this,
I think, would be helpful for further determining DCI- and CI-groups.

Problem 4. *Assume that $\mathrm{Aut}X$ and B are two DCI- or CI-groups
with $(|A|, |B|) = 1$. Under what conditions is $A \times B$ still DCI or CI? As
the first step, what can we say for abelian case?*

Since DCI- or CI-groups are very rare, Xin-Gui Fang and the author
defined and studied so-called m-DCI- and m-CI-groups in [F1], [F2], [FW],
[FX1], [FX2] and [FX2p].

Definition 2.2 *A finite group G is called an m-DCI-group if any
subset S with $1 \notin S$ and $|S| \leq m$ is a CI-subset.*

Definition 2.3 *A finite group G is called an m-CI-group if any sub-
set S with $1 \notin S$, $S^{-1} = S$ and $|S| \leq m$ is a CI-subset.*

These concepts are useful for studying isomorphisms of Cayley (di)graphs of small valency.

If $m = 1$, then by definition G is 1-DCI if and only if Aut$G$ acts transitively on every subset of elements having the same order. (Such groups are called T-groups.) So the first task faced by us is to determine all finite T-groups. This problem is not easy. For a p-group G, p odd, E.E. Shult [S] proved that G is a T-group if and only if G is homocyclic; that is, G is a direct product of a finite number of cyclic p-groups of the same order. In particular, G must be Abelian. But for 2-groups, this is not true. The quaternion group Q of order 8 is a counterexample. We have the following

Conjecture 5 *Non-Abelian T-2-group are Suzuki 2-groups or the quaternion group Q.*

For Suzuki 2-groups, the reader is refered to [HB, VII, §7]. Zhang proved in [Zh] that a Suzuki 2-group G is a T-group if $|G| = |Z(G)|^2$.

In the literature there are several papers devoted to the so-called 2-automorphic 2-groups. (See [Gr] for instance.) The concept of T-2-groups is stronger than that of 2-automorphic 2-groups. So it would be easier to determine T-2-groups than to determine 2-automorphic ones.

If a T-group is not nilpotent, W. Gaschütz and T. Yen [GY] proved that the p-length of a p-solvable T-group, for odd p, is equal to 1. It is not difficult to check that A_5 is a T-group. This was the only knowledge we had before Zhang's work [Zh]. Zhang gave an explicit description for T-groups in [Zh], which is the most important result for T-groups so far. It is difficult to formulate his result here, I only mention the following theorem of his.

Theorem 2.4

(1) *If G is a T-group of odd order, then G has a normal nilpotent Hall subgroup H such that all Sylow subgroups of H are homocyclic and the factor group G/H is cyclic.*

(2) *If G is an insoluble T-group, then G is isomorphic to one of the following groups: $PSL(2,5)$, $PSL(2,7)$, $PSL(2,8)$, $PSL(2,9)$, $SL(2,5)$, $SL(2,7)$, $SL(2,9)$, and a semi-direct product of B by $SL(2,5)$, where B is abelian and for any p divides $|B|$, $p = 11, 19, 29$ or 59, all Sylow p-subgroup are homocyclic of rank 2, and the center of this semi-direct product is identity.*

In virtue of Zhang's work, it should not be very difficult to give a classification for T-groups which are not 2-groups.

Coming back to m-DCI- or m-CI-groups, we have some significant results for Abelian groups; namely, in [F1], [F2], [FW], [FX1] and [FX2p] we proved the following

Theorem 2.5 *Let G be a finite abelian group, p any prime divisor of $|G|$, and let G_p be the Sylow p-subgroup of G. Then*

(1) *G is 1-DCI if and only if G_p is homocyclic for any p;*

(2) *G is 2-DCI if and only if G_p is homocyclic for any odd p and G_2 is cyclic or elementary Abelian;*

(3) *G is 3-DCI if and only if G_p is homocyclic for any $p > 3$, G_3 is cyclic or elementary Abelian, and G_2 is elementary Abelian or $G_2 \cong Z_4$;*

(4) *G is 1-CI if and only if G_2 is homocyclic; and*

(5) *G is m-CI, $2 \leq m \leq 5$, if and only if G is 2-DCI.*

As a by-product, we proved a conjecture posed by F. Boesch and R. Tindell in [BT] which claimed, in our terminology, that for any n, the cyclic group Z_n is 4-CI. As another by-product, we proved an interesting theorem about abelian groups. To formulate this we give the following

Definition 2.6 *A finite group G is called homogeneous, if for any two isomorphic subgroups H and K, and for any isomorphism $\sigma : H \to K$, σ can be extended to an automorphism α of G.*

Obviously, homogeneous groups are 1-DCI, and we proved

Theorem 2.7 *A finite Abelian group G is homogeneous if and only if G is 1-DCI; that is for any p, the Sylow p-subgroup of G is homocyclic.*

The determination of non-Abelian homogeneous groups is much more difficult, but it should be easier than determining T-groups. So we propose the following

Problem 6. *Determine all finite homogeneous groups.*

Let G be a finite group. An interesting question is to ask if there exists a CI-subset in G which is not $G - \{1\}$ or $\emptyset$; such CI-subset is called nontrivial. In [LPX] the authors gave a positive answer to this question; namely they proved

Theorem 2.8 *Any finite group of order greater than 2 has a nontrivial CI-subset.*

Since if S is a CI-subset then $G - S - \{1\}$ is also CI, so, for any finite group of order greater than 2, we can find a nontrivial CI-subset which generates the group.

§2.2 Classifications of Cayley Digraphs and Cayley Graphs

Algebraists like classifications. Since they set feet on graph theory, they have been trying very hard to classify various kinds of graphs, such as Cayley graphs, vertex-transitive graphs, arc-transitive graphs, distance-transitive graphs. They *have had* good success in this attempt and obtained a lot of beautiful results, although the pure graph theorists do not like them.

In this subsection, I do not want to give an exposition of all these efforts, I only want to say a few words about our results in classifying Cayley graphs.

If G is a DCI-group, then, by definition, the classification of Cayley digraphs of G can be completed by determining the orbits of the action of Aut$G$ on the set of all subsets not containing 1 of G. For example, classifying Cayley digraphs of Z_p, (graph theorists would prefer to call them circulant digraphs of order p), is simply an arithmetic problem, because we know that Z_p is a DCI-group and Aut$G$ is isomorphic to Z_{p-1}. But for other groups, even for cyclic ones, the problem is not so easy. Two kinds of difficulties could occur. One is that nontrivial isomorphisms between Cayley digraphs have to be determined, and we lack effective methods to do this. The other is that there exist isomorphisms between Cayley digraphs of different groups. Both difficulties are very serious, so we have succeeded very little so far.

If the group is nonabelian, solving this problem is very difficult even for the case of valency 3 case. For example, in order to determine Cayley (undirected) graphs of valency 3 with at most 120 vertices and with a regular full automorphism group, H.S.M. Coxeter, R. Frucht and D.L. Powers wrote a book [CFP], in which they listed 440 such graphs. No general methods and results have appeared so far.

However, for Abelian groups our situation is better. At least we can determine the Cayley (di)graphs of small valency for them. Chang-Nian Dong and the author are preparing a paper which will give a classification of Cayley graph of valency at most 4 for cyclic groups.

Another thing which is worth to mention is that in our workshop, Lu-Yan Wang is trying to classify the Cayley graphs of valency three for nilpotent groups, and he has achieved a lot. A paper is in preparation.

The most difficult thing in classifying Cayley graphs of small valency is to solve the isomorphism problem. Since there are very few groups which are m-DCI or m-CI, I think the following concept would be helpful.

Definition 2.9   *We call a finite group G weakly m-DCI (or weakly m-CI), if every subset S of G with $1 \notin S$, $|S| \leq m$, $\langle S \rangle = G$ (and $S^{-1} = S$)*

is CI.

Weakly m-DCI- (or m-CI-)groups are quite different from m-DCI- (or m-CI-)ones. For example, every finite p-group G is weakly m-DCI provided $m < p$ (see [B2]); but even for $m = 1$, G being 1-DCI would imply G being Abelian (for $p > 2$).

For the purpose of classifying Cayley (di)graphs, this weaker concept is more useful because in solving the isomorphism problem for Cayley digraphs, we only need to consider the connected ones.

(Note that the concept of "weakly DCI- (or CI-)groups" does not make sense, because it is equivalent to that of DCI- (or CI-)groups.)

Now we would like to propose the following

Problem 7. *Determine weakly m-DCI- and weakly m-CI-groups for small m. First do it for Abelian groups.*

Problem 8. *Is every minimal generating subset for a finite group a CI-subset?*

I think Problem 8 is very difficult. Besides any m-generating set of a p-group ($m < p$) is CI, we only know that is true for 2-generating sets of 2-generator abelian groups.

Another attempt in classifying Cayley graphs we would like to mention is about arc-transitive circulants. A complete classification seems to be very hard to obtain. However, we have achieved the 2-arc-transitive case; namely, Alspach, Marušič and the author [AMX] proved the following

Theorem 2.10 *A connected, 2-arc-transitive circulant of order n, $n \geq 3$, is one of the following graphs:*

(1) *the complete graph K_n, which is 2-arc-transitive but not 3-arc-transitive;*

(2) *the complete bipartite graph $K_{n/2,n/2}$, which is 3-arc-transitive but not 4-arc-transitive;*

(3) *the complete bipartite graph $K_{n/2,n/2}$ minus a 1-factor, $n/2$ odd, which is 2-arc-transitive but not 3-arc-transitive; and*

(4) *the cycle C_n of length n, which is k-arc-transitive for all $k \geq 0$.*

§2.3 Hamiltonian Property of Cayley Graphs and Vertex-Transitive Graphs

Let $X = (V, E)$ be a connected graph. A Hamiltonian cycle in X is a cycle $(v_1, v_2, \cdots, v_n, v_1)$ where $V = \{v_1, v_2, \cdots, v_n\}$ and $v_1 v_2$, $v_2 v_3 \cdots$, $v_n v_1$ are edges of X. A Hamiltonian path in X is a path $(v_1, v_2, \cdots, v_n)$ where $V = \{v_1, v_2, \cdots, v_n\}$ and $v_1 v_2$, $v_2 v_3 \cdots$, $v_{n-1} v_n$ are edges of X.

Studying the Hamiltonian property of graphs is a favourite problem for graph theorists, and there is a problem of this kind which is also interesting to group theorists. That is the Lovász problem.

As is well known, up to now we have only found four non-trivial connected vertex-transitive graphs which do not have Hamiltonian cycles; they are the Petersen graph, Coxeter graph, and the graphs obtained from each of these two graphs by replacing each vertex with a triangle and joining the vertices in a natural way. However, all of these four graphs have Hamiltonian paths. So in 1970 Lovász proposed the following conjecture.

Conjecture 9. *Every connected vertex-transitive graph has a Hamiltonian path.*

Furthermore, it is noted that all of the above four graphs are not Cayley graphs. So several authors made the following

Conjecture 10. *Every connected Cayley graph of a finite group has a Hamiltonian cycle.*

Both of these two conjectures are still open. Most results obtained so far about the second conjecture were surveyed in [Al], [Wit1] and [WG]. For example we know that is true for Abelian groups and for nonabelian groups whose derived subgroup is cyclic of prime power order. Besides, in 1989, Alspach and Zhang [AZ] proved that every cubic Cayley graph of a dihedral group has a Hamilton cycle. The best result in this direction is due to Witte [Wit2] who obtained in 1986 the following theorem, in the proof of which some group-theoretic technology was successfully used.

Theorem 2.11 *Every connected Cayley digraph of a finite p-group has a directed Hamiltonian cycle.*

Some authors wanted to generalize Witte's result to vertex-transitive digraphs of prime-power order, but very little has been achieved. However, the author proved the following theorem in [X2].

Theorem 2.12 *Every connected digraph with a primitive automorphism group acting on the vertex set has a directed Hamiltonian cycle.*

Actually the above theorem is only an immediate consequence of Theorem 2.11 and the following group-theoretic result proved in [X2].

Theorem 2.13 *Every primitive group of prime-power degree has a regular subgroup.*

This theorem is proved using the O'Nan-Scott theorem for primitive groups, (see [LPS] for instance), and a classification of maximal subgroups of prime-power indices of simple groups in [Gu]. So it relies on the finite simple group classification.

Another result which is worth to mention is that Wang and Xu [WX] proved every connected, so-called quasi-Abelian Cayley graph has a Hamiltonian cycle.

Let G be a finite group, and S a subset of G with $1 \notin S$ and $S^{-1} = S$. If S is a union of several entire conjugate classes of G, then the Cayley graph $X = \text{Cay}(G, S)$ of G with respect to S is called *quasi-Abelian*.

First, we observed that

Lemma 2.14 *Let $X = \text{Cay}(G, S)$ be the Cayley graph on a finite group G with respect to S. Then the following are equivalent.*

(1) *X is quasi-abelian.*

(2) *$\text{Aut}X \geq R(G)L(G)$.*

(3) *$\text{Aut}X \geq R(G)\text{Inn}(G)$, where $\text{Inn}(G)$ is the subgroup of $\text{Aut}X$ induced by the group inner automorphisms of G.*

Using this lemma and some graph-theoretic technology we proved

Theorem 2.15 *Let $X = \text{Cay}(G, S)$ be a connected quasi-Abelian Cayley graph on a finite group G and $|G| > 2$. Then X is Hamiltonian.*

As an example of quasi-abelian Cayley graphs, and also as the motivation of the research in [WX], we mention the definition of Parsons graphs and three conjectures about them posed by Zaks in [Z].

Let F_q be a finite field with q elements where q is a prime power. Let d be an even positive integer. For any $b \in F_q$, we define the Parsons graph $T_b(d, q)$ as follows:

(1) The vertex set $V = V(T_b(d, q))$ is the set of all $d \times d$ matrices A over F_q with $\det A = 1$, i.e., $V = SL(d, q)$, the special linear group over F_q.

(2) The edge set $E = E(T_b(d, q))$ is the set of all pairs (A, B) of elements in $V(T_b(d, q))$ satisfying $\det(A - B) = b$.

Joseph Zaks raised in [Z] the following

Conjectures 2.16   (1) *For all $d \geq 4$, or $d = 2$ and $q \geq 3, T_b(d, q)$ has a 1-factor.*

(2) *For all $d \geq 4$, or $d = 2$ and $q \geq 3, T_b(d, q)$ is Hamiltonian.*

(3) *Every Parsons graph, except for $T_1(2, 2)$, is connected.*

In [WX] the authors also proved all the three conjectures above.

A graph is called *Hamiltonian decomposable*, if its edge set is a disjoint union of several Hamiltonian cycles. Now we would like to suggest the following

Problem 11.   *Are any connected quasi-Abelian Cayley graphs Hamiltonian decomposable?*

All results we surveyed so far in this subsection is about Conjecture 10. As for Conjecture 9 there are still fewer results. I only want to mention Marušič's work. He proved this conjecture for vertex-transitive graphs of orders p, $2p$, $3p$, $4p$, $5p$, p^2, p^3 and $2p^2$. See [Ma1], [MP] and the references in them.

§3. Classifications of Some Special Classes of Arc-Transitive Graphs

In this section we shall give an exposition of some results obtained in our workshop about classifications of some special classes of arc-transitive graphs.

Let X be a simple undirected graph. Recall that X is said to be arc-transitive, if $\operatorname{Aut}X$ acts arc-transitively on the vertex set of X; that is $\operatorname{Aut}X$ acts arc-transitively on the set of ordered adjacent pairs of vertices of X. Let G be a subgroup of $\operatorname{Aut}X$. X is said to be G-arc-transitive if G acts arc-transitively on the vertex set of X. It is easy to see that X is G-arc-transitive if and only if G acts transitively on $V(X)$ and for any vertex $v \in V(X)$, the stabilizer G_v of v in G is transitive on the neighborhood $X_1(v)$ of v, which is the set of vertices adjacent to v.

§3.1 Classifying arc-transitive graphs of order a product of two distinct primes

The earliest work in this direction was done by C.Y. Chao. In 1971 he classified all arc-transitive graphs with a prime number of vertices, see [Ch]. In 1972 J. L. Berggren [Ber] simplified Chao's proof of this classification theorem. Since then several authors tried to classify arc-transitive graphs of order $2p$, p a prime, but no such attempts had succeeded until 1987 when Y.Cheng and J.Oxley [CO] determined all such graphs. (Actually they classified so-called "weakly symmetric" graphs of order $2p$, that is, vertex- and edge-transitive graphs; they proved that all weakly symmetric graphs of order $2p$ are also arc-transitive.)

It took 16 years from p to $2p$ because in group theory the classification of the primitive permutation groups of degree $2p$ took a very long time. It was completed only in 1985 when M.W. Liebeck and J. Saxl [LS] classified all primitive groups of degree mp with $m < p$. The proof of their result relied heavily on the finite simple group classification.

In virtue of this result of Liebeck and Saxl it is not difficult to classify the arc-transitive graphs of order mp, $m < p$, with a primitive group acting on it; we did this for $3p$ in [WX1], and for kp where k and p are distinct primes in [PX], and finally for mp, $m < p$, in [LWWX]. We shall give an exposition for these pieces of work in subsection 3.2. But for imprimitive graphs, (the graphs whose automorphism groups are imprimitive on their vertex sets), we need other tools. As the first attemp, Wang and the author classified the arc-transitive graphs of order $3p$, see [WX1]. Later, Praeger joined us and we finished the classification of the arc-transitive graphs of order kp where k and p are two distinct primes, see [PWX].

Recall that if G acts imprimitively on an arc-transitive graph X and $\overline{X} = \{B_0, B_1, \cdots, B_{n-1}\}$ is a complete block system of G, we defined the *quotient graph* of X corresponding to $\overline{X}$ in §1. This graph is also denoted by $\overline{X}$. G induces an action on $\overline{X}$ naturally. Assume that the kernel of this action is K. Set $\overline{G} = G/K$. Then $\overline{G}$ acts on $\overline{X}$ faithfully. A lemma due to Praeger [Pr2] is basic for quotient graphs.

Lemma 3.1 *Assume that X is G-arc-transitive. Then*

(1) $\overline{X}$ *is $\overline{G}$-arc-transitive;*

(2) *If X is connected, then so is $\overline{X}$;*

(3) *If the induced subgraph by B_i has an edge, then B_i is a union of some connected components of X.*

Now we assume that X is an arc-transitive graph of order kp where $k < p$ are primes. By this lemma, if X is connected, then there is no edge in each induced graph B_i, and the quotient graph $\overline{X}$ has order k or p, and hence is known. So what we have to do is to lift the graph $\overline{X}$ to X, preserving the arc-transitivity.

Here I do not want to give a complete formulation of our classification theorem for arc-transitive graphs of order kp, because the explicit description of the graphs we found needs a lot of space. I only want to say some words about our method and propose an open problem.

Consider the kernel K of the action of G on $\overline{X}$. We distinguish two cases where K acts on each block B_i unfaithfully and faithfully. In the former case, it is easy to prove that X is just a lexicographic product of $\overline{X}$ by an empty graph or a graph obtained easily from this lexicographic product; but in the latter case, to determine X is rather difficult. In most cases we need to determine a sort of covering graphs of a smaller arc-transitive graph. Let $X = (V, E)$ be an arc-transitive graph and W a k-element set. A k-fold covering graph Y of X is a graph whose vertex set is $V \times W$ and if $vu \in E$ then the induced graph between two blocks

$\{(v, w) \mid w \in W\}$ and $\{(u, w) \mid w \in W\}$ is a perfect maching. We need to determine all arc-transitive k-fold covering graphs of a given graph X. This problem is very difficult in general. But in our situation k is a prime and X is often the complete graph K_p. Even in this case, we need the simple group classification theorem, especially the classification of doubly-transitive groups which is a consequence of the simple group classification. We also need methods from geometry, linear groups and the representation theory of finite groups. To know more details, the reader has to read the original paper [WX1] and [PWX].

Now we suggest the following

Problem 12   *Determine all arc-transitive k-fold covering graphs of a given arc-transitive graph X. First do the case where X is a complete graph.*

Nearly at the same time as we did the above-mentioned work, Marušič and Scapellato independently classified vertex transitive graphs of order a product of two distinct primes, which they called pq-graphs. They proved that a vertex-transitive graph on kp $(p > k)$ vertices is either a (k, p)-metacirculant defined in [AP1] or one of a certain family of graphs with automorphism group containing $SL(2, q)$ acting transitively on vertices, where $q = 2^{2^s}$, $p = q + 1$ is a Fermat prime, and k divides $q - 1$. However, they did not determined, for a (k, p)-metacirculant graph X, whether its automorphism group contains a subgroup G which is arc-transitive on X and has a block system consisting of p blocks of size k. Nor is it clear when the graphs of order kp defined in [MP1] have full automorphism group larger than $SL(2, q)$. But in [PWX] we give a straightforward analysis, for the subclass of vertex-transitive graphs of order kp which are arc-transitive, of imprimitive graphs of order kp which includes the necessary analyses of metacirculants and the graphs associated with $SL(2, q)$ to solve these problems. But for (k, p)-metacirculant graphs X, nobody solves the isomorphism problem. So we suggest the following

Problem 13   *Find conditions under which two (k, p)-metacirculant graphs are isomorphic.*

After finishing the classification of arc-transitive pq-graphs, Lu-Yan Wang, Ru-Ji Wang, Praeger and Xu tried to do the same thing for arc-transitive graphs of order $4p$, and recently Marušič joined us to do this, a paper is in preparation ([MPWWX]). The method we used here is the same as in cases $3p$ and pq, but some new phenomena happened.

Another piece of work we would mention is that Hui-Ling Li, Ru-Ji

Wang and Xu are doing the case of p^2, and we have obtained a classification for primitive groups of degree p^2, and also obtained a classification for arc-transitive graphs of that order; a paper is in preparation.

Next, we would suggest the following

Problem 14   *Classify arc-transitive graphs of orders $2pq$, p^2q and pqr, where p, q and r are three distinct primes.*

As for this problem, Ru-Ji Wang is trying to do the case of $6p$, and he has got a classification of arc-transitive graphs of order $6p$ which have a soluble automorphism group, see [W].

§3.2 Classifying vertex-primitive arc-transitive graphs

In this section we shall describe bisic strategy and basic steps for classifying the vertex-primitive arc-transitive (di)graphs of order kp where k and p are distinct primes and $k < p$.

As mentioned above, M.W. Liebeck and J. Saxl [LS] classified all primitive groups of degree mp with $m < p$ in 1985. So we can extracted all primitive groups of degree kp from Liebeck and Saxl's paper and list them in the following table 1:

Table 1. Simply Primitive groups G of degree kp
p and k are primes, and $3 \leq k < p$

$T = \mathrm{soc}(G)$	k	p	action	comment
A_p	$\frac{p-1}{2}$	p	pairs	$p \geq 7$
A_{p+1}	$\frac{p+1}{2}$	p	pairs	$p \geq 5$
A_7	5	7	triples	
$PSL(3,2)$	3	7	incident point-line pairs	G has a graph automorphism
$PSL(n,q), n = 4 \text{ or } 5,$	5	7 or 31	2-spaces or $(n-2)$-spaces	
$\Omega^{\pm}(2d,2), d \geq 4,$	$2^{d-1} \pm 1$	$2^d \mp 1$	singular 1-spaces	
$PSp(4, 2^{2^t}), t > 0$	$2^{2^t} + 1$	$2^{2^{t+1}} + 1$	1-spaces or totally isotropic 2-spaces	
$PSL(2, k^2), k$ odd prime	k	$\frac{k^2+1}{2}$	cosets of $PGL(2,k)$	
$PSL(2, 19)$	3	19	cosets of A_5	
$PSL(2, 29)$	7	29	cosets of A_5	
$PSL(2, 59)$	29	59	cosets of A_5	
$PSL(2, 61)$	31	61	cosets of A_5	
$G = PGL(2, 11)$	5	11	cosets of S_4	
$PSL(2, 23)$	11	23	cosets of S_4	
$PSL(2, p), p \geq 11$	$\frac{p \mp 1}{2}$	p	cosets of $D_{p \pm 1}$	
M_{11}	5	11		
M_{23}	11	23		
M_{22}	7	11		

For each group G we shall do the following:

(1) finding all orbital digraphs for G, they are G-vertex-primitive, (noting by Proposition 1.14 that all such digraphs are connected);

(2) determining which orbital digraph are undirected;

(3) determining the full automorphism group for each orbital digraph;

(4) determining isomorphisms between the orbital digraphs you found.

Doing (1) and (2) is equivalent to (1') determining all suborbits, which are G_v-orbits for any $v \in V$, and (2') determining which suborbits are self-paired, (a suborbit is said to be *self-paired* if the corresponding orbital is self-paired.)

When doing this, for different groups we use different methods. Some groups have geometric structure, then we can use geometric considerations. For example, assume that $PSp(4, q) \leq G \leq P\Gamma Sp(4, q)$, that is G is the 4-dimensional symplectic groups acting on totally isotropic 1-spaces or 2-spaces. Note that these two representations are interchanged by an automorphism of the symplectic groups, so they are equivalent and it is sufficient to consider only the action on 1-spaces. Thus we may assume that G acts on the set Ω of 1-spaces of the 4-dimensional symplectic space $V = V(4, q)$ over $GF(q)$, where $q = 2^{2^t}$, $t > 0$, and $k = q + 1$, and $p = q^2 + 1$ are Fermat primes. An easy calculation will give that $|\Omega| = (q+1)(q^2+1) = kp$, G has rank 3, and the subdegrees of G are 1, $q+q^2$ and q^3, and that suborbits are self-paired. In fact, taking $\alpha = \langle e \rangle \in \Omega$, and letting $V = V_1 \oplus V_2$ be a decomposition of V as a direct sum of two hyperbolic planes $V_1 = \langle e_1 = e, f_1 \rangle$ and $V_2 = \langle e_2, f_2 \rangle$, the orbits of G_α in Ω are $\{\alpha\}$, $\Sigma_1 = \{\langle x \rangle \in \Omega \mid \langle x \rangle \perp \langle e \rangle, \langle x \rangle \neq \langle e \rangle\}$, and $\Sigma_2 = \{\langle x \rangle \in \Omega \mid \langle x \rangle \not\perp \langle e \rangle\}$. Now $\langle x \rangle \in \Sigma_1$ if and only if $x = a_1 e_1 + a_2 e_2 + b_2 f_2$, where $(a_2, b_2) \neq (0, 0)$. There are $q(q^2 - 1)$ choices for the coefficients a_1, a_2, b_2, so $|\Sigma_1| = q(q+1)$ and $|\Sigma_2| = q^3$.

If some sporadic groups are involved, we can use known information about these groups, to say that in the Atlas [CCNPW]. For example, in the last line of Table I the Mathieu group M_{22} appears. In this case $M_{22} \leq G \leq \mathrm{Aut} M_{22}$ and $|G : H| = 77$, G acts on the cosets of H. Then, by the Atlas, the action of G on the cosets of H has rank 3, with subdegrees 1, 16 and 60, and of course all suborbits are self-paired.

If it is difficult to use geometric consideration and the Atlas, we have to use purely group-theoretic methods. In our situation, it is in the case where $PSL(2, p) \leq G \leq P\Gamma L(2, p)$. In this case the following elementary lemmas quoted from [PX] will be useful in analyzing the suborbit structure of G. The first result is due to Manning [Ma, Theorem XIV].

Lemma 3.2   *Let G be a transitive group on V, let $H = G_v$ for some $v \in V$, and let $K \leq H$. If the set of G-conjugates of K which are contained in H form t conjugacy classes $C_1, C_2, ..., C_t$ of H, then K fixes $\Sigma_{i=1}^t |N_G(K_i) : N_H(K_i)|$ points of V, where $K_i \in C_i$ for $1 \leq i \leq t$. In particular, if $t = 1$, that is if every G-conjugate of K in H is conjugate to K in H, then K fixes $|N_G(K) : N_H(K)|$ points of V.*

The next lemma is just a special case where the stabilizer subgroup H of the primitive groups G is isomorphic to A_5. In this case all subgroups of H of the same order are conjugate in H, and, applying Lemma 3.2 to the action of H on its orbits, we have the following technical result. (We denote the set of fixed points in V of a subgroup K of $\text{Sym}(V)$ by $\text{Fix}_V(K)$.)

Lemma 3.3   ([PX, Lemma 2.2]) *Let G and H be as in Lemma 3.2 and suppose that $H = A_5$. Let $K_1, ..., K_7$ be seven subgroups of H satisfying $K_1 \cong A_4$, $K_2 \cong D_{10}$, $K_3 \cong D_6$, $K_4 \cong Z_5$, $K_5 \cong Z_3$, $K_6 \cong D_4$ and $K_7 \cong Z_2$. Let k_i be the number of points in V fixed by K_i, for $i = 1, 2, \cdots, 7$. Then G has 1 suborbit of length 1, $k_1 - 1$ suborbits of length 5, $k_2 - 1$ suborbits of length 6, $k_3 - 1$ suborbits of length 10, $\frac{k_4 - k_2}{2}$ suborbits of length 12, $\frac{k_5 - 2k_1 - k_3 + 2}{2}$ suborbits of length 20, $\frac{k_6 - k_1}{3}$ suborbits of length 15, $\frac{k_7 - 2k_2 - 2k_3 - k_6 + 4}{2}$ suborbits of length 30, and all the other suborbits have length 60.*

The next lemma will be used to show that certain suborbits in primitive groups are self-paired.

Lemma 3.4   ([PX, Lemma 2.3]) *Let $D = \langle a, b \rangle \cong D_{2n}$, $n \geq 2$, be a permutation group on $V = \{1, 2, \cdots, n\}$, where $a = (1, 2, \cdots, n)$ and $b = (1)(2, n)(3, n-1) \cdots (i, n+2-i) \cdots$. Then the nontrivial orbitals of D are $\Delta_i = (1, i)^D = (1, n+2-i)^D$, for $2 \leq i \leq \frac{n+2}{2}$. Each of these orbitals is self-paired. Moreover, for all points i, j, with $i \neq j$, there is an involution in D which interchanges i and j.*

Deciding whether or not a suborbit is self-paired is most difficult in the case of *regular suborbits*, that is those suborbits of length equal to the order of the point stabilizer. The next lemma allows us to determine, for some groups, the number of self-paired regular suborbits, given complete information about all suborbits of smaller length.

Lemma 3.5   ([PX, Lemma 2.4]) *Let G be a transitive group on V and let $H = G_v$ for some $v \in V$. Suppose that G has one conjugacy class of involutions, and that each involution has N cycles of length 2. For a nontrivial self-paired orbital Δ and a pair $(v, u) \in \Delta$, let $\text{inv}(\Delta)$ be the* *number of involutions in G with a 2-cycle (v, u). Then*

$$N = \frac{c}{2|H|} \sum_{\Delta = \Delta'} |\Delta(v)| \; inv(\Delta),$$

where c is the order of the centralizer of an involution.

We shall give an example to show how to use the above lemmas to calculate suborbit structure for primitive groups later.

When determining the full automorphism group of a graph and the isomorphisms between graphs, a result about the 2-closures of primitive groups due to Liebeck, Praeger and Saxl [LPS1] is very useful.

Let G be a transitive group on V. G has a natural action on $V \times V$. The orbitals of G are Δ_0, $\Delta_1,...,\Delta_{r-1}$. We call

$$G^{(2)} = \{g \in G \mid \Delta_i^g = \Delta_i, \; i = 0, 1, ..., r-1\}$$

the 2-closure of G. If $G = G^{(2)}$, we say that G is 2-closed. It is easy to see that any full automorphism group of a digraph is 2-closed. So, if there is no subgroups between $G^{(2)}$ and the symmetric group $\mathrm{Sym}(V)$, then the automorphism group of the orbital digraph must be $G^{(2)}$ A similar argument can be used in deciding if two orbital digraphs are isomorphic.

Now we give an example.

Example 3.6   *Let $G = PSL(2, 59)$, acting on the set $V = [G : H]$ of right cosets of a subgroup $H \cong A_5$. Then*

(a) *$|V| = 1711 = 29 \cdot 59$, and G has rank 38, and the suborbits of G have lengths 1, 6, 10, 2 of length 12, 4 of length 20, 5 of length 30, and 24 of length 60. All the suborbits of G of length less than 60 and 16 of the suborbits of length 60 are self-paired.*

(b) *The 29 nontrivial (self-paired) orbital graphs are pairwise non-isomorphic. Also the 8 non-self-paired orbital digraphs are pairwise non-isomorphic. All have automorphism group $PSL(2, 59)$. For each non-self-paired suborbit $\Delta(v)$ of length 60, the graph on V with edge set $\{\{v, u\} \mid (v, u) \in \Delta \cup \Delta'\}$ has automorphism group $PSL(2, 59)$ and is edge transitive but not arc-transitive; there are 4 such graphs.*

Proof. We have $|V| = |G|/|A_5| = 29 \cdot 59 = 1711$. Let $K_i, 1 \leq i \leq 7$ be seven subgroups of $H \cong A_5$, as in Lemma 3.3. We have the following table:

K_i	$K_1 \cong A_4$	$K_2 \cong D_{10}$	$K_3 \cong D_6$	$K_4 \cong Z_5$	$K_5 \cong Z_3$	$K_6 \cong D_4$	$K_7 \cong Z_2$
$N_H(K_i)$	A_4	D_{10}	D_6	D_{10}	D_6	A_4	D_4
$N_G(K_i)$	A_4	D_{20}	D_{12}	D_{60}	D_{60}	A_4	D_{60}
$|\mathrm{Fix}_V(K_i)|$	1	2	2	6	10	1	15

By Lemma 3.3, the suborbits of G have lengths 1, 6, 10, 2 of length 12, 4 of length 20, 5 of length 30, and 24 of length 60, and so G has rank 38.

Next we show that all the nontrivial suborbits of length less than 60 are self-paired. Here we only give a proof for those orbital digraphs of length 20. Let Δ be such an orbital, that is $|\Delta(v)| = 20$. Let $u \in \Delta(v)$. Then $G_{vu} = Z_3$, and $N_G(G_{vu}) \cong D_{30}$ acts on $\mathrm{Fix}_V(G_{vu})$, a set of size 5, as D_{10}. By Lemma 3.4, some element of $N_G(G_{vu})$ interchanges v and u. So Δ is self-paired. For orbitals of other length, the proof is similar and omitted.

Now we use Lemma 3.5 to determine the number of self-paired suborbits of length 60. In the following table $y_{|\Delta(v)|}$ denotes the number of self-paired suborbits of length $|\Delta(v)|$ and $\mathrm{inv}(\Delta)$ is as in Lemma 3.5.

$|\Delta(v)|$	$y_{|\Delta(v)|}$	G_{vu}	$G_{\{v,u\}}$	$\mathrm{inv}(\Delta)$
6	1	D_{10}	D_{20}	6
10	1	D_6	D_{12}	4
12	2	Z_5	D_{10}	5
20	4	Z_3	D_6	3
30	5	Z_2	D_4	2
60	y	1	Z_2	1

By Lemma 3.3 every involution has 15 fixed points, so every involution has $N = (1711 - 15)/2 = 848$ cycles of length 2. Then by Lemma 3.5,

$$848 = \frac{1}{2}\{6 \cdot 1 \cdot 6 + 10 \cdot 1 \cdot 4 + 12 \cdot 2 \cdot 5 + 20 \cdot 4 \cdot 3 + 30 \cdot 5 \cdot 2 + 60 \cdot y \cdot 1\},$$

so $y = 16$.

The assertions that the automorphism group of every orbital graph and digraph is $PSL(2,59)$ follows from the facts that the 2-closure of $PSL(2,59)$ (acting on the cosets of A_5) is itself and that $PSL(2,59)$ is maximal in the symmetric group.

Next we show that the orbital graphs are pairwise non-isomorphic. If σ is an isomorphism between two of the orbital graphs, Δ_1 and Δ_2, then σ lies in the normalizer in $\mathrm{Sym}(V)$ of their common automorphism group $A = PSL(2,59)$. However, by the maximality of A the group A is self-normalizing in $\mathrm{Sym}(V)$ so $\sigma \in A$, whence $\Delta_1 = \Delta_2$. $\square$

§3.3 Classifying arc-transitive graphs of small valency

Other work I would like to mention is an earlier paper of Praeger and Xu (see [PX1]). We gave a characterization theorem for arc-transitive graphs of valency $2p$, p a prime, the automorphism groups of which have non-semiregular normal abelian p-subgroups. Originally, this paper came from an attempt to classify 4-valent G-arc-transitive graphs where G is solvable. In this case, G has at least one nontrivial Abelian normal subgroup. We found that it was necessary to distinguish the following two essentially different cases: (1) G has a non-semiregular Abelian normal subgroup; and (2) all of the abelian normal subgroups of G are semiregular. We classified all 4-valent arc-transitive graphs in Case 1, and when we tried to do the same thing for Case 2. We found that we needed to generalize this theorem to valency $2p$. But here I would like to state our theorem only for the 4-valent case.

First, we define a family of graphs $C(r, t)$, where r and t are positive integers and $r \geq 3$.

Definition 3.7   (a) *The graph $C(r, 1)$ has vertex set $Z_r \times Z_2$ and its edges are defined by*

$$(i, x) \sim (i + 1, y), \ \forall i \in Z_r, \ \forall x, y \in Z_2.$$

(b) *The graph $C(r, 2)$ has vertex set $Z_r \times Z_2 \times Z_2$ and its edges are defined by*

$$(i, x, y) \sim (i + 1, y, z), \ \forall i \in Z_r, \ \forall x, y, z \in Z_2.$$

(c) *For $t \geq 3$ we define the graph $C(r, t)$ to have vertex set $Z_r \times Z_2^t$ and edges defined by*

$$(i, x_1, ..., x_t) \sim (i + 1, y_1, ..., y_t)$$

if and only if (i) $(i, x_1, ..., x_{t-2})$ and $(i + 1, y_1, ..., y_{t-2})$ are adjacent in $C(r, t - 2)$, (ii) $x_t = y_{t-2}$, and (iii) if $t \geq 4$, $y_{t-1} = x_{t-3}$, and if $t = 3$, $y_2 = x_1$.

We proved

Theorem 3.8   (1) *$C(r, t)$ is arc-transitive if and only if $r \geq t + 1$, and is vertex-transitive if and only if $r \geq t$;*

(2) *Suppose that X is a connected G-arc-transitive graph of valency 4, and that G has an Abelian normal subgroup M which is not semiregular on the vertices of X. Then $X \cong C(r, t)$ for some $r \geq \max(3, t+1)$, $t \geq 1$.*

This is the first step of classifying 4-valent arc-transitive graphs. For the second step we should do the case where G has nontrivial Abelian normal subgroups and all of them are semiregular. Recently, I was informed that A. Gardiner and C. E. Praeger have finished this case, see [GP, GP1]. So the only remaining case is that G has no nontrivial Abelian normal subgroups. This would be the most difficult case in this problem. Speaking precisely, we propose the following

Problem 15. *Let X be a arc-transitive graph of valency 4, and $A = \mathrm{Aut}X$. Assume that A has no nontrivial Abelian normal subgroups. Determine X.*

The family of graphs $C(r,t)$ provides a new class of examples of 4-valent G-arc-transitive graphs for which the order $|G_v|$ of a vertex stabilizer can be arbitrarily large (see [W3]). This is related to the famous Weiss conjecture for arc-transitive graphs. In order to state this conjecture we need the following

Definition 3.9 *A arc-transitive graph X is called G-locally primitive, if for any vertex $v \in V(X)$ the stabilizer G_v acts primitively on the neighborhood $X_1(v)$ of v.*

The Weiss conjecture is

Conjecture 16. *There exists a function $f(d)$ defined on the set of natural numbers $d \geq 3$ such that, for every G-locally primitive arc-transitive graph of valency d, the order of a vertex stabilizer $|G_v|$ is less than $f(d)$.*

This is an analogue of Sims' conjecture for arc-transitive graphs. Sims' conjecture is for primitive groups. It says that there exists a function $f(d)$ defined on the set of natural numbers $d \geq 3$ such that, for any primitive group G which has a suborbit of length d, the order of a one point stabilizer is less than $f(d)$. Sims' conjecture has been proved in [CPSS] using the finite simple group classification, but the Weiss conjecture is still open.

Our graph $C(r,t)$ is not locally primitive, the order of a one vertex stabilizer of its automorphism group equals 2^{r-t}, which can be arbitrarily large when r increases (for a fixed value of t).

For 4-valent arc-transitive graphs the Weiss conjecture is true (see [G2] and [G3]); $3^6 \cdot 2^4$ is an upper bound of the order of a one vertex stabilizer. I would like to propose the following

Problem 17. $\cdot$ *Let X be a 4-valent G-arc-transitive graph with $|G_v| > 3^6 \cdot 2^4$, where G_v is the one vertex stabilizer. Is X isomorphic to a $C(r,t)$ for some r and t?*

Another interesting thing is that if we regard our graph $C(r,t)$ as an oriented digraph with directed edges $((i,\vec{x}),(i+1,\vec{y}))$, where $\vec{x} =$

$(x_1, x_2, ..., x_t)$, $\vec{y} = (y_1, y_2, ..., y_t)$, $(i, \vec{x})$ and $(i+1, \vec{y})$ are adjacent in $C(r, t)$. Then this digraph is $(r - t)$-arc transitive, but not $(r - t + 1)$-arc transitive. When r is large, $C(r, t)$ is a nontrivial s-arc transitive digraph for arbitrarily large s; but for nontrivial undirected s-arc transitive graphs, we have $s \leq 7$.

§4. A Few Words For 1/2-Transitive Graphs

Let $X = (V, E)$ be a simple undirected graph. X is said to be *1/2-transitive*, if it is vertex-transitive and edge-transitive, but not arc-transitive.

W.T. Tutte was the first person who considered 1/2-transitive graphs. In his book [Tut3] he proved

Theorem 4.1   *If a graph X is vertex-transitive and edge-transitive, and if it has odd degree, then X is arc-transitive.*

This theorem means that there are no 1/2-transirive graphs of odd degree. He asked whether there are such graphs of even degree.

In 1970, I.Z. Bouwer [Bo] gave an affirmative answer to Tutte's question; he constructed a 1/2-transitive graph of degree n for each even number $n \geq 4$. The smallest graph in his family has order 54 and degree 4.

In 1981, D.F. Holt [H] found another 1/2-transitive graph which has degree 4 and order 27. He asked

Question 1.   Are there any 1/2-transitive graphs of order less than 27?

Question 2.   How many 1/2-transitive graphs of order 27 and degree 4 are there up to isomorphism?

D.A. Holton observed that all 1/2-transitive graphs found by Bouwer and Holt have imprimitive automorphism groups. In [Ho] he asked the following question in 1982.

Holton's Question.   Are there any 1/2-transitive graphs whose automorphism groups are vertex-primitive?

What we had talked about so far was the status of the 1/2-transitive graph theory before 1990.

Since 1990 several authors have done much work on 1/2-transitive graphs; they are B. Alspach, M. Conder, D. Marušič, L. Nowitz, C.E. Praeger, D.E. Taylor and M.Y. Xu. They have found more examples of 1/2-transitive graphs, and they answered the three questions above and proposed several new problems on 1/2-transitive graphs, some of which are still open until now.

For these results, the reader is refered to a survey article [X4] and research papers [AMN], [AX], [MWX], [TX] and [X3].

References

[A]   A.Ádám, Research problem 2-10, *J. Combin. Theory* **2** (1967), 393.

[Al]   B.Alspach, The search for long paths and cycles in vertex-transitive graphs and digraphs, in *Combinatorial Mathematics VIII*, Lecture Notes in Math. 884, Ed. K.L. McAvaney, Springer-Verlag, Berlin, 1981, pp. 14-21.

[AMN]   B. Alspach, D. Marušič and L. Nowitz, Constructing graphs which are 1/2-transitive, *J. Austral. Math. Soc.*, **56** (1994), 1–12.

[AMX]   B.Alspach, D. Maruši c and Ming-Yao Xu, Classification of 2-arc-transitive circulants, *Journal of Algebraic Combinatorics*, 5 (1996), 83–86.

[AP]   B.Alspach and T.D.Parsons, Isomorphisms of circulant graphs and digraphs, *Discrete Math.* **25** (1979), 97-108.

[AP1]   B.Alspach and T.D.Parsons, A construction for vertex-transitive graphs. *Canad. J. Math.*, 1982, 34:307–318.

[AZ]   B.Alspach and C.Q.Zhang, Hamilton cycles in cubic Cayley graphs on dihedral groups, *Ars Combinatoria* **28** (1989), 101-108.

[AX]   B. Alspach and Ming-Yao Xu, 1/2-transitive graphs of order 3p, *Journal of Algebraic Combinatorics*, 3 (1994), 347–355.

[B1]   L.Babai, On the abstract group of automorphisms, *Combinatorics* (Ed. H. N. V. Temperley), London Math. Soc. Lecture Note Series 52, 1981; 1-40.

[B2]   L.Babai, Isomorphism problem for a class of point-symmetric structures, *Acta Math. Acad. Sci. Hungar.* **29** (1977), 329-336.

[BF1]   L.Babai and P.Frankl, Isomorphisms of Cayley graphs I, Colloq. Math. Soc. J. Bolyai, 18. *Combinatorics*, Keszthely, 1976; North-holland, Amsterdam, 1978; 35-52.

[BF2]   L.Babai and P.Frankl, Isomorphisms of Cayley graphs II, *Acta Math. Acad. Sci. Hungar.* **34** (1979), 177-183.

[Ber]   J.L.Berggren, An algebraic characterization of symmetric graphs with p points, *Bull. Aus. Math. Soc.* **7** (1972), 131-134.

[BT]   F. Boesch and R. Tindell, Circulants and their connectivities, *J. Graph Theory* **8** (1984), 487-499.

[Bo]   I.Z. Bouwer, Vertex and edge-transitive but not 1-transitive graphs. *Canad. Math. Bull.*, **13** (1970), 231-237.

[BCN]   A.E. Brower, A.M. Cohen and A. Neumaier, *Distance Regular Graphs*, Springer-Verlag, Berlin, 1989.

[C1]   P.J. Cameron, Finite permutation groups and finite simple groups, *Bull. London Math. Soc.* **13** (1981), 1-22.

[C2]   P.J.Cameron, Automorphism groups of graphs, *Selected Topics in Graph Theory II*, (Eds. L.W. Beineke and R.J. Wilson), Academic Press, 1983; 89-127.

[C3] P.J.Cameron, Suborbits in primitive groups, *Combinatorics*, Part 3, pp. 98-129. (Edited by M. Hall Jr. and J. H. van Lint), Mathematical Centre Tracts **57**, Amsterdam, 1974.

[CPSS] P.J.Cameron, C.E.Praeger, J.Saxl and G.M.Seitz, On the Sims' conjecture and distance-transitive graphs, *Bull. London Math. Soc.* **15** (1983), 499-506.

[Ch] C.Y.Chao, On the classification of symmetric graphs with a prime number of vertices, *Trans. Amer. Math. Soc.* **158** (1971), 247-256.

[CO] Y.Cheng and J.Oxley, On weakly symmetric graphs of order twice a prime, *J. Combin. Theory Ser.* **B 42** (1987), 196-211.

[CCNPW] J.H. Conway, R.T. Curtis, S.P. Norton, R.A. Parker and R.A.Wilson, *An Atlas of Finite Groups,* Clarendon Press, Oxford, 1985.

[CFP] H.S.M. Coxeter, R. Frucht and D.L. Powers, *Zero-symmetric Graphs,* Academic Press, New York, 1981.

[D] D.Z.Djoković, Isomorphism problem for a special class of graphs, *Acta Math. Acad. Sci. Hungar.* **21** (1970), 267-270.

[EM] V.N.Egorov and A.I.Markov, On Adam's conjecture for graphs with circulant adjacency matrices (Russian), *Doklady Akad. Nauk SSSR* **249** (1979), 529-532.

[ET] B.Elspas and J.Turner, Graphs with circulant adjacency matrices, *J. Combin. Theory* **9** (1970), 297-307.

[F1] Xin-Gui Fang, A characterization of finite abelian 2-DCI-groups (Chinese), *J. of Math. (Wuhan)* **8** (1988), 315-317.

[F2] Xin-Gui Fang, Abelian 3-DCI-groups, *Ars Combin,* **32**(1991), 263–267.

[FW] Xin-Gui Fang and Min Wang, Isomorphisms of Cayley graphs of valency m (≤ 5) for a finite commutative group (Chinese), *Chinese Ann. Math. Ser. A,* **13** (1992), suppl., 7–14.

[FX1] Xin-Gui Fang and Ming-Yao Xu, Abelian 3-DCI-groups of odd order, *Ars Combinatoria* **28** (1989), 247-251.

[FX2] Xin-Gui Fang and Ming-Yao Xu, On isomorphisms of Cayley graphs of small valency (Research Communications) (in Chinese), *Kexue Tongbao* **37** (1992), 283.

[FX2p] Xin-Gui Fang and Ming-Yao Xu, On isomorphisms of Cayley graphs of small valency, *Adv. in Math.,* Algebra Colloq., **1**(1994), 67–76.

[G1] A.D.Gardiner, Symmetry conditions in graphs, *Surveys in Combinatorics* (Ed. B. Bollobas), London Math. Soc. Lecture Note Series 38, 1979; 22-43. MR 81e:05081.

[G2] A.Gardiner, Arc transitivity in graphs, *Quart. J. Math. Oxford* (2)**24** (1973), 399-407.

[G3] A.Gardiner, Arc transitivity in graphs II, *Quart. J. Math. Oxford* (2)**25** (1974), 163-167.

[GP] A.Gardiner and C.E.Praeger, On 4-valent symmetric graphs I, preprint, 1990.

[GP1] A.Gardiner and C.E.Praeger, On 4-valent symmetric graphs II, preprint, 1991.

[GY] W.Gaschütz and T.Yen, Groups with an automorphism group which is transitive on the elements of prime order, *Math. Z.* **86** (1964), 123-127.

[Gfd] Ya.Yu.Gol'fand, Description of subrings of $V(Sp_1 \times Sp_2 \times \cdots \times Sp_m)$ (in Russian), *Investigations in the Algebraic Theory of Combinatorial Objects,* Proc. Seminar Inst. System Studies, Moscow, 1985; 65-76.

[Go] C.D.Godsil, On Cayley graph isomorphisms, *Ars Combin.* **15** (1983), 231-246.

[Gr] F.Gross, 2-automorphic 2-groups, *J. Algebra* **40** (1976), 348-353.

[Gu] R.M.Guralnick, Subgroups of prime-power index in a simple group, *J. Algebra* **81** (1983), 304-311.

[H] F.Harary, *Graph Theory*, Addison-Wesley, Reading, Mass., 1969.

[Hi1] D. G. Higman, Intersection matrices for finite permutation groups, *J. Algebra* **6** (1967), 22-42.

[Hi2] G.Higman, Suzuki 2-groups, *Illinois J. Math.* **7** (1963), 79-96.

[Ho] D.F. Holt, A graph which is edge transitive but not arc-transitive, *J. Graph Theory*, **5** (1981), 201-204.

[Hol] D.Holton, Research Problem 9, *Discrete Math.* **38** (1982), 125.

[Hu] B.Huppert, *Endliche Gruppen I*, Springer-Verlag, 1967.

[HB] B.Huppert and N. Blackburn, *Finite Groups II*, Springer-Verlag, 1982.

[J] G.A.Jones, Abelian subgroups of simply primitive groups of degree p^3, *Quart. J. Math. Oxford* (2), **30** (1979), 53-76.

[KP] M.H.Klin and R.Pöschl, The König problem, the isomorphism problem for cyclic graphs and the method of Schur rings, Colloq. Math. Soc. J. Bolyai, 25. *Algebraic Methods in Graph Theory*, Szeged, 1978; North-Holland, Amsterdam, 1981; 405-434.

[LPX] Cai-Heng Li, C.E. Praeger and Ming-Yao Xu, On m-DCI-groups, preprint, 1992.

[LWWX] Hui-Ling Li, Jie Wang, Lu-Yan Wang and Ming-Yao Xu, Vertex primitive graphs of order containing a large prime factor, *Communications in Algebra*, **22**(1994), 3449-3477.

[LPS] M.W.Liebeck, C.E.Praeger and J.Saxl, On the O'Nan-Scott theorem for finite primitive permutation groups, *J. Austral. Math. Soc.* (A)**44** (1988), 389-396.

[LPS1] M.W.Liebeck, C.E.Praeger and J.Saxl, On the 2-closures of finite permutation groups, *J. London Math. Soc.* (2) **37** (1988), 241-252.

[LS] M.W.Liebeck and J.Saxl, Primitive permutation groups containing an element of large prime order, *J. London Math. Soc.* (2) **31** (1985), 237-249.

[Lo1] P.Lorimer, Vertex transitive graphs: symmetric graphs of prime valency, *J. Graph Theory* **8** (1984), 55-68.

[Ma] W.A. Manning, On the order of primitive groups III, *Trans. Amer. Math. Soc.* **19** (1918), 127-142.

[MP] D.Marušič and T.D.Parsons, Hamiltonian paths in vertex-symmetric graphs of order $5p$, *Discrete Math.* **42** (1982), 227-242.

[Ma] D.Marušič, Vertex-transitive graphs and digraphs of order p^k, *Ann. of Discrete Math.* **27** (1985), 115-128.

[MS1] D. Marušič and R. Scapellato, Characterizing vertex-transitive pq-graphs, preprint, 1990.

[MS2] D. Marušič and R. Scapellato, Imprimitive representations of $SL(2, 2^k)$, preprint, 1990.

[Mc1] B.D.McKay, Transitive graphs with fewer than twenty vertices, *Math. Comp.*, **33** (1979), 1101-1121.

[Mc2] B.D.McKay, unpublished.

[Neu] P.M.Neumann, Finite permutation groups, edge-coloured graphs and matrices, *Topics in Group Theory and Computation*, (Proc. of a summer school at University College, Galway, 1973); Edited by M. P. J. Curran, Academic Press, 1977. Chap. 5, pp.82-118.

[P1] P.P.Pàlfy, On regular pronormal subgroups of symmetric groups, *Acta Math. Acad. Sci. Hungar.* **34** (1979), 287-292.

[P2]   P.P.Pàlfy, Isomorphism problem for relational structures with a cyclic automorphism, *Europ. J. Combin.* **8** (1987), 35-43.

[Pr1] C.E.Praeger, Symmetric graphs and the classification of the finite simple groups, *Groups–Korea 1983*, Proc. of Conference on Combinatorial Group Theory, Kyoungju, Korea, 1983. Lecture Notes in Math. No. 1098, Springer Verlag, 1984. pp.99-110.

[Pr2] C.E.Praeger, Imprimitive symmetric graphs, *Ars Combin.* **19A** (1985), 149-163.

[Pr3] C.E.Praeger, On automorphism groups of imprimitive symmetric graphs, *Ars Combin.* **23A** (1987), 207-224.

[Pr4] C.E.Praeger, Finite vertex transitive graphs and primitive permutation groups, in *Coding Theory, Design Theory, Group Theory*, (Burlington, VT, 1990), Wiley-Intersci. Publ., Wiley, New York, 1993. pp.51–65.

[PSY] C.E. Praeger, J. Saxl and K. Yokoyama, Distance transitive graphs and finite simple groups, *Proc. London Math. Soc.*, (3)**55** (1987), 1–21.

[MPWWX] D. Marušič, C.E.Praeger, Ru-Ji Wang, Lu-Yan Wang and Ming-Yao Xu, A classification of symmetric graphs of order $4p$, (in preparation).

[PWX] C.E.Praeger, Ru-Ji Wang and Ming-Yao Xu, Symmetric graphs of order a product of two distinct primes, *J. Combin. Theory Ser. B*, (1993), 299–318.

[PX1] C.E.Praeger and Ming-Yao Xu, A characterization of a class of symmetric graphs of twice prime valency, *European J. Combin.* **10** (1989), 91-102.

[PX2] C.E.Praeger and Ming-Yao Xu, Vertex primitive graphs of order a product of two distinct primes, *J. Combin. Theory Ser. B*, **59**(1993), 245–266.

[Sa]   G.O.Sabidussi, Vertex-transitive graphs, *Monatsh. Math.* **68** (1964), 426-438.

[S]   E.E.Shult, On finite automorphic algebras, *Illinois J. Math.* **13** (1969), 625-653.

[Si]   C.C.Sims, Graphs and finite permutation groups, *Math. Z.* **95** (1967), 76-86.

[TX]   D.E. Taylor and Ming-Yao Xu, Vertex-primitive 1/2-transitive graphs, *J. Austral. Math. Soc.*, **57**(1994), 113–124.

[Tu]   J.Turner, Point-symmetric graphs with a prime number of points, *J. Combin. Theory* **3** (1967), 136-145.

[Tut1] W. T. Tutte, A family of cubical graphs, *Proc. Camb. Phil. Soc.* **43** (1947), 459-474. **MR 9** p.7.

[Tut2] W. T. Tutte, On the symmetry of cubic graphs, *Canad. J. Math.* **11** (1959), 621-624. **MR 22**#679.

[Tut3] W.T. Tutte, *Connectivity in Graphs*, Toronto University Press, 1966.

[W]   Ru-Ji Wang, A classification of symmetric graphs of order $6p$ with a soluble automorphism group, Preprint.

[WX]   Jun Wang and Ming-Yao Xu, A class of Hamiltonian Cayley graphs and Parsons graphs, submitted to Europ. J. Combin.

[WX1] Ru-Ji Wang and Ming-Yao Xu, A classification of symmetric graphs of order $3p$, *J. Combin. Theory Ser. B*, (1993), 197–216.

[W1]   R.M.Weiss, s-transitive graphs, *Colloq. Math. Soc. J. Bolyai*, 25. *Algebraic Methods in Graph Theory*, Szeged, 1978; North-Holland, Amsterdam, 1981; 827-847.

[W2]   R.M.Weiss, The non-existence of 8-transitive graphs, *Combinatorica* **1** (1981), 309-311.

[W3]   R.M.Weiss, An application of p-factorization methods to symmetric graphs, *Math. Proc. Camb. Phil. Soc.* **85** (1979), 43-48.

[Wi1] H.Wielandt, *Finite Permutation Groups*, Academic Press, New York, 1964.

[Wi2] H.Wielandt, *Permutation Groups Through Invariant Relations And Invariant Functions*, Ohio State University, Columbus, 1969.

[Wit1] D.Witte, On hamiltonian circuits in Cayley diagrams, *Discrete Math.***38** (1982), 99-108.

[Wit2] D.Witte, Cayley diagrams of prime-power order are hamiltonian, *J. Combin. Theory* **40B** (1986), 107-112.

[WG] D.Witte and J.A.Gallian, A survey: hamiltonian cycles in Cayley graphs, *Discrete Math.***51** (1984), 293-304.

[Wo] W.J. Wong, Determination of a class of primitive permutation groups, *Math. Z.* **99** (1967), 235-246.

[X1] Ming-Yao Xu, DCI- and CI-groups of order p^3, *Adv. in Math. (China)* **17** (1988), 427-428.

[X1p] Ming-Yao Xu, On isomorphisms of Cayley digraphs and graphs of groups of order p^3, (submitted).

[X2] Ming-Yao Xu, Vertex-primitive digraphs of order p^k are hamiltonian, *Discrete Math.*, **128** (1994), 415 – 417.

[X3] Ming-Yao Xu, Half-transitive graphs of prime-cube order, *Journal of Algebraic Combinatorics* **1** (1992), 275-282.

[X4] Ming-Yao Xu, Some new results on 1/2-transitive graphs, Adv. in Math., **23** (1994), 505 – 516.

[Y] H.P.Yap, *Some Topics in Graph Theory*, London Math. Soc. Lecture Note Series 108, 1986. Chap. 3. Symmetries of graphs, 88-155.

[Z] Joseph Zaks, Parsons graphs of matrices, *Discrete Math.* **78** (1989), 187-193.

[Zh] J.P.Zhang, On finite groups all of whose elements of the same order are conjugate in their automorphism groups, J. Algebra, **153** (1992), 22 – 36.

Some New Developments in the Study
of Differential Geometry
# of Homogeneous Spaces

Yan Zhida   Hou Zixin

Nankai University, Tianjin

Here we give a brief introduction to the main work in the study of differential geometry of homogeneous spaces done in recent years in China.

1. Locally and Globally Classifications of Noncompact Symmetric Spaces

Pseudo-Riemannian symmetric spaces are the extension of Riemannian Symmetric spaces. For definition, they are homogeneous spaces G/H, where G is a connected Lie group acting effectively over G/H, and there is an involutive isomophism σ of G such that $(H_\sigma)_0 \subset H \subset H_\sigma$, where H_σ stands for the subgroup of G which consists of all σ-invariant elements in G and $(H_\sigma)_0$ is the identity component of H_σ. In the middle of 1950s, M. Berger gave a locally classification to irreducible symmetric spaces by counting the spaces one by one. Later, Yan Zhida gave a general result to this locally classification problem by using the classification theory of noncompact semisimple real Lie algebra given by Yan self [1]. Furthermore, Yan Zhida studied the so called symmetric spaces, gave a complete answer to the classification problem of Berger type locally symmetric spaces. Yang Qi counted this locally symmetric spaces explicitly by using the above mentioned result [3] (see also Yang Zhenxiang's [4]).

The globally classification is essentially a matter of calculating the fundamental groups. It is reduced to the problem of finding the centers of noncompact semisimple Lie groups. This problem has been solved by Goto-Kobayashi and Sirota-Solodovnikow almost at the same time. Jiang Caikun gave a method to calculate the fundamental groups for all symmetric spaces in [5] by giving a corresponding relation between the fundamental groups of the noncompact symmetric spaces and fundamental groups of compact Riemannian globally symmetric spaces. The latter were known by E. Cartan. Jiang Caikun used the Satake diagrams and the diagrams in the above mentioned classifications theory of noncompact semisimple Lie algebras.

2. The Study of Homogeneous Complex Manifolds and Kählerian Manifolds

The condition for an oriented even dimensional real manifold has complex structures and when it does have, to classify these complex structures are naturally two essential problems in the theory of complex manifolds. The necessary and sufficient condition that an almost complex structure has an associated complex structure is that it satisfies integrability condition given by Newlander and Nirenberg and it is of fundamental importance. J. Koszul and other mathematicians turned this condition into a more algebraic condition for a Lie algebra in the case of a homogeneous manifold. Inspired by S. Murakami, we studied the complex structure s on an important class of homogeneous manifolds. Let G be a connected semisimple Lie group, L be its compact subgroup and the centralizer of a torus in G. Hou Zixin constructed a one to one corresponding relation between the G-invariant complex structures on G/L and some elements which satisfy certain conditions in the Weyl group of G^c, characterized the invariant complex structures on G/L in a new way [6], this characterization can be expressed simply by using Dynkin diagrams. Based on this result, Hou Zixin and Zou Yiming gave a holomophically equivalent classification for the invariant complex structures on G/L for the case when G is compact and semisimple, i.e. G/L is the well known D-spaces. The main classification theorem is as follows: For a given invariant complex structure J on G/L, there is an associated pair (Π, Π_0), where Π is a simple root system of G, Π_0 is a root base of L which is compatible with Π, and two complex structures J and J' are holomorphically equivalent if and only if there exists an isomorphism of the root system of G which maps (Π, Π_0) to (Π', Π'_0). They gave a complete classification to this case [6,8], [10]. Japanese mathematician M. Nishiyama also got the similar result (see Osaka J. Math. 21(1984), 39-58). It remains a problem to classify the invariant complex structures on G/L when G is noncompact. Even the isomorphism group of G/L is a Lie group or not is unknown yet. Besides, Hou Zixin studied the Chern characteristic classes of D-spaces, gave a very simple formula for the counting of the first Chern characteristic classes by using the diagrams of complex structures [8].

We should point out that the importance of this type of homogeneous complex manifolds being such that for a homogeneous Kählerian manifold acted by a reductive Lie group, when expressed as a coset space G/L, L must be the centralizer of a torus in G. On the other hand, there is a corresponding Kählerian metric which makes a D-space into a homogeneous Kählerian manifold. Thus, naturally to ask, when G is noncompact, is there a corresponding Kählerian metric for an invariant complex structure on G/L (someone calls it D_1-space)? What is necessary and sufficient condition? Based on the necessary and sufficient condition given by S. Murakami and Morakami's study on complex structures, Hou Zixin gave a new and completely algebraic condition [7]. Based on all these work, Xu Yicao proceeded to prove an important result: Suppose G is noncompact and acts effectively over G/L. Then the isomorphism group of G/L is G. He gave a classification to and, realized all homogeneous Kählerian manifolds which are acted effectively by reductive Lie groups [25].

Parallelly, the theory of homogeneous paracomplex and parakähler mannifolds has been developed in China. As an analogue, by the definition, an almost paracomplex structure I on a manifold M^{2n} is a (1,1), tensor field on M^{2n} satisfies the following conditions.

(i) $I_p^2 = id$ on $T_p(M)$ for all $p \in M$,

(ii) The dimensions of (± 1)-spaces are the same.

It is similar to the complex case that the definition of integrablity of almost paracomplex structure can be given by the vanished Nuijenhuis tensor. Let (M, g) be a pseudo-Riemannian manifold and I be a paracomplex structure on M. If

$$g(IX, Y) + g(X, IY) = 0, \quad \forall\, X, Y \in \mathfrak{X}(M)$$

then (M, I, g) is called parahermitian manifold. Put $w(X, Y) = g(IX, Y)$, which is a α-form on M, if $dw = 0$, (M, I, g) is called a parakähler manifold. S. Kaneyuki gave the fundmental theory of homogeneous paracomplex and parakähler manifolds. Hou Zixin and Deng Shaoqian developed the theory. Hou discussed the G-invariant parakähler metrics on

the semisimple homogeneous manifolds G/H. He gave a parametrization
of these metrics for G being a connected split semisimple Lie group [20].
Deng gave the first example of non-symmetric diplorization on Lie algebra, which corresponds parakähler structure on homogeneous manifolds
[21]. Hou and Deng discussed the diplorizations on compact Lie algebras
and proved that compact homogeneous parakähler manifolds must be tori
[22].

Hou Zixin and Zhao Dun gave a new concept–semicomplex structure,
which is a generalization of complex and paracomplex structure and has
both of mathematical and physical backgrounds. They gave some very
important examples of homogeneous semikähler manifolds [23].

3. The Study of Minimal Immersions of Symmetric Spaces

The study of minimal submanifolds is an active branch in differential geometry. Even the study of minimal submanifolds in n-sphere S^n
is rather difficult. Using the representation theory of Lie group and Lie
algebra as a tool the study of the minimal immersions of homogeneous
manifolds developed rapidly in recent years. By a theorem given by Takahashi, which says, for an isometric immersion $x : M \to E^n$, where M is
a m-dimensional connected Riemannian manifold, if $\Delta x = -\lambda x$, $\lambda \neq 0$,
then $\lambda > 0$, and x is a minimal immersion from M into rS^{n-1} with
$r = (m/\lambda)^{1/2}$, the minimal immersion problem of a Riemannian space
into a sphere was made related to the spectrum theory of the Laplace
operator Δ. In the case of a compact homogeneous space, the spectrum
theory is closely related to the so called class one representations of Lie
groups. Thus, there exists the posibility of using the representation theory
of Lie group and Lie algebra to study the minimal immersions. We did
some work on this subject in recent years.

In 1970s, based on the theorem given by Takahashi, M. do-Carmo and
N. Wallach studied the minimal immersions from spheres into spheres,
gave a nice description to these minimal immersions. Zou Yiming generalized their results to all rank one compact Remannian symmetric spaces.
The main difficulty in using the same method to discuss the minimal
immersions of all compact symmetric spaces is how to decompose the representation spaces. He gave some new methods. And, Zou Yiming also
gave a method to count the degree of a standard minimal immersion of
a compact symmetric space into a sphere, and counted the degrees of all

standard minimal immersions of rank one symmetric spaces into spheres [13]. K. Mashimo, a Japanese mathematician, also counted the degrees of the standard minimal immersions of rank one symmetric spaces. Their works were independent.

The so called zonal spherical functions are closely related to the class one representations. J. A. Wolf gave a necessary and sufficient condition to determine whether all zonal spherical functions on a given compact symmetric space are real value functions or not. But not all compact symmetric spaces satisfy this condition. Zou Yiming gave a complete answer to this problem in [12] by using the results on the representation theory given by Yan Zhida and Zhang Dagan [18]

On the other hand we also studied the following problem: What is the necessary and sufficient condition that the image of a standard minimal immersion of an irreducible compact symmetric spaces has a symmetric Riemannian structure? When is an immersion induced by a class one representation an imbedding? Zhang Jianbo [14] and Hu Yi [15] gave the answer to these problems.

4. Spectrum Theory and Some Other Results

The spectrum theory of the Laplace-Beltrami operator of a compact Riemannian symmetric space M, especially the study of the spectrum of $C^\infty(\wedge^p M)$ $(p = 0, 1, \cdots, \dim M)$ had many results in the recent years. The problem reduced to the decomposition of the left regular unitary representations into the irreducible representations of a subgroup. Yan Zhida gave an effective method and counted all spectrums for rank one compact symmetric spaces [2]. Later, Wang Ming studied rank two compact Lie groups, especially the spectrum of G_2 [16]. Together with the results given by some foreign mathematicians, the spectrum theory is well done for rank ≤ 2 compact symmetric spaces. A development is the result given by Japanese mathematician Kaneda, he counted the spectrum of $C^\infty(\wedge^1 M)$ for all simply connected irreducible compact symmetric spaces (see J. Math., Kyoto Univ. 23(1984), 369-395, and 24(1985), 141-162).

Besides, we also studied the algebra problem that closely related to differential geometry. As is well known, the Weyl groups of the semisimple complex or compact semisimple Lie algebras play an important role in the study of homogeneous spaces. And the Weyl groups of noncompact semisimple Lie algebras have been used in the study of geometry, but until

Hou Zixin and Zhang Zhixue gave an explicit expression for this group [9], there had been no explicit expression. And, Kang Wei counted the Weyl groups for all exceptional simple Lie algebras [17]. Thus this theory was completed. They also discussed the structures of the so-called Cartan groups, [19].

References

[1] Yan Zhida, Sur les espaces symetriques noncompacts Scientia Sinica, 15(1965), 31-38.

[2] Yan Zhida, The spectrum of rank one compact symmetric spaces, Proc. of 1980 Beijing Symposium on Differential Geometry and Differential Equations.

[3] Yang Qi, The classification of symmetric spaces, Chinese Ann. of Math., 5(1984), 425-436.

[4] Yang Zenxiang, On the classification of noncompact locally symmetric spaces, Acta Math. Sinica, No.3(1978).

[5] Jing Caikun, On the fundamental groups of noncompact symmetric spaces, Chinese Annals of Math., 1988.

[6] Hou Zixin, On invariant complex structures on certain important homogeneous spaces, Proc. of 1981 Shanghai Symposium on Differential Geometry and Differential Equations.

[7] Hou Zixin, On invariant Kähler structures on noncompact homogeneous complex manifolds, Chinese Annals of Math., 6:3(1985).

[8] Hou Zixin, D-spaces and its characteristic classes, Chinese Ann. of Math., 6:5(1985).

[9] Hou Zixin and Zhang Zhixue, Structures of Weyl groups of real semisimple Lie algebras, Scientia Sinica A, 28(1985), 311-319.

[10] Zou Yiming, The classification of the complex structures on one class of simply connected compact homogeneous spaces, Kexue Tongbao, 29(1984), 999-1003.

[11] Zou Yiming, Minimal immersions of rank 1 compact symmetric spaces into spheres, Scientia Sinica, 28(1985), 263-272.

[12] Zou Yiming, On zonal spherical functions of compact symmetric spaces, Chinese Ann. of Math. Ser. **A6** (1985), No. 3, 263-269.

[13] Zou Yiming, On degrees of minimal immersions of compact Riemannian symmetric spaces into spheres, Lecture Notes in Math. 1255 (1987), Springer.

[14] Zhang Jianbo, The minimal immersions of symmetric spaces into spheres, Lecture Notes in Math. 1255 (1987), Springer.

[15] Hu Yi, Imbedding problem of compact symmetric spaces, Lecture Notes in Mathematics, Springer, 1255(1987).

[16] Wang Ming, The spectrum theory of rank 2 compact Lie groups, Acta Math. Sinica, 27(1984), 613-623.

[17] Kang Wei, The Weyl groups of the exceptional simple real algebras, Chinese Annals of Mathematics, Ser. **A9** (1988), No. 4, 408-397.

[18] Yan Zhida and Zhang Dagan, A method of classification of real irreducible representations of real semisimple Lie algebras, Scientia Sinica, 25(1983), 14-24.

[19] Hou Zixin and Kang Wei, Structures of Cartan groups of real simple Lie algebras, SEA Math. Bull., 15:1(1991).

[20] Hou Zixin, The parakähler metrics on semisimple homogeneous manifolds, Chinese . Science Bulletin, 31:18(1992), 1504-1507.

[21] Hou Zixin, The parakähler metrics on semisimple homogeneous manifolds, Chinese Science Bulletin, 31:18(1992), 1504-1507.

[22] Deng Shaoqian and S. Kaneyuki, An example of non-symmetric diplorization on Lie algebra. (to appear)

[23] Hou Zixin and Deng Shaoqian, On diplorization on compact Lie algebras. (to appear)

[24] Hou Zixin and Zhao Dun, Electromagnetic tensor fields and semicomplex structures, Proc. XXI Diff. Geo. Methods in Theoretical Physics Tianjin, 1993, 343-346.

[25] Xu Yicao, A classification of a class of homogeneous Kählerian manifolds, Scientia Sinica Ser.A, 29(1986), 449-463.